2004

KRIPKE

Kripke

Names, Necessity, and Identity

Christopher Hughes

CLARENDON PRESS · OXFORD

OXFORD
UNIVERSITY PRESS

Great Clarendon Street, Oxford OX2 6DP

Oxford University Press is a department of the University of Oxford.
It furthers the University's objective of excellence in research, scholarship,
and education by publishing worldwide in

Oxford New York

Auckland Bangkok Buenos Aires Cape Town Chennai
Dar es Salaam Delhi Hong Kong Istanbul Karachi Kolkata
Kuala Lumpur Madrid Melbourne Mexico City Mumbai Nairobi
São Paulo Shanghai Taipei Tokyo Toronto

Oxford is a registered trade mark of Oxford University Press
in the UK and in certain other countries

Published in the United States
by Oxford University Press Inc., New York

© Christopher Hughes 2004

British Library Cataloguing in Publication Data

Data available

Library of Congress Cataloging in Publication Data

Data available

ISBN 0–19–824107–0

1 3 5 7 9 10 8 6 4 2

Typeset by Newgen Imaging Systems (P) Ltd., Chennai, India
Printed in Great Britain
on acid-free paper by
Biddles Ltd., King's Lynn, Norfolk

To Marta

PREFACE

Why write a book on Kripke? Well, Kripke is one of the most influential analytic philosophers of the twentieth century; his best-known work (*Naming and Necessity*) is arguably the single most important contribution to metaphysics and the philosophy of language in the last fifty years. And writing a book on a philosopher is an excellent way to get a finer-grained, broader, and deeper understanding of his (or her) thought.

But why read this book on Kripke? I hope it will be useful. Although much has been written on Kripke, most of it has consisted of discussion and criticism of this or that particular view of Kripke (say, on reference, or on the intersubstitutability or otherwise of names in belief contexts, or on the identity of mental states with physical states). This book is an attempt to provide something of an overview of the central themes of Kripke's metaphysics and philosophy of language. As well as expounding Kripke, I juxtapose what I take to be the most important criticisms of Kripke's views with those views. My aim is to put the reader in a better position to arrive at an overall judgement concerning how well those views stand up to the varied criticisms that have been made of them. In addition, the reader will see, I am not averse to putting in my own two cents' worth.

As most readers of this book will be aware, different Kripkean ideas have differing degrees of controversiality. Certain central Kripkean views—especially in the philosophy of language—are, if not uncontroversial, as close to uncontroversial as any interesting views in analytic philosophy. Most analytic philosophers think Kripke has shown that the descriptivist account of the reference of proper names targeted in *Naming and Necessity* is hopeless; that causal relations between a user's use of a proper name and an initial act of reference-fixing typically play a crucial role in explaining why, when a speaker uses this particular name, it refers to this particular thing; that proper names are rigid, and that identity statements involving only proper names are accordingly necessarily true or necessarily false; that we cannot simply assume that analyticity = apriority = necessity, and syntheticity = aposteriority = contingency; that the distinction between accidental and essential properties is not a distinction foisted upon us by Aristotle, but one that has what Kripke calls 'intuitive content'. Other Kripkean views are somewhere between moderately controversial and highly controversial. A good many analytic philosophers don't agree with Kripke that the identification of mental states with physical states has severely counterintuitive modal consequences. Not a few analytic philosophers think that, *pace* Kripke, we need a counterpart-theoretic account of modal predication to solve certain logical difficulties. Few analytic philosophers find Kripke's alleged examples of the contingent a priori convincing.

Like most people, I find some Kripkean ideas, and some Kripkean arguments, more compelling than others. Sometimes I think that Kripke has made a thorough, exceptionally clear, and completely convincing case for a claim; sometimes I do not. So, when I wrote this book, I found that in some cases Kripke's treatment of a question left me with almost nothing to object to, and almost nothing to add; in other cases, quite a lot.

This left me with three options. I could have written a book whose aim was simply to expound some central themes in Kripke's metaphysics and philosophy of language. The drawback of this option was that, as the reader of an earlier version of this book said, it is nearly impossible to improve on Kripke's exposition. (And, as the reader said, I don't.) I could instead have written a book in which I addressed only those themes in Kripke in which I thought there was much to object to, or much to clarify, or much to add to—a book I might have called something like *Variations on Kripkean Themes*, or *Reflections on Kripke*. But my hope is that some readers will find it useful to have a book that doesn't limit itself to those bits of Kripke's thought that are to some extent obscure, or incomplete, or controversial.

I have accordingly ended up with a book that is less uniform than it might otherwise have been. Some sections—especially in the first two chapters—consist mainly of exposition, together with a bit of discussion of why certain objections that have been made to Kripke's view are ineffectual, and of why certain extant interpretations of Kripke's position are erroneous. In other sections—especially, but not exclusively, in the last two chapters—I spend much more time, not only discussing extant objections to Kripke's arguments, but also developing objections of my own, or exploring possible lines of defence of a view that Kripke seems to me to have left undefended.

In one way, the non-uniformity of this book may give it a less pleasing shape than it would otherwise have had. In another way, though, its non-uniformity seems appropriate, given the structure of *Naming and Necessity* (and 'Identity and Necessity'). As I have said, some of Kripke's ideas (especially, though not exclusively, in metaphysics) are highly controversial; others (especially, though not exclusively, in the philosophy of language) are highly uncontroversial. Also, some of Kripke's views (especially ones set out in the first two lectures of *Naming and Necessity*) are set out in great detail. Others (especially ones that appear in the third lecture of *Naming and Necessity*, and towards the end of 'Identity and Necessity') are presented in a much more compressed fashion. For example, Kripke's discussion of descriptivist theories of reference is far more leisurely, and less compressed, than his discussion of various versions of the identity theory. (I take it that this reflects the fact that *Naming and Necessity* and 'Identity and Necessity' were given as talks, and, as we all know, there is never enough time in a talk or series of talks for us to cover all the things we had intended to cover.) Given that Kripke treats some questions in a more leisurely and exhaustive fashion than he treats other questions, it seems natural that a commentator will sometimes have more to say, and sometimes have less to say, about what Kripke has to say on those questions. In some cases I will have less to say because Kripke has less to say. This happens, for example, when Kripke simply registers his view on a certain question, and says that he leaves the defence of that view to another time and place. In other cases I have more to say because Kripke has less to say. For example, if Kripke gives an argument for a conclusion that looks like an enthymeme, I may spend a good bit of time considering what sort of premises (if any) that are both Kripkean and plausible will turn the enthymeme into a valid argument.

If, as I hope, this book will be useful, to whom exactly might it be useful? Those who are new to philosophy would probably find it rather hard going. But undergraduates with a good bit of philosophy under their belt should find the book helpful, though they may sometimes wish for more forest and fewer trees. Graduate students would, I hope, find it instructive. (Depending on their background in logic, both undergraduates and graduates may find the section on Kripke's modal logic challenging, although I tried to make it straightforward and to presuppose very little background in logic.) Because I have tried to write the book in such a way as not to exclude upper-level undergraduates or graduate students, professional philosophers—especially metaphysicians and philosophers of language—may find some of the exposition surplus to requirements. I should like to think, though, that they too will find some food for thought here. My discussion of Kripke on, say, names of kinds, or the essentiality of origin, or trans-world identity, or the relation of persons to bodies, will, I hope, make a contribution to an ongoing debate started by Kripke.

Many different people have helped me in many different ways with this book. Thanks are due to all my colleagues at King's College London, especially to Keith Hossack, M. M. McCabe (a department head *quo maior non cogitari possit*), Mark Sainsbury, and Gabriel Segal. Thanks also go to Andrea Bottani, Massimiliano Carrara, Pierdaniele Giaretta, Carl Ginet, Michele Marsonet, Enrico Martini, Mario Mignucci, Ernesto Napoli, Carlo Penco, Achille Varzi, and Nicla Vassallo, as well as to Marta, Laura, and Amanda. Finally, I am grateful to two anonymous referees whose helpful and encouraging suggestions significantly changed (and, I hope, improved) this book.

CONTENTS

ABBREVIATIONS AND CONVENTIONS

'APAB'   'A Puzzle About Belief', in A. Margalit (ed.), *Meaning and Use* (Dordrecht: D. Reidel, 1979)

'I & N'   'Identity and Necessity', in M. Munitz (ed.), *Identity and Individuation* (New York: New York University Press, 1971)

'N & N'   'Naming and Necessity', in D. Davidson and G. Harman (eds.), *Semantics of Natural Language* (Dordrecht: D. Reidel, 1972)

N & N   *Naming and Necessity*, slightly rev. edn., with new introd. (Oxford: Basil Blackwell, 1980)

In cases where confusion will not result, I use variables $(x, y, \ldots)$, individual constants $(a, b, \ldots)$, and formulae autonymously (that is, as names of themselves).

1

Names

1, DESCRIPTIVIST THEORIES OF PROPER NAMES

> At least if one is not familiar with the philosophical literature about this matter, one naively feels something like the following about proper names...if someone says 'Cicero was an orator', then he uses the name 'Cicero' in that statement simply to pick out a certain object...It would, therefore, seem that the function of names is *simply* to refer, and not to describe the objects so named...
>
> (Kripke, 'Identity and Necessity')

If names simply refer to their referents, it would seem, then names directly refer to their referents. In particular, names do not refer to their referents by specifying a condition which their referent uniquely satisfies. By contrast, if definite descriptions refer, they refer to their referents indirectly, by specifying a condition which their referent uniquely satisfies. If, for example, 'the man who saved Venice from the Genoese' refers to Vettore Pisani, it does so by specifying a condition (saving Venice from the Genoese) that only Vettore Pisani satisfies. If some other man had saved Venice from the Genoese, then 'the man who saved Venice from the Genoese' would not have referred to Pisani.[1]

So on the 'naive' view of proper names, they turn out to be very different from definite descriptions. Nevertheless, Kripke holds, 'the classical tradition in modern logic' has assimilated proper names to definite descriptions: both Russell and Frege thought that proper names were abbreviated or disguised definite descriptions.[2]

---

[1] I have said that definite descriptions refer indirectly if at all, because it is a matter of dispute whether (any) definite descriptions refer. Some philosophers think that definite descriptions should be treated as (non-referring) quantifier phrases: 'the F' should be assimilated to 'some F' and 'each F', rather than to proper names and demonstrative phrases such as 'this F'. (See e.g. G. Evans, 'Reference and Contingency', *The Monist*, 62 (1979), 161–89, esp. 169–70.) If definite descriptions do not refer, the gap between them and proper names (construed as directly referential) will be at least as wide as if definite descriptions refer indirectly.

[2] See Kripke, 'Naming and Necessity', in D. Davidson and G. Harman (eds.), *Semantics of Natural Language* (Dordrecht: D. Reidel, 1972), 255; hereafter 'N & N'. Mark Sainsbury has raised doubts about

Given that proper names don't look like definite descriptions, why should anyone suppose that that is what they are? According to Kripke, some powerful considerations favour the supposition. First, it enables us to say, in a quite general way, what a speaker refers to when she uses a proper name, and why she refers to it. When a speaker uses a name, she refers to the satisfier of the definite description the name abbreviates or disguises (or she refers to nothing at all, if that definite description is improper). And she refers to the individual satisfying the definite description, rather than to anything else, just because it is the individual satisfying that description.

Secondly, whether or not a given identity statement involving proper names is true is often an empirical matter: it was, for example, an empirical discovery that Hesperus is Phosphorus. But what was the content of that discovery? What did astronomers come to know when they discovered that Hesperus is Phosphorus? Arguably, that the celestial body visible here in the evening is the celestial body visible there in the morning. If, however, the discovery that Hesperus is Phosphorus is the discovery that the celestial body visible here in the evening is the celestial body visible there in the evening, it seems natural to suppose that 'Hesperus' abbreviates or disguises—or at least is synonymous with— 'the celestial body visible here in the evening' (and likewise for Phosphorus).

Also, whatever the discovery that Hesperus is Phosphorus is, it is not the discovery that Hesperus is Hesperus: as Frege emphasized, it appears that someone who already knew that Hesperus is Hesperus could come to know that Hesperus is Phosphorus. Now there is no problem about how this is possible, if 'Hesperus' is short for 'the celestial body visible here in the evening' and 'Phosphorus' is short for 'the celestial body visible there in the morning'. We know that co-referential definite descriptions are not in general intersubstitutable *salva veritate* in belief contexts, and Russell's standard account of the semantics of definite descriptions gives us no reason to think that definite descriptions should be intersubstitutable in such contexts.[3] By contrast, as Kripke has noted, if names refer without describing, it is hard to see why co-referential proper names should not be intersubstitutable within belief contexts.[4]

---

whether either Frege or Russell held that names are (abbreviated or disguised) descriptions: see his 'Philosophical Logic', in A. Grayling (ed.), *Philosophy: A Guide Through the Subject* (Oxford: Oxford University Press, 1995), 74–6. For some citations of Russell that suggest he *did* hold the view attributed to him by Kripke, see K. Donnellan, 'Proper Names and Identifying Descriptions', in Davidson and Harman (eds.), *Semantics of Natural Language*, 357 *et passim*. (Russell characterizes 'Romulus' as a 'sort of truncated description', and says that 'when I say, e.g., "Homer existed", I am meaning by "Homer" some description, say "the author of the Homeric Poems"'.)

[3] On Russell's account, a statement of the form 'The F is G' may be analysed as: (*a*) there is exactly one F, and (*b*) whatever is F is G. Given this analysis, one will not be able to infer 'S believes the H is the G' from 'S believes that the F is G' and 'The F is the H'; see B. Russell, 'On Denoting', *Mind*, 14 (1905), 479–93.

[4] See Kripke, 'A Puzzle About Belief', in A. Margalit (ed.), *Meaning and Use* (Dordrecht: D. Reidel, 1979), 104–5; hereafter 'APAB'.

Support for the assimilation of proper names to definite descriptions also seems to come from proper-name-involving existential statements. Suppose someone asserts, say, that God does not exist. It is at least initially plausible that she is asserting something along the lines of: there is no such thing as the omnipotent creator of the world.[5] Again, some astronomers used to believe there was a planet between Mercury and the Sun—one they called Vulcan. It seems plausible to suppose that when such astronomers asserted the existence of Vulcan, they were asserting that there was such a thing as the planet between Mercury and the Sun.

Existential statements involving proper names appear to support the account of names in question in another way, not explicitly discussed by Kripke in *Naming and Necessity*. Since they are sometimes false, this suggests that there are empty names (such as 'Santa Claus' or 'Vulcan' (the supposed planet between Mercury and the Sun)). There is no difficulty about how this could be, if names are abbreviated or disguised descriptions: an empty name will be an improper description (that is, one not satisfied by exactly one thing). If, on the other hand, a proper name is not a definite description—if indeed it does not describe at all, only refer—it is unclear how something that really is a name could fail to refer.

If a proper name is an abbreviated or disguised definite description, we can ask which definite description it abbreviates or disguises. In some cases, there is arguably a particular definite description that is naturally thought of as the proper name unabbreviated or undisguised. Someone might hold, for example, that 'o' abbreviates or disguises 'the smallest natural number' (rather than, say, 'the number you cannot divide anything by' or 'the identity element for addition'); that 'ø' abbreviates or disguises 'the null set' (rather than 'the set that is a subset of every set'), and so on.[6] More often, though, it seems arbitrary to single out a particular definite description as the one a name abbreviates or disguises. Which particular definite description is abbreviated or disguised by the proper name of the person now reading these words?

Some philosophers have responded to this difficulty as follows: rather than saying that a proper name abbreviates or disguises a single definite description, we should say that a proper name is associated with a (possibly vague or shifty) cluster or family of (definite or indefinite) descriptions. A name is not—or at least, need not be, and typically will not be—synonymous with any one of the descriptions in the family associated with it; it refers to the

---

[5] This example is a bit tricky, in view of the fact that 'God' (or at least 'god') can be used as a sortal, rather than as a proper name; for example, in the question 'How many gods are there?' Be that as it may, it looks as though 'God' can also be used as a proper name; for example, in 'God is eternal'. (Compare: 'Thursday' can be used as a sortal ('How many Thursdays are there in September?') and also, it seems, as a proper name ('Thursday is Thanksgiving').)

[6] Actually, expressions like 'ø' suggest a certain difficulty for the claim that proper names abbreviate definite descriptions. It seems natural to say that 'ø' is not a name, precisely because it is just an abbreviation for the definite description 'the null set'.

(unique) satisfier of most, or of a weighted most, of those descriptions (the unique exemplifier of most, or of a weighted most, of a certain set of properties). In the absence of a unique satisfier or exemplifier, the name does not refer.[7]

The aim here is to offer a more flexible, 'open-textured' version of the view that names are abbreviated or disguised definite descriptions. There must accordingly be more to the view than just the claim that for every (non-empty) proper name there is a corresponding true identity statement of the form *n = the unique satisfier of most (or of a weighted most) of the descriptions in F*, where *n* is the name in question and F is a family of descriptions. After all, someone who thinks names are simply and directly referential could accept this last claim. The same goes for the stronger claim that for every (non-empty) proper name there is a corresponding *necessarily* true identity statement of the form *n = the unique satisfier of most (or of a weighted most) of the descriptions in F*. Proponents of the modified definite-description account of proper names have accordingly generally supposed that, where *n* is a name and F is the associated family of descriptions, it is necessarily true, analytic, and knowable a priori that (*a*) if *n* exists, then *n* is the unique satisfier of most (or of a weighted most) of the descriptions in F, and (*b*) if exactly one thing satisfies most (or a weighted most) of the descriptions in F, then *n* does. To put it another way, they have generally supposed that it is necessary, analytic, and a priori that a thing is *n* if and only if it is the unique satisfier of most (or of a weighted most) of the descriptions in F. They have accordingly typically supposed that there are very tight connections between a proper name *n* and the description 'the satisfier of most (or of a weighted most) of the descriptions in F'—connections tight enough that we may think of the definite description as providing or expressing the sense of the proper name, whether we think of the sense of a name as a conceptual representation of a referent employed by a competent user of that name, or think of it as what determines a name's reference, or think of it as a name's cognitive or informational value.[8]

Suppose that on a theory of proper names, where *n* is an arbitrarily chosen proper name, there is always a definite description—either an ordinary definite description, or one of the sort 'the thing with most (or a weighted most) of such-and-such properties'—free from proper names, demonstratives, or indexicals, such that it is necessary, analytic, and knowable a priori that a thing is *n* if and only if it is the satisfier of that definite description.[9] Then we may

---

[7] See J. Searle, 'Proper Names', *Mind*, 67 (1958), 166–73, and L. Linsky, *Names and Descriptions* (Chicago: University of Chicago Press, 1977).

[8] For a discussion of various ways of understanding the notion of sense, see 'N & N', 277, and N. Salmon, *Reference and Essence* (Oxford: Basil Blackwell, 1982), 9–14.

[9] Demonstrative expressions include 'this', 'that', 'he', and 'she'; indexical expressions include pronouns such as 'I', adverbs such as 'here' and 'now', and adjectives such as 'present' and 'actual'. For a discussion of demonstratives, indexicals, and the difference between them, see D. Kaplan, 'Demonstratives', MS, 1977,

call that theory of proper names a *(pure) descriptivist* theory of proper names. (Pure) descriptivist theories of proper names are the principal target of Lecture II of *Naming and Necessity*.

Before discussing Kripke's arguments against pure descriptivism, it may be worth saying something about why that theory, as I have characterized it, requires that, where *n* is an arbitrarily chosen proper name, there be a 'purely qualitative' (that is, name-free, demonstrative-free, and indexical-free) definite description such that it is necessary, analytic, and knowable a priori that something is *n* if and only if it satisfies that description. It is presumably necessary, analytic, and knowable a priori that someone is Socrates if and only if he is the individual who is the same as Socrates. So if 'the individual that is the same as *n*' is allowed as a *substituend* for 'the F', someone who thinks that names are purely referential can agree that, for an arbitarily chosen name *n*, there is always a description 'the F' such that it is necessary, analytic, and a priori that a thing is *n* if and only if it is the F. In order to get a theory that contrasts with the naive view of proper names as purely referential, we need to suppose that descriptions like 'the individual who is the same as *n*' are debarred as *substituends* for 'the F'.

Of course, one might suppose that, for a proper name *n*, there is always some associated *n*-free definite description such that it is necessary, analytic, and a priori that a thing is *n* if and only if it satisfies that definite description, without requiring that the associated description be name-free. And the examples of associated (analysis-providing) descriptions offered by descriptivists have in fact typically included proper names. For example, according to Russell, the name 'Romulus' is a kind of truncated version of 'the person who killed Remus, founded Rome, and so on'.[10]

I take it, though, that a wholehearted descriptivist would have to regard 'the person who killed Remus, founded Rome, and so on' as a promissory note, which could ultimately be cashed in for a description that does not contain 'Remus', 'Rome', or any other proper name. For suppose the descriptivist denies this, and says that no name-free definite description is such that it is necessary, analytic, and a priori that someone is Romulus if and only if that someone satisfies that definite description. Suppose, for simplicity, that she holds that 'Romulus' refers to the satisfier of the definite description 'the founder of Rome' (and is 'analytically equivalent' to the definite description 'the founder of Rome'). We can ask her what 'Rome' refers to. She might provide an account of the reference of 'Rome' which does not involve an associated (analysis-providing) definite description. Or she might not provide any account of the reference of 'Rome', on the grounds that she does not know

---

7–10; a bit of the relevant material is also included in Kaplan's 'Thoughts on Demonstratives', in P. Yourgrau (ed.), *Demonstratives* (Oxford: Oxford University Press, 1990).

[10] Russell, 'Lectures on Logical Atomism', in R. C. Marsh (ed.), *Logic and Knowledge* (London: Allen & Unwin, 1956), 243.

what a good account would be like. Either way, it seems, she is a descriptivist not about proper names (as such), but only about some proper names ('Romulus', but not 'Rome'). So let us suppose that, when asked what 'Rome' refers to, and how, she says that 'Rome' refers to the G, where 'the G' is the definite description associated with (the analysis-providing definite description for) 'Rome'. If, as our hypothetical descriptivist supposes, there is no name-free way of saying what 'Romulus' refers to, then the same must be true of the definite description 'the G'. (Otherwise there would after all be a name-free way of saying what 'Romulus' refers to—namely, the founder of the G.) Suppose that 'the G' is 'the city on the Tiber'. We can now ask the descriptivist what 'the Tiber' refers to, and how. If she keeps on long enough answering this sort of question in a descriptivist way, by providing a (name-including) analysing definite description, then, assuming that a proper name cannot appear in its own analysis-providing definite description, and given that there are only finitely many proper names in English, she will eventually come round in a circle. As Kripke has emphasized, this would vitiate her account ('N & N', 292). It is as though you were to tell me that the man who deserved it got the prize he won. I ask you, 'Which man?', and you answer, 'The one who won the prize.' I say, 'But which prize is that?', and you answer 'The one the man deserved to win.' We haven't got any further forward.

Another possible theory of proper names would hold that, for each name, there is an (analysis-providing) name-free definite description, but allow that that definite description might have an ineliminably demonstrative or indexical character, in that it contained demonstratives or indexicals and could not be 'cashed in for' any other demonstrative- and indexical-free definite description. Such a theory is antithetical to the spirit of pure descriptivism. For the pure descriptivist, a proper name, as it were, attains singular reference via its association with a uniquely satisfied pure description. A description is *pure* if it describes and all it does is describe. Intuitively, '... is *this*' is not a pure description: to say that something is *this* is not to describe it all. (If asked to describe my wife, I could hardly reply: she is *this* (demonstrating her).) Nor is '... is *this K*' a pure description. Even if saying that something is *this K* involves describing it as a K, it also involves identifying it (rather than describing it) as this.[11] Of course, if there were a way of trading in expressions such as 'this' and 'this K' for synonymous definite descriptions built up from purely qualitative predicates, and free from names, demonstratives, and indexicals, the intuitions I am appealing to would be mistaken. But on the theory under consideration, there are some demonstrative expressions (or indexical expressions) that cannot be traded in for (synonymous) purely qualitative, demonstrative- and indexical-free definite descriptions—namely, the demonstratives or

---

[11] 'I', 'here', and 'now' are not purely descriptive either. Although they do not require an associated demonstration to attain reference, in other ways they appear to resemble demonstrative phrases built up from a demonstrative and a sortal predicate. Thus 'I' refers to this person (the one speaking), 'here' to this place (the place of utterance), and 'now' to this time (the time of utterance).

indexicals that appear in the *analysantia* of names whose *analysantia* have an ineliminably demonstrative or indexical character. If there are such names, we can if we want say that they refer via their association with a uniquely satisfied description, but the description will do more than just describe.

I do not mean to suggest here that a theory on which names can be analysed by definite descriptions with an ineliminably indexical or demonstrative character is uninteresting or unworkable. As we shall see, some philosophers have responded to Kripke's attack on descriptivism by proposing theories of just this kind. But such theories, whatever their merits, are not (what I call) purely descriptivist, and they are not Kripke's target in *Naming and Necessity*. This comes out clearly when Kripke discusses whether his criticisms refute descriptivism or only a particular formulation of it:

> Have I been unfair to the description theory? Here I have stated it very precisely . . . So then it's easy to refute. Maybe if I tried to state mine with sufficient precision in the form of six or seven or eight theses, it would also turn out that when you examine the theses one by one, they will all be false. That might even be so, but the difference is this. What I think the examples I've given show is not simply that there's some technical error here or some mistake there, but that the whole picture given by this theory of how reference is determined seems to be wrong from the fundamentals. It seems to be wrong to think that we give ourselves some properties which somehow *qualitatively* uniquely pick out an object and determine our reference in that manner. ('N & N', 300; my emphasis)

Kripke offers three sorts of arguments against pure descriptivism. Following Salmon, we may call these the modal, epistemological, and semantical arguments.[12]

Suppose the proper name 'Hesperus' abbreviates or can be 'analysed as' 'the celestial body visible over there in the evening'. Then it will be a necessary truth that if Hesperus exists (if something is Hesperus), it is visible over there in the evening. But this seems wrong: if Hesperus had collided with a meteor, it might not have been visible at all (from earth) in the evening, or it might have been visible (from earth) in the evening somewhere else ('N & N', 276). Also, if 'Hesperus' just abbreviates 'the celestial body visible over there in the evening', then it will be a necessary truth that if exactly one celestial body is visible over there in the evening, then Hesperus is. Again, though, this does not seem to be a necessary truth, since some other celestial body might have been visible in the evening (from earth) in just the place where Hesperus is actually visible (ibid.).

Michael Dummett has objected that this type of argument has very little force.[13] It is no good, he says, arguing that 'St Anne' does not mean 'the mother of Mary' on the grounds that St Anne might never have been Mary's (or anyone else's) mother. A statement such as

(1) The mother of Mary could not but have been a parent

---

[12] See Salmon, *Reference and Essence*, 23–31.
[13] Cf. M. Dummett, *Frege: Philosophy of Language*, 2nd edn. (London: Duckworth, 1981), 112–16.

has a true reading, and a false reading. (On the true reading, it is equivalent to *It could not have been true that: the mother of Mary was not a parent*. On the false reading, it is equivalent to *It is true of the mother of Mary that she could not but have been a parent*.) In just the same way, Dummett maintains, a statement such as

(2) St Anne could not but have been a parent

has a true reading, and a false reading:

Even though there is an intuitive sense in which it is quite correct to say, 'St. Anne might never have become a parent', there is an equally clear sense in which we may rightly say, 'St. Anne cannot but have been a parent', provided always that this is understood as meaning that, if there was such a woman as St. Anne, then she can only have been a parent.[14]

If Dummett is right in maintaining that (1) and (2) exhibit the same sort of ambiguity, that would certainly make trouble for modal arguments against descriptivist theories of proper names. But, like Kripke, I cannot believe that (2) is ambiguous in the way Dummett supposes: 'It could not have been true that: St Anne was never a parent', 'It could never have happened that: someone was St Anne, but not a parent' and their congeners all sound false.[15] (Also, counterfactuals such as 'If St Anne hadn't been anyone's parent, she wouldn't have been St Anne' seem (unambiguously) false, in contradistinction to 'If the mother of Mary hadn't been anyone's parent, she wouldn't have been the mother of Mary.'[16]) I find it puzzling, and somewhat worrisome, that Dummett takes a sentence which I think has no true reading to have one. Why would Dummett see a reading of (2) that just isn't there? Well, perhaps his intuitions are corrupted by a prior commitment to descriptivism. But then, how do I know that my own intuitions aren't corrupted by exposure to Kripke, Putnam, *et alii*? I'm not sure, but I take comfort from the fact that if I am blind to a certain reading of (2), my form of blindness is very widespread.

It might be (indeed, it has been) said, by way of defending descriptivism from the modal argument under discussion, that although names are synonymous with definite descriptions, there is a standing convention that a sentence like '*n* might have been F' is to be understood as equivalent to 'The G is such that it might not have been F', and not to 'It might have been that: the G is F' (where 'the G' is the description synonymous with *n*). The existence of such a convention for names, and the absence of a similar ('wide scope') convention for unabbreviated definite descriptions, would explain why we are wont to think 'The mother of Mary could not but have been a parent' has a true reading and a false

[14] Cf. Dummett, *Frege: Philosophy of Language*, 113.

[15] Cf. Kripke, *Naming and Necessity*, slightly rev. edn., with new introd. (Oxford: Basil Blackwell, 1980), 13; hereafter *N & N*.

[16] I ignore complications arising from the fact that Dummett's example involves 'St Anne' rather than 'Anne'. (I am inclined to think that—on at least one way of understanding it—'St Anne' (or 'Saint Anne') is not really a proper name (though it contains one), any more than 'Countess Dysart' is a proper name.)

reading, and wont to think 'St Anne could not but have been a parent' has no true reading. The reason this defence seems ineffectual is that, on the face of it, there is no way of understanding the *non-modal* sentence 'St Anne was a parent' on which it expresses a necessary truth, while there is a way of understanding the non-modal sentence 'The mother of Mary was a parent' on which it expresses a necessary truth. This contrast cannot be explained in terms of a convention governing the way names and modal operators interact within a sentence (see *N & N*, 10–14).

So the considerations Kripke sets out in *Naming and Necessity* seem to show that the proper name 'Hesperus' cannot be 'analysed by' (is not analytically equivalent to) the definite description 'the celestial body visible over there in the evening'. It doesn't in any obvious way follow that there is anything wrong with pure descriptivism as such. The pure descriptivist might say:

> To refute pure descriptivism, Kripke would have to show that, for 'Hesperus' or some other proper name, there is no definite description that provides the sort of analysis of that name that pure descriptivism implies the existence of. And he certainly has not done that by showing that one hypothesis about what the 'analysis-providing description' for 'Hesperus' is mistaken.

True, Kripke's remarks about 'Hesperus' and 'the celestial body visible over there in the evening' do not (and are not intended to) constitute a logically watertight argument against a (pure) descriptivist account of the reference of proper names. In the 'Naming and Necessity' lectures Kripke does not aim to set out such an argument; instead he aims to persuade his audience that the (pure) descriptivist 'picture' of how proper names refer is mistaken. I take it, though, that Kripke would say that, if we reflect on the failure of 'the celestial body visible over there in the evening' to provide the sort of 'analysis-providing description' the pure descriptivist needs, this will bring to light the unattractiveness of pure descriptivism as such. And this seems right.

Remember that, for the pure descriptivist, there is a (purely) qualitative definite description such that necessarily a thing is Hesperus if and only if it satisfies that description. We can challenge the descriptivist to say what such a description might be. If we do, the pure descriptivist might say: 'the celestial body visible over there in the evening'. Here we may, appealing to Kripke, object that it is not a necessary truth that a thing is Hesperus if and only if it is the celestial body visible over there in the evening: something could have been Hesperus without being (a celestial body) visible over there in the evening, and something could have been (the only celestial body) visible over there in the evening without being Hesperus.

A pure descriptivist might respond to this objection by saying:

> The analysis-providing description for 'Hesperus' is 'the celestial body visible over there in the evening', but this should be construed as a 'rigid'

definite description. That is, it should be understood as equivalent to 'the thing that is in actual fact the celestial body visible over there in the evening'. As long as we understand the analysis-providing description in this way, the modal difficulties alleged by Kripke do not arise. We can grant that something might have been Hesperus without being (a celestial body) visible over there in the evening, and that something might have been (the only celestial body) visible over there in the evening without being Hesperus, and still insist that necessarily a thing is Hesperus if and only if it is the celestial body visible over there in the evening—that is, if and only if it is *the thing that is in actual fact* the celestial body visible over there in the evening.

This response seems inadequate. It may be true that necessarily something is Hesperus if and only if it is the thing that is in actual fact the celestial body visible over there in the evening. But 'the thing that is in actual fact the celestial body visible over there in the evening' is not an indexical-free definite description, inasmuch as it explicitly involves the indexical 'actual'. And, if we take 'the celestial body visible over there in the evening' to be elliptical for, or at any rate equivalent to, 'the thing that is in actual fact the celestial body visible over there in the evening', the former, like the latter, will involve an indexical element (even if it involves it implicitly rather than explicitly). We have seen that, for the pure descriptivist, there must be some name-free, demonstrative-free, indexical-free definite description that necessarily a thing satisfies if and only if it is Hesperus. When we challenge the pure descriptivist to provide a purely qualitative definite description that necessarily a thing satisfies if and only if it is Hesperus, we are challenging the pure descriptivist to come up with a name-free, demonstrative-free, and indexical-free definite description that necessarily a thing satisfies if and only if it is Hesperus. She hasn't met that challenge if the definite description she offers us is (explicitly or implicitly) indexical-involving.

In fact, the definite description 'the celestial body visible over there in the evening' is not demonstrative- and indexical-free, whether or not it is construed rigidly, simply in virtue of containing 'over there'. On those grounds alone, it cannot be the sort of purely qualitative analysis-providing description the pure descriptivist needs there to be (the sort of purely qualitative analysis-providing description she is challenged to provide).

In light of this, a pure descriptivist might say, the considerations Kripke adduces concerning 'Hesperus' and 'the celestial body visible over there in the evening' don't raise any difficulties for pure descriptivism, inasmuch as no pure descriptivist with her wits about her would suggest that 'the celestial body visible over there in the evening' is the analysis-providing description for 'Hesperus'.

But this just brings out how unlikely it seems that the pure descriptivist will be able to meet the challenge to come up with an analysis-providing description for 'Hesperus' that is name-free, demonstrative-free, and indexical-free (n-d-i-free, as I shall say for brevity's sake). I have no clue what the n-d-i-free definite description might be such that necessarily a thing is Hesperus if and only if it satisfies that description. Indeed, I have no clue what n-d-i-free definite description might be such that *actually* a thing is Hesperus if and only if it satisfies that description. For I don't know of any purely qualitative property that Hesperus has and nothing else has: any property that I can think of that I know Hesperus and only Hesperus has I can specify only via names, demonstratives, or indexicals. So I have no idea what purely qualitative definite description, if any, might be such that necessarily a thing is Hesperus if and only if it satisfies that description, and I doubt the pure descriptivist does either.

There might nevertheless be some purely qualitative property that Hesperus and only Hesperus has, or even some purely qualitative property that Hesperus actually has and necessarily nothing different from Hesperus has. For reasons I shall set out later on in this section, I suspect that there isn't any purely qualitative property that Hesperus and necessarily nothing different from Hesperus has. In fact, for reasons I shall touch upon in Chapter 3, I suspect that there isn't even any purely qualitative property that Hesperus and only Hesperus has. But suppose there is some purely qualitative property that Hesperus and only Hesperus has. It still seems a safe bet that among the qualitative properties of Hesperus there is no purely qualitative property such that we are in a position to know a priori that, if Hesperus has it, nothing else does. If some quality of Hesperus is not had by anything else anywhere in the universe, the fact, that if Hesperus has it, nothing else does, is ascertainable only empirically. If, for example, no other planet has Hesperus' exact shape or Hesperus' exact volume, it is only through empirical inquiry that we can establish that, if Hesperus has that shape or that volume, nothing else anywhere in the universe does.

For a pure descriptivist, though, there is an n-d-i-free definite description that any competent user of the name 'Hesperus', simply in virtue of being a competent user of that name, is in a position to know a priori is satisfied by a thing if and only if that thing is Hesperus. (Again, the definite description might be of the ordinary sort, or might involve having most, or a weighted most, of a set of (purely qualitative) properties.) This implies that there is a purely qualitative property that the competent user of the name 'Hesperus' is in a position to know a priori Hesperus has only if nothing else does—to wit, the purely qualitative property a thing must be the unique bearer of in order to be the unique satisfier of the n-d-i-free definite description that a competent user of the name 'Hesperus' is in a position to know a priori is satisfied by a thing if and only if that thing is Hesperus. Suppose, for example, that the n-d-i-free definite description that any competent user of the name 'Hesperus' is in a position to know a priori is satisfied by a thing if and only if it is Hesperus were 'the round thing'. Then the competent user of the name

'Hesperus' would be in a position to know a priori that Hesperus is round only if nothing else is round. But this is the kind of thing that nobody could know a priori. Now 'the round thing' is an n-d-i-free definite description that no pure descriptivist would offer as the analysis of the name 'Hesperus'. But that is not the point. The point is that, if there is an n-d-i-free analysis-providing description for the name 'Hesperus', then among Hesperus' purely qualitative properties is one that any competent user of the name 'Hesperus' is in a position to know a priori Hesperus has only if nothing else does. This seems incredible to me.

To sum up: let us say that 't' and 't''' are *intensionally equivalent* if necessarily something is t if and only if it is t'. And let us say that 't' and 't''' are *epistemically equivalent* (for a person) just in case that person is in a position to know a priori that something is t if and only if it is t'. Pure descriptivism requires that there be some 'purely qualitative' (n-d-i-free) definite description that is both intensionally and epistemically equivalent to the name 'Hesperus'. For the reasons adduced by Kripke, 'the celestial body visible over there in the evening' cannot fit the bill. Nor does it look as though any other description could: even if some purely qualitative (n-d-i-free) definite description is intensionally equivalent to 'Hesperus' (because Hesperus has a (purely) qualitative essence), that definite description will not be epistemically equivalent to 'Hesperus', lest there be qualities one could know a priori Hesperus has only if nothing else does.

Kripke's epistemological arguments against pure descriptivism are structurally similar to his modal arguments. Suppose, for example, that the analysis-providing description for the name 'Gödel' is 'the man who discovered the incompleteness of arithmetic'. Then *if anyone is Gödel, he (uniquely) discovered the incompleteness of arithmetic* and *if anyone (uniquely) discovered the incompleteness of arithmetic, Gödel did* will be truths that any competent user of the name 'Gödel' is in a position to know a priori. But even though we are competent users of the name 'Gödel', we are in no position to know those truths a priori; we could perfectly well find out that the discovery of the incompleteness of arithmetic was misattributed to Gödel and is due to a different man, Schmidt. 'The man who discovered the incompleteness of arithmetic' accordingly cannot be the analysis-providing definite description for 'Gödel', lest truths that some competent users of the name 'Gödel' could know only a posteriori be misclassified as truths that any competent user of the name 'Gödel' could know a priori ('N & N', 294–6).

Again, the pure descriptivist could respond that this shows only that 'the man who discovered the incompleteness of arithmetic' is the wrong 'analysis' of 'Gödel'. And again, we can challenge the pure descriptivist to say what a better one might be. The pure descriptivist is committed to saying that there is some purely qualitative analysis-providing definite description that is both intensionally equivalent to 'Gödel' and epistemically equivalent to 'Gödel' for any competent user of the name 'Gödel'. But why suppose there is?

As Kripke points out, a competent user of a name needn't be in possession of any non-trivial identifying description of the referent of that name. A competent user of the name 'Peano', if asked who Peano was, may not be able to give a better answer than 'the guy who discovered the Peano postulates'. If he is then asked, 'What are the Peano postulates?', he may be unable to give a better answer than 'some postulates Peano discovered'. In that case, he won't have a non-trivial identifying description of Peano. Suppose, though, that a competent user of the name 'Peano' can in fact give an independent characterization of the Peano postulates. She still does not have a (non-trivial, genuinely) identifying description of Peano, because the postulates whose discovery is commonly misattributed to Peano were actually discovered by Dedekind ('N & N', 294–5). Also, if enough of what is attributed to Jonah in the Bible is legend, it may be that no one is in a position to give a (non-trivial, genuinely) identifying description of Jonah ('N & N', 282). Even when there isn't a problem about sifting out misattributions, we may simply not have enough information about a historical figure to enable us to provide a (non-trivial) identifying description of him. As David Kaplan reports, the entry under 'Rameses VIII' in the *Concise Biographical Dictionary* is 'One of a number of ancient pharaohs about whom nothing is known.'[17] And even when there is knowledge that would allow a competent user to offer a non-trivial identifying description of the referent of a name, if the user had it, that user needn't actually have it. Going back to our original example, although the competent user of the name 'Gödel' may be able to give a non-trivial identifying description of Gödel, he needn't be able to.

Even if a competent user of the name 'Gödel' is not in possession of any non-trivial identifying description, he may still be in possession of what we might call a trivially identifying description of Gödel. Waiving problems about the multireferentiality or unireferentiality of names, he may know, for example, that Gödel is the individual called 'Gödel'. And precisely because it is trivial that 'Gödel is the individual called "Gödel"', it might be said, he is in a position to know a priori that someone is Gödel if and only if someone is the individual called 'Gödel'. So, it might be said, there is a definite description that, unlike 'the man who discovered the incompleteness of arithmetic', is epistemically equivalent to the name 'Gödel' for any competent user of that name.

But 'the individual who is called "Gödel"' (unlike, perhaps 'the discoverer of the incompleteness of arithmetic') is obviously not an n-d-i-free definite description. So it won't help the pure descriptivist make the case that there is a purely qualitative definite description that is both intensionally equivalent to 'Gödel' and epistemically equivalent to 'Gödel', for any competent user of that name. Given that a competent user of the name 'Gödel' may be bereft of any interesting identifying description of Gödel—may indeed be unable to say anything more about Gödel than that he was a famous mathematical logician—it

---

[17] D. Kaplan, 'Bob and Carol and Ted Alice', MS, 1978, n. 9.

is very hard to see how one could suppose that for any competent user of the name 'Gödel', there is bound to be a purely qualitative description that the user is in a position to know a priori is satisfied by someone if and only if that someone is Gödel.

Perhaps there are, or at any rate could be, cases in which some competent user of a name is in a position to provide an n-d-i-free definite description that is epistemically equivalent to that name (for that user). Suppose that someone was and took herself to be the only conscious being there is, and introduced the name 'Conscia' via the following ceremony: *Conscia* shall be the (only) conscious being. It is at least arguable that for her the name 'Conscia' and the n-d-i-free definite description 'the (only) conscious being' are epistemically equivalent. (For more on the relevant issues, see Chapter 2, section 2.)

If that is indeed the case, then there will be some quality of Conscia (namely, *being conscious*) that Conscia can know a priori Conscia has only if nothing else does. The case will accordingly be different from the case of Hesperus discussed earlier, where there is no quality of Hesperus that any competent user of the name 'Hesperus' is in a position to know a priori Hesperus has only if nothing else does.

This does not mean, though, that the pure descriptivist will be able to come up with an analysis-providing n-d-i-free definite description for 'Conscia'. In order to do so, she would have to provide an n-d-i-free description that not only is epistemically equivalent to 'Conscia' (for any competent user of that name), but also is intensionally equivalent to 'Conscia'. Even if 'the conscious being' is epistemically equivalent to 'Conscia' (for its introducer), it is not intensionally equivalent to 'Conscia'. Someone could have been Conscia without being the only conscious being (since Conscia might have had a conscious twin). And someone could have been the only conscious being without being Conscia (because the conscious twin that Conscia didn't actually have might have 'taken the place' of Conscia in the world, and thus might have been the only conscious being). If, however, 'the conscious being' is not both intensionally equivalent to 'Conscia' and epistemically equivalent to 'Conscia' (for any competent user of that name), it is hard to see what other n-d-i-free description might be both intensionally and epistemically equivalent to 'Conscia'.

The point here is a kind of mirror image of the one made with respect to 'Hesperus'. There we saw that, even if there is some n-d-i-free description that is intensionally equivalent to 'Hesperus', it is hard to believe that any such description will be epistemically equivalent to 'Hesperus' for every (or indeed, for any) competent user of the name 'Hesperus'. Here we see that, even if there is some n-d-i-free description that is epistemically equivalent to 'Conscia' (for some competent user of that name), it is hard to believe that any such description will be intensionally equivalent to 'Conscia'. In both cases the pure descriptivist appears unable to defend the existence of the kind of analysis-providing definite description of a proper name that pure descriptivism commits her to.

The third sort of argument Kripke offers against pure descriptivism is a semantic one. Suppose that 'Peano' means 'the mathematician who discovered such-and-such postulates (the ones whose discovery is usually misattributed to Peano)'. Then 'Peano' refers to Dedekind, and it is true after all that Peano discovered the Peano postulates (since 'Peano' refers to Dedekind, and Dedekind discovered the (postulates usually called the) Peano postulates). This seems wrong ('N & N', 295). A similar argument is offered by Donnellan:[18] Suppose that 'Thales' means 'the Greek philosopher who held that everything is water'. Now suppose that the man referred to by Aristotle and Herodotus when they used the Greek counterpart of 'Thales'—the man to whom our uses of the name 'Thales' may be traced back—never actually held that everything is water. Suppose, finally, that an obscure Greek hermit–philosopher whom Aristotle and Herodotus had never heard of—a philosopher who is in no way causally connected with our uses of the name 'Thales'—did (uniquely) hold that everything is water. Then when we use the name 'Thales', we refer to the hermit–philosopher. But even if the above story is true, 'Thales' surely does not refer to the hermit–philosopher.

The semantic argument against descriptivism should not be confused with the modal argument. The modal argument turns on the contingency of certain truths that would be necessary if descriptivism were true. For example, it is a contingent fact that, if someone was St Anne, then she had children, but it would be necessary if 'St Anne' meant 'the mother of Mary'. The semantic argument turns on the claim that certain truths about reference are *not* contingent, or at least not contingent in the way they would be if descriptivism were true. For example, that 'Peano' refers to this particular person is not contingent upon his having discovered a certain set of postulates, although it would be so if 'Peano' meant 'the person who discovered such-and-such postulates'.

The descriptivist can respond to the semantic arguments of Kripke (and Donnellan) by saying that all they show is that 'the man who held that everything is water' is not what 'Thales' means, and 'the discoverer of such-and-such postulates' is not what 'Peano' means. The correct analysis of the name 'Peano', she can insist, will be given by a purely qualitative definite description that is 'semantically equivalent' (analytically equivalent) to 'Peano', as well as intensionally equivalent and epistemically equivalent, to 'Peano'. But we have already seen the limitations of this response: it is very doubtful that any purely qualitative definite description is equivalent to 'Peano' in all of those ways. Because the considerations should be familiar to the reader, I shall not rehearse them here.

Before concluding the discussion of pure descriptivism, I shall sketch an argument against that view that Kripke does not advance (or explicitly discuss).

---

[18] See Donnellan, 'Proper Names and Identifying Descriptions', 373–5.

Near Belluno there is a mountain called Mount Faverghera. Could it have happened that something different from Mount Faverghera had some of the qualitative properties Mount Faverghera actually has? Certainly: lots of mountains different from Mount Faverghera actually have some of the qualitative properties Mount Faverghera actually has. Could it have happened that something different from Mount Faverghera had all of the qualitative properties Mount Faverghera actually has? It is hard to see why not. Just as something else could have had Mount Faverghera's size, or its shape, something else could have had its size and shape. And something else could have had its size, shape, and mass. It looks as though we could go on this way indefinitely: and the limit of the series would be a possible state of affairs in which something else had all of Mount Faverghera's qualitative properties. It is very hard to see how one could motivate the claim that, if we considered states of affairs in which something else has more and more of Mount Faverghera's qualitative properties, we would eventually cross a line on the far side of which we were considering impossibilities rather than possibilities.[19]

Note that this argument does not depend on a sorites fallacy. We can consider a series of states of affairs in which a man who isn't actually bald has one fewer hairs than he actually has, then two fewer, three fewer, and so on. It is impossible to say where exactly in this series the man becomes bald. But it is obvious that subtracting hairs long enough will, as it were, eventually take you from an unbald man to a bald man. It is anything but obvious that conjoining qualitative properties of Mount Faverghera long enough will eventually take you from a property that something else could have had to a property that nothing else could have had.[20]

Here someone might object: suppose something did have all the qualitative properties that Mount Faverghera actually has. Why shouldn't we suppose that it just is Mount Faverghera? Well, consider a possible world in which Mount Faverghera exists and has a perfect qualitative duplicate in New Zealand. We shouldn't suppose that the New Zealand mountain is Mount Faverghera, because the New Zealand mountain is on the other side of the world from Mount Faverghera. Or consider a possible world in which Mount

---

[19] Suppose you thought that individuals were sets or mereological aggregates of qualitative properties, and that qualitative properties were universals (in Armstrong's sense). Sets with different members (or aggregates with different parts) must be distinct, and sets with the same members (or aggregates with the same parts) must be identical. So, you might say, it is possible for something different from Mount Faverghera to have all but one of the qualitative properties (universals) that Mount Faverghera actually has; but it is impossible for something different from Mount Faverghera to have all of the qualitative properties or universals Mount Faverghera actually has. The line between the possible and the impossible would be crossed at the last possible moment. The difficulty is in motivating the account of individuals and properties under discussion. See D. Zimmerman, 'Distinct Indiscernibles and the Bundle Theory', *Mind*, 106 (1997), 305–9, and C. Hughes, 'Bundle Theory from *A to B*', *Mind*, 108 (1999), 149–56.

[20] For more discussion, see R. Adams, 'Primitive Thisness and Primitive Identity', *Journal of Philosophy*, 76 (1979), 5–26, and C. Hughes, 'Omniscience, Negative Existentials, and Cosmic Luck', *Religious Studies*, 34 (1998), 375–401.

Faverghera is destroyed, and thousands of years later, by an amazing coincidence, a perfect qualitative duplicate of it comes into existence (either in New Zealand, or in the Veneto). We shouldn't suppose that the duplicate is Mount Faverghera, because Mount Faverghera and the duplicate don't exist at any of the same times.

If something else could have had all of Mount Faverghera's qualitative properties, then pure descriptivism is false. For the pure descriptivist, there is some purely qualitative condition C such that necessarily: Mount Faverghera exists if and only if C is uniquely satisfied. If, however, something else could have had all of the qualitative properties of Mount Faverghera, then it might have happened that Mount Faverghera had a perfect qualitative duplicate in New Zealand, or in another epoch. In that case Mount Faverghera would have existed, but C would not have been uniquely satisfied. Also, it might have happened that Mount Faverghera did not exist but a perfect duplicate of it did and was the only thing to satisfy C. In that case C would have been uniquely satisfied, but Mount Faverghera would not have existed.

The argument just sketched is pleasingly general. Because it is not aimed at any particular analysis of 'Mount Faverghera', the (pure) descriptivist cannot blame the difficulties it brings to light on the (alleged) *analysans* of 'Mount Faverghera', as opposed to (pure) descriptivism as such. As far as I can see, the only way the (pure) descriptivist can block the argument is to insist, *more Leibniziano*, that only Mount Faverghera could have been exactly like Mount Faverghera.

At the risk of belabouring a point already made: note that, for the pure descriptivist, 'If Mount Faverghera exists, nothing else is just like it' is not only a necessary truth, but also one that can be known a priori, simply by reflecting on the meanings of its terms. This last claim is surely incredible. Suppose you present me with a mass of meticulously documented evidence supporting the claim that, amazingly enough, there is another mountain in New Zealand *just* like Mount Faverghera. I could hardly dismiss your evidence by noting that 'Another mountain in New Zealand is just like Mount Faverghera' is synonymous with something whose falsity may be determined a priori— that is, 'Some mountain in New Zealand which is not the (only) thing with such-and-such qualitative properties has all the same qualitative properties as the (only) thing having such-and-such qualitative properties.'[21]

Why doesn't Kripke avail himself of what might be called duplicability arguments against descriptivism? Perhaps because he does not want to pronounce

---

[21] I would be confident that no mountain in New Zealand is *exactly* like Mount Faverghera, even before looking at your evidence. But the grounds of my confidence would be empirical rather than a priori, and they would be macro-object-specific. Suppose I introduced a name 'Phi' that referred to a particular subatomic particle of a certain kind, or to a particular photon. I would be much less surprised to learn that something else on earth is just like Phi than I would be to learn that something else on earth is just like Mount Faverghera. But the pure descriptivist would have to say that 'Something else is just like Mount Faverghera' and 'Something else is just like Phi' are equally excludable on a priori semantic grounds.

on whether there are, for example, a set of purely qualitative conditions that are necessary and sufficient for, say, being Richard Nixon ('N & N', 268).

Suppose that pure descriptivism is a failed account of the reference of proper names. What sort of fallback position is available to the descriptivist? As we have seen, she might hold on to the idea that a proper name abbreviates, or is synonymous with, or is analysed by, a definite description, and give up the requirement that the definite description must be purely qualitative. As long as the impure descriptivist allows as *analysantia* non-qualitative definite descriptions of the sort, 'the individual who is actually F', she need not worry about Kripke's modal arguments. (Presuming that Aristotle is the most famous individual who taught Alexander philosophy, 'If Aristotle existed, then he is actually the most famous individual who taught Alexander philosophy' and 'If anyone is actually the most famous individual who taught Alexander philosophy, then Aristotle is' are necessary truths (on the right (rigidifying) reading of 'actually'.[22]) But she does still need to worry about Kripke's epistemological arguments. Even if we do not restrict our attention to purely qualitative descriptions, it is not clear that there is any (interesting) definite description 'the F' such that any competent user of the name 'Aristotle' will be in a position to know a priori that, if Aristotle exists, then Aristotle is the F, or that if anything is the F, Aristotle is. (Examples of *non*-interesting definite descriptions would be 'the individual who is identical to Aristotle' and 'the individual to whom "Aristotle" refers'.) It is not, for example, true that any competent user of the name 'Aristotle' is in a position to know a priori that, if Aristotle exists, then he is actually the most famous individual who taught Alexander philosophy, or to know a priori that, if someone is actually the most famous individual who taught Alexander philosophy, then Aristotle is.[23] Moreover, as we shall see in the next section, it may be that a (non-empty) name denotes its referent with respect to every possible world—including worlds in which its referent does not exist. But, it is plausible to suppose, no definite description of the form 'the F' denotes anything with respect to a possible world where nothing is uniquely F. If all this is right, then not even the definite description 'the individual who is identical to Aristotle' is synonymous with the proper name 'Aristotle'.

Alternatively, a descriptivist might give up the claim that names are analysable by any (qualitative or non-qualitative) definite descriptions, and hold on to the idea that names don't simply refer; they also describe. (As we have already seen, impure demonstratives such as 'this man' or 'this yellow thing' arguably describe; so the view that names are descriptive does not in any obvious way entail the idea that they are analysable by definite descriptions.)

This kind of impure descriptivism amounts to the denial of what Kripke calls the 'naive' view of names. While Kripke does not find this sort of weak

---

[22] See A. Plantinga, 'The Boethian Compromise', *American Philosophical Quarterly*, 15 (1978), 129–38.
[23] Salmon makes this point in *Reference and Essence*, 27.

descriptivism as initially plausible as the naive view, he does not attempt to refute it in *Naming and Necessity* (or, indeed, in any of his other works). That is because, as we shall see, he finds the 'naive' view of names attractive but problematic.[24]

2, RIGIDITY AND INTERSUBSTITUTABILITY

Although Kripke argues against a certain kind of theory of proper names, he does not attempt to replace that theory with a better one ('N & N', 300–3). But he does endorse certain theses about how proper names refer, and offer a certain picture of how reference is transmitted. In the next two sections I shall concentrate on the theses.

One of the principal (positive) theses of *Naming and Necessity* is that proper names are rigid designators. In *Naming and Necessity* and 'Identity and Necessity' Kripke defines a *rigid designator* as a term that designates the same object in any possible world in which that object exists ('N & N', 269–70; 'I & N', 145–6). If a term rigidly designates an object that exists in every possible world, Kripke says, we may call the term *strongly rigid* (ibid.).

How do we determine whether or not a term—say, 'Nixon'—is rigid? In *Naming and Necessity* and 'Identity and Necessity' Kripke suggests the following test for the rigidity of a term x: ask yourself whether something other than x might have been x. Or ask yourself whether x might have been other than (different from) the thing that is in fact x. The idea would seem to be that if the answer is no, then x is rigid; if the answer is yes, then x is non-rigid:

> One of the intuitive theses I will maintain in these talks is that *names* are rigid designators. Certainly they seem to satisfy the intuitive test mentioned above: although someone other than the U.S. President in 1970 might have been the U.S. President in 1970 (e.g. Humphrey might have), no one other than Nixon might have been Nixon. ('N & N', 270)

> We have a simple, intuitive test for [rigidity] ... We can say that the inventor of bifocals might have been someone other than the man who *in fact* invented bifocals. We cannot say, though, that the square root of 81 might have been a different number than it in fact is... If we apply this intuitive test to proper names, such as for example 'Richard Nixon', they would seem intuitively to come out to be rigid designators. ('I & N', 148–9)

In light of these remarks, it is not surprising that some philosophers have thought that, for Kripke, a rigid designator is just a designator that could not have designated anything except its actual referent.[25] But this cannot be right. Suppose that Abel's origin is essential to him, so that no one but Adam could

[24] Cf. *N & N*, introd. 20–1, and 'APAB'.

[25] See e.g. C. McGinn, 'Rigid Designation and Semantic Value', *Philosophical Quarterly*, 32 (1982), 97. McGinn offers 'the father of Abel' as an example of a rigid designator.

have been Abel's father. Then 'the father of Abel' could not have designated anything other than its actual referent. But it is not rigid, since it does not designate Adam in all the worlds in which Adam exists. (It fails to designate Adam in worlds in which Adam is childless).[26]

Let us call a designator *inflexible* if it couldn't have referred to anything different from the thing it actually refers to. Any rigid designator will be inflexible, but the converse does not hold. The intuitive tests described above appear to test for inflexibility. If we want to test for the rigidity of *x*, it seems we should ask whether *x* could have existed without being *x*. On that test 'Nixon' comes out rigid, but neither 'the number of the planets' nor 'the father of Abel' do.[27]

Given that 'Nixon' designates the same thing (Nixon) in every world where Nixon exists, what does it designate in worlds where he does not exist? There would seem to be two possibilities: Nixon, or nothing. If 'Nixon' designates Nixon even in Nixonless worlds, then 'Nixon' denotes Nixon in every possible world. In that case, 'Nixon' is as rigid as a term can get, and, in particular, as rigid as strongly rigid designators such as '0' and 'the (positive) square root of 81'. So Kripke's terminology suggests that 'Nixon' fails to designate in Nixonless worlds. Indeed, in 'Identity and Necessity' he appears to say that in a (possible) situation in which the referent of a rigid designator does not exist, the designator has no referent:

> When I use the notion of a rigid designator, I do not imply that the object referred to necessarily exists. All I mean is that in any possible world where the object in question *does* exist, in any situation where the object *would* exist, we use the designator in question to designate that object. In a situation where the object does not exist, then we should say the designator has no referent… ('I & N', 146)

Why might someone suppose (why might Kripke suppose) that 'Nixon' fails to designate anything in a possible world where Nixon does not exist? Kripke does not say; but the following is one possibility: designation is a relation between a term and an individual. A term cannot stand in a relation to something if that thing does not so much as exist. So, necessarily, a term designates an individual only if that individual exists. If, however, it is a necessary truth that 'Nixon' designates only if Nixon exists, then if Nixon had not existed, 'Nixon' would not have designated him. So, it might seem, in a counterfactual alternative in which Nixon does not exist, 'Nixon' does not designate him.

The trouble with the (last sentence of) this line of reasoning is that it blurs the distinction between two questions. We can ask whether a term would have designated that individual had that individual not existed. We can also ask whether a term (actually) designates an individual *with respect to* a possible

---

[26] Or, to avoid appeal to the essentiality of origin: 'The individual who is both (identical to) Adam and the father of Abel' does not designate anyone other than Adam in any possible world, but does not designate Adam in possible worlds in which Adam exists but is not the father of Abel. (In such worlds, it designates nobody at all.)

[27] Something very like this point is made in J. Almog, 'Naming Without Necessity', *Journal of Philosophy*, 20 (1986), n. 10.

world in which that individual does not exist. In other words, we can ask whether a term designates an individual when we speak of a possible world in which that individual does not exist. The answer to the second question can be affirmative even if the answer to the first one is not. So we can say that a proper name designates the same thing with respect to every possible world (even worlds in which that thing does not exist), in much the way that a proper name designates the same thing with respect to every time (even times at which the thing no longer exists, or did not yet exist). Indeed, we can say that a sentence such as 'Nixon does not exist' expresses a truth with respect to a possible world (or with respect to a time) just in case what 'Nixon' designates with respect to that world (or time) does not exist at that world (or time).[28]

Moreover, it seems congenial to Kripke's anti-descriptivism, and to the view of names Kripke considers pre-theoretically plausible, to suppose that a proper name designates the same individual with respect to every possible world. A non-rigid definite description designates whomever or whatever uniquely satisfies a certain condition. So there is no way of saying what such a description designates with respect to a given possible world, without, as it were, looking at that world to see what (if anything) uniquely satisfies the relevant condition there. If names are not disguised or abbreviated (non-rigid) descriptions—if moreover, their function is *simply* to refer—there is no obvious motivation for, as it were, leaving open the question of what a name designates, pending the choice of a possible world (or time).[29]

If designation with respect to a world is understood in the way just sketched, Kripke would not want to deny that (non-empty) proper names designate the same thing with respect to every possible world. For he notes that, when someone says, 'Suppose Hitler had never been born', 'Hitler' refers rigidly to something that would not exist in the counterfactual situation described ('N & N', 290). This remark suggests a 'Kaplanite' construal of rigidity, but it is not tied in with the definition of rigidity offered elsewhere in *Naming and Necessity*. In the preface to the second edition of *Naming and Necessity*, Kripke cites the relevant passage, and endorses the view that a proper name rigidly designates its referent 'even when we speak of counterfactual situations where that referent would not have existed' (*N & N*, preface, n. 21).[30]

Kripke's remarks on rigidity in the preface suggest an account on which there are three sorts of inflexible designators. A designator like 'the father of Abel' (or—if you have doubts about the essentiality of origin—'the individual that is both Carlo Zeno and lucky') is inflexible, but not rigid. A designator like 'the individual identical to Adam' is what we might call *quasi-rigid* or

---

[28] See Kaplan, 'Bob and Carol and Ted and Alice', app. x, for a very clear statement of the point.

[29] Cf. Salmon, *Reference and Essence*, 38–9 and n. 35.

[30] Even the passage from 'Identity and Necessity' cited above is perhaps less clearly antithetical to a Kaplanite construal of rigidity than might be supposed. I only cited part of the last sentence, and the full sentence reads: 'In a situation where the object does not exist, then we should say that the designator has no referent *and that the object in question so designated does not exist*' (my emphasis).

barely rigid, in that it designates Adam in all possible worlds except Adamless ones. And a designator like 'Adam' is *fully rigid*, in that it designates Adam in all possible worlds.

In his discussion of rigidity at note 21 Kripke introduces a distinction between de jure rigidity and de facto rigidity. A designator is de jure rigid if its referent is stipulated to be a single object, whether we are speaking of the actual world or a merely possible one; a designator is (merely) de facto rigid if it is a definite description of the form 'the F' that happens to contain a predicate 'F' that in each possible world is true of one and the same thing.

The distinction between de jure and de facto rigidity is not the distinction between full rigidity and bare or quasi rigidity, since the de facto rigid designator 'the smallest prime' is fully rigid (assuming that the number 2 is a necessary individual).[31] But in the footnote under discussion Kripke appears to say that de jure rigidity entails (what I have been calling) full rigidity.[32] Nor is the distinction between de jure rigid designators and de facto rigid designators the distinction between proper names and (rigid) definite descriptions, since some demonstratives or indexicals ('this', 'I') are, I take it, de jure rigid. But Kripke's characterization of de facto rigidity at least suggests that all and only de facto rigid designators are rigid definite descriptions. And Kripke holds that (all) proper names are de jure rigid.[33] It is, as it were, the job of a name to designate rigidly, and any designator of an actual or hypothetical language that is not rigid is so unlike the things we ordinarily call names that it should not be called a name (*N & N*, preface, n. 5). So it appears that, for Kripke, de jure rigid designators comprise names, demonstratives, and indexicals, while de facto rigid designators comprise (rigid) definite descriptions.

It is interesting to compare Kripke's characterization of de jure rigidity with Kaplan's (quite similar) characterization of direct referentiality:

I intend to use '*directly referential*' for an expression whose referent, once determined is fixed for all possible circumstances...For me the intuitive characterization is not that of an expression which *turns out* to designate the same object in all possible circumstances, but an expression whose semantical rules provide *directly* that the referent in all possible circumstances is fixed to be the actual referent.[34]

Kripke characterizes rigidity (and de jure rigidity) modally. We might, though, characterize rigidity more broadly, and regard what Kripke characterizes as a particular kind of rigidity. That is, we might say that a designator is (fully) rigid just in case it designates the same thing at every 'index of evaluation' (and quasi- or

---

[31] Also, Kripke says that a merely de facto rigid designator 'happens to use a predicate that in each possible world is true of one and the same unique object' (*N & N*). If a predicate can be true of an object in a world only if that object exists in that world, then only fully (and strongly) rigid designators are de facto rigid.

[32] 'Since names are rigid *de jure*...I say that a proper name rigidly designates its referent even when we speak of counterfactual situations where that referent would not have existed.'

[33] 'Clearly my thesis about names is that they are rigid *de jure*, but in the monograph I am content with the weaker assertion of rigidity' (*N & N*).   [34] Kaplan, 'Demonstratives', 12.

barely rigid just in case it designates the same thing at every 'index of evaluation' at which that thing exists). An index of evaluation might be a possible world, but it might also be a particular time. It might even be a place, if a term such as 'the spring of 2001' were thought of as designating different periods of time at different hemispheres. Actually, although there are temporally flexible designators, it is not clear that there are locally flexible designators. On some ways of filling in the context——*was not always (will not always be)*——, we get a truth; it is doubtful that the same goes for the context——*isn't*——*everywhere*. (It sounds at least odd, for example, to say (in the northern hemisphere), 'The spring of 2001 isn't the spring of 2001 in Australia.') Be that as it may, it seems that there could be languages with locally flexible designators and 'local tenses'. If there were, such languages could contain both locally rigid and locally non-rigid designators.

Why distinguish, say, temporal rigidity from modal rigidity? Well, a designator such as '7' is both temporally and modally rigid, and a designator such as 'my favourite food' would appear to be temporally as well as modally non-rigid. (Just as there is a true reading of *My favourite food might not have been my favourite food*, there seems to be a true reading of *My favourite food was not always my favourite food*).[35] But a designator such as 'the number you thought of at noon on 24 June 2002' is temporally rigid, but not modally rigid (compare the acceptable *The number you thought of at noon on 24 June 2002 might not have been the number you thought of at noon on 24 June 2002* with the unacceptable *The number you thought of at noon on 24 June 2002 wasn't always the number you thought of at noon on 24 June 2002*).

Kripke's principal positive semantic thesis about proper names is that they are de jure rigid designators. At this point it may be worth discussing a number of theses about proper names that are not Kripke's but have often been attributed to him.

First, Kripke does not hold (and never held) that there is nothing more to being a name than being a rigid designator, or being a de jure rigid designator. I would not belabour this point if Joseph Almog had not offered the following interpretation of Kripke: In *Naming and Necessity* Kripke proposed a 'modally oriented' account of names, on which a (genuine) name just was a rigid designator. He subsequently came to see that such an account would misclassify definite descriptions such as 'the only even prime' as names. So he offered a revised account, on which names are identified with de jure rigid designators—designators whose rigidity is a result of semantic stipulation, rather than a consequence of certain modal truths. Almog notes that

In fairness to Kripke, he can deny . . . that he intended the notion of 'rigid designator' to serve as an explication of the intuitive idea of 'genuine name'. Instead, he might say, he

---

[35] Compare *The Pope has always been Polish* with *(From the dawn of the republic) the President has always been the commander-in-chief of the army*. The former comes out true only if 'the Pope' is taken as temporally rigid; the latter only if 'the President' is taken as temporally non-rigid (the (current) President wasn't the commander-in-chief at the dawn of the republic, since he hadn't been born yet).

introduced rigidity as part of his attempt to refute the Frege–Russell description theory. How so? By showing that names are rigid, and the descriptions they supposedly abbreviate, are not. This interpretation of Kripke's work, even if it made sense, would have to be weakened (considerably) to get off the ground. Obviously, there are many rigid descriptions, sometimes even 'easy-to-come-by' transforms of ordinary descriptions, and names could abbreviate *them*. The view would now be that Kripke attempted to refute only the claim that the *original* descriptions of Frege and Russell 'give the meaning' of names. I do not know quite what to say of this interpretation. It seems (to me) incredibly weak: the description theory may turn out to be true after all, names do refer by way of (rigid) senses, the only trouble with the theory was this or that description used by Frege (Russell) to 'give the meaning' ... Such an interpretation of Kripke's work seems to me literally incredible.[36]

There is much to take issue with here. To start with, as I have tried to show, the descriptivism Kripke targets implies that for each name there is a *purely qualitative* definite description that is intensionally equivalent to it, and it is by no means clear that for each name there is a corresponding purely qualitative intensionally equivalent definite description.

Almog is right to suppose that, even if names are rigid, they might still refer by way of (rigid) senses. But, for the description theory, as Kripke characterizes it, to be true, it would have to be not just necessarily true, but also knowable a priori by a competent user of the name *n* that if *n* exists, then *n* is the F, and if exactly one thing is the F, *n* is (cf. again 'N & N', 285). While it is a doddle for a (sufficiently reflective) user of the name *n* to come up with a (rigidified) definite description that is intensionally equivalent to *n* (of the form *the actual* F, where *n* is uniquely F), Kripke has argued convincingly that there may well be no definite description 'the F' such that the competent user of the name *n* is in a position to know a priori that, if *n* exists, then *n* is the F, or that if exactly one thing is the F, then *n* is—assuming we exclude (as descriptivists want to do) non-qualitative definite descriptions that are trivially intensionally equivalent to *n*, such as 'the individual identical to *n*' or the like.

So it is simply not true that, on the interpretation of Kripke Almog rejects, descriptivism as such is left untouched, and only particular hypotheses about which descriptions provide the meaning of proper names are discredited.[37]

Furthermore, to my knowledge Kripke never says, or even suggests, that there is nothing more to being a name than being a rigid (or de jure rigid) designator. As we have seen, he often says that names *are* rigid (and, in the preface to the second edition of *Naming and Necessity*, says that names are de jure rigid). This, however, no more implies that 'namehood' (nominality?) is rigidity (or de jure rigidity) than 'horses are animals' implies that equinity is animality. Also, Kripke explicitly says that, as he uses the term, 'name' applies only to those things which in ordinary parlance would be called proper names ('N & N', 254).

---

[36] Almog, 'Naming Without Necessity', n. 14.

[37] In any case, as we shall see, Kripke does not want to definitely exclude the possibility that (some) names do refer via something like a rigid sense; see 'APAB' n. 43.

Since at the time Kripke gave the 'Naming and Necessity' lectures, he thought that some non-names (e.g. 'the positive root of 81') were rigid, he could hardly have thought at that time that names just were rigid designators.

In fairness to Almog, what he attributes to Kripke is not the thesis that all and only names (in the ordinary sense of the term) are rigid (or de jure rigid) designators, but the thesis that all and only 'genuine names' or 'genuine naming devices' are rigid (or de jure rigid) designators. That said, Kripke nowhere draws a distinction between names in the ordinary sense (which are never definite descriptions, or demonstratives, or indexicals), and 'genuine names' (which sometimes are).

Finally, suppose that, at the time of the composition of *Naming and Necessity* and 'Identity and Necessity' Kripke thought that namehood and rigidity came to the same thing. In that case he surely would have flatly rejected the 'naive view' of proper names, on which all a name does is refer. (It is obvious that some rigid definite descriptions describe their referent and don't simply refer to it.) And Kripke does not in fact reject the naive view of names in either *Naming and Necessity* or 'Identity and Necessity'.

Indeed, many philosophers have taken *Naming and Necessity* to be a defence of a naive (or 'Millian') account of names.[38] In a certain sense, it is. Kripke clearly finds that account natural and initially plausible (cf. the passage cited at the beginning of this chapter), and the first two lectures of *Naming and Necessity* are a sustained attack on what Kripke considers the best-known and best-worked-out alternative to the Millian account. Moreover, in at least one place Kripke says explicitly that Mill was right to maintain that proper names (unlike definite descriptions) have denotation but not connotation:

Mill...held that although some 'singular names', the definite descriptions, have both denotation and connotation, others, the genuine proper names, had denotation but no connotation. Mill further maintained that 'general names', or general terms, had connotation.... The modern logical tradition, as represented by Frege and Russell, disputed Mill on the issue of singular names, but endorsed him on that of general names. Thus *all* terms, both singular and general, have a 'connotation' or Fregean sense.... The present view, directly reversing Frege and Russell, *endorses* Mill's view of *singular* terms, but *disputes* his view of *general* terms. ('N & N', 327)

Suppose we thought of the Millian thesis that names lack connotation as equivalent to the naive (and Millian) view that all names do is refer. Then we might naturally read the above passage as an endorsement of the naive view. It appears, though, that in that passage Kripke is equating a Millian connotation with a Fregean sense. If connotation is so understood, the thesis that names lack connotation does not entail the naive (and Millian) view that names are purely referential.

[38] See e.g. C. Travis, 'Are Belief Ascriptions Opaque?', *Proceedings of the Aristotelian Society*, 97 (1993), 97–8, where Travis speaks of the 'Mill–Kripke view of names'. See also J. Katz, 'Names Without Bearers', *Philosophical Review*, 103 (1994), 787–825, where Katz characterizes Kripke as a Millian about proper names (i.e. one who claims that names lack sense and linguistic meaning).

In any case, Kripke elsewhere stops short of endorsing the no-connotation thesis for proper names: at 'N & N', 322, he avers that Mill was 'more-or-less right' in holding that proper names, unlike definite descriptions, are non-connotative. Also, in note 58 of *Naming and Necessity* Kripke considers the view that each name is associated with a sortal that is (in some sense) a part of the meaning of that name. A view of this sort—which attributes descriptive senses to names, but not (necessarily) ones that determine their referents—is clearly inconsistent with the naive and Millian account of names: but Kripke says he does not need to take a position on its correctness. *Naming and Necessity* articulates and defends an account of names that is, as Kripke says, closer to the Millian one than to a Fregean one; it neither endorses nor rejects a purely Millian account.[39]

Kripke is also sometimes thought to have asserted, or at any rate to be committed to, the thesis that co-referential proper names are intersubstitutable *salva veritate* in doxastic and epistemic as well as modal contexts. One can see why someone who took Kripke to be committed to the view that names are purely referential might take him to be committed to such intersubstitutability. If the meaning of a name simply is its referent, then sentences such as 'Cicero is bald' and 'Tully is bald' should mean the same thing. (After all, the meaning of a whole sentence is determined by the meanings of its parts and the way those parts are put together.) If 'Cicero is bald' and 'Tully is bald' mean the same thing, it seems, then the proposition the first sentence expresses (*that Cicero is bald*) should be the same proposition as the proposition the second sentence expresses (*that Tully is bald*). (Compare: If 'Cicero is bald' means the same thing as 'Cicerone è calvo', then the proposition that 'Cicero is bald' expresses (in English) is the same proposition as the proposition that 'Cicerone è calvo' expresses (in Italian). If *that Cicero is bald* and *that Tully is bald* are the same proposition, then it is impossible to stand in the believing relation to either without standing in the believing relation to the other. So, it would seem, someone believes that Cicero is bald if and only if she believes that Tully is bald; cf. 'APAB', 241–2.)

As we have seen, though, Kripke is tempted by, but not committed to, the claim that names are purely referential. He is accordingly not committed to the intersubstitutability *salva veritate* of co-referential terms in doxastic or epistemic contexts.[40] Because Kripke's views on intersubstitutability have evolved, it may be worth saying something about that evolution.

In *Naming and Necessity* and 'Identity and Necessity' Kripke does not—so far as I know—explicitly assert that co-referential names are not (always) intersubstitutable *salva veritate* in contexts involving knowledge and belief. On the other hand, Kripke grants that we could discover that Hesperus and

---

[39] See Kripke, 'APAB', esp. sect. I.

[40] Kripke is committed to (and endorses) the intersubstitutability of co-referential proper names in modal contexts (cf. 'APAB', sect. I).

Phosphorus are different planets: it is very unlikely, but not (epistemically) impossible, that astronomers have misidentified Hesperus with Phosphorus ('I & N', 143). And Kripke would presumably deny that we could discover that Phosphorus and Phosphorus are different planets. So it is natural to attribute to the Kripke of 'Identity and Necessity' the view that 'Hesperus' and 'Phosphorus' are co-referential names that are non-intersubstitutable in the (epistemic) context *we could discover that——is not Phosphorus*. Similarly, in *Naming and Necessity*, Kripke emphasizes that Hesperus and Phosphorus might have turned out to be different planets, and would presumably deny that Hesperus and Hesperus might have turned out to be different planets. So it is natural to regard *Naming and Necessity*, along with 'Identity and Necessity', as implicitly, if not explicitly, committed to the denial of the inter-substitutability *salva veritate* of proper names in doxastic and epistemic con-texts.[41] If this is right, there is a certain tension in *Naming and Necessity* and 'Identity and Necessity': both works show sympathy for (if not commitment to) a purely referential account of names, and acceptance of a non-intersub-stitutability thesis apparently uncongenial to that account.

In 'A Puzzle About Belief' Kripke discusses the relation between the inter-substitutability thesis, Millianism about names, and the account of names offered in *Naming and Necessity*. He grants that, if all names do is refer, then it looks as though co-referential proper names should be interchangeable in doxastic and epistemic contexts. And he grants that the apparent failure of such interchangeability is a difficulty, not just for Millianism, but for 'the Millian spirit' of his account of names in *Naming and Necessity* ('APAB', n. 10).

Faced with this difficulty, one might draw a number of conclusions. One would be that names are not purely referential: they must have something like a Fregean sense, even if that sense is rigid and not purely qualitative, and even if that sense is not sufficient to determine a referent. Another would be that the apparent non-interchangeability of co-referential names in doxastic and epistemic contexts is an illusion.[42] Yet another would be that, since names are purely referential and co-referential names are not interchangeable in doxas-tic or epistemic contexts, Millianism does not after all entail (what I shall call, for the sake of brevity) the intersubstitutability thesis. Interestingly, Kripke draws none of these conclusions. Indeed, he thinks that it would be unwise to

---

[41] See *N & N*, introd., 20, where Kripke says that 'Hesperus is Phosphorus' can be used to raise an empirical issue, while 'Hesperus is Hesperus' cannot.

[42] Before *Naming and Necessity* it was often thought that 'Hesperus' and 'Phosphorus' were not interchangeable *salva veritate* in the modal context, *Necessarily,——is Phosphorus*. Kripke convinced most people that such non-interchangeability was merely apparent. Mightn't the same hold for the non-interchangeability of 'Hesperus' and 'Phosphorus' in doxastic and epistemic contexts? Perhaps. But note that, in explaining away the apparent non-interchangeability of 'Hesperus' and 'Phosphorus' in a context such as *It could have been that:——is not Phosphorus*, Kripke seems to presuppose that 'Hesperus' and 'Phosphorus' are *genuinely* non-interchangeable in the (epistemic) context *It could have turned out that——is not Phosphorus*.

draw *any* significant theoretical conclusions about proper names from the apparent failure of the intersubstitutability thesis ('APAB', 271).

Why so? Because (Kripke thinks) we can get the same kind of paradoxical conclusions that follow from the intersubstitutability thesis from some other very plausible principles, and a very plausible claim, without appealing to that thesis.

The principles are:

>(DQ) If a normal speaker of English, upon reflection, sincerely assents to '*p*', then he believes that *p*

and

>(TR) If a sentence of one language expresses a truth in that language, then any translation of it into any other language also expresses a truth (in that other language).

('DQ' stands for 'disquotation' and 'TR' for 'translation'.)

The claim is:

>(C) 'London is pretty' is a translation of 'Londres est jolie'.

Suppose, Kripke says, that Pierre is a normal speaker of French. Pierre has heard of a certain city—one that we call 'London' and he calls 'Londres'—and believes that it is pretty. He consequently sincerely and reflectively assents to 'Londres est jolie'. If DQ is true, so is the counterpart principle for French. So the following sentence of French expresses a truth: 'Pierre croit que Londres est jolie.' In that case, by TR and C, 'Pierre believes that London is pretty' expresses a truth. By 'semantic descent', we may conclude that Pierre believes that London is pretty.

Suppose, Kripke continues, that Pierre later moves to (an unattractive part of) London. Since Pierre arrives in London not knowing any English, and since there are no French speakers in his part of London, he becomes a normal speaker of English by 'the direct method' of language acquisition. One of the things he learns, on his way to becoming a normal speaker of English, is that the city in which he is now living is called 'London'. Since he believes that the city in which he is now living is ugly, he does not, upon reflection, sincerely assent to 'London is pretty'. In fact, he sincerely and reflectively assents to 'London is not pretty'. So, by DQ, Pierre believes that London is not pretty. Suppose, though, that Pierre still sincerely and reflectively assents to 'Londres est jolie': he takes it for granted that the city called 'Londres' and the city called 'London' are different cities. Then, it would seem, Pierre believes both that London is pretty and that London is not pretty. If so, he has contradictory beliefs. But, Kripke says, this seems wrong: if you have contradictory beliefs, logical acumen could reveal this fact to you, and no amount of logical acumen could reveal to Pierre that he has contradictory beliefs.

So, Kripke concludes, we have a puzzle. DQ and TR look unexceptionable. But from those principles, together with (the apparently unimpeachable) C, we get the (intuitively unacceptable) consequence that Pierre believes both that London is pretty and that London is not pretty. Moreover, given C, TR, and the plausible

> (SDQ) A normal English speaker who is not reticent will be disposed to sincere reflective assent to 'p' if and only if he believes that p

('SDQ' stands for 'strengthened disquotation'), it follows that Pierre both does and does not believe that London is pretty! (Given Pierre's indisposition to assent sincerely and reflectively to 'London is pretty', we get that he does not believe that London is pretty; given his disposition to assent sincerely and reflectively to 'Londres est jolie', the (French counterpart of) SDQ, and TR, we get that he *does* believe that London is pretty.)

What does all this have to do with intersubstitutability? As Kripke sees it, the falsity of the intersubstitutability thesis must be argued for, not assumed. The following is a plausible line of argument: suppose that Jones (sincerely, reflectively) assents to both 'Cicero is bald' and 'Tully is not bald'. Then (by DQ) he believes that Cicero is bald and that Tully is not bald. If the intersubstitutability thesis were true, it would follow that Jones has contradictory beliefs: he would believe that Cicero is bald and that Cicero is not bald. But, intuitively, Jones does not have contradictory beliefs; so the intersubstitutability thesis is false.

Kripke objects: Given DQ and the intersubstitutability thesis, we get the counter-intuitive result that Jones has contradictory beliefs. But given DQ and the (apparently unassailable) ancillary premisses TR and C, we get the equally counter-intuitive result that Pierre has contradictory beliefs. So it is no good arguing from DQ and the premiss that Jones does not have contradictory beliefs to the falsity of the intersubstitutability thesis.

It may be worth stressing that Kripke is *not* arguing here that, upon reflection, we can hold on to intersubstitutability (and DQ, TR, and C), and live with the initially counter-intuitive consequence that Jones and Pierre have contradictory beliefs. While I would describe that consequence as 'initially counter-intuitive', Kripke goes much further: the intersubstitutability thesis, 'when combined with our normal disquotational judgments of belief, leads to straightforward absurdities. But...the "same" absurdities can be derived by replacing the interchangeability principle with our normal practices of translation and disquotation...' ('APAB', 255).

As Kripke sees it, DQ and TR are apparently self-evident, and C is obvious. Yet we can derive 'straightforward absurdities' from DQ, TR, and C. He concludes that Pierre's case is one in which our normal practices of belief attribution are 'placed under the greatest strain' and 'may even break down'. And the same goes for Jones's case ('APAB', 271). In light of that strain, we should regard the intersubstitutability thesis (like DQ) as problematic, but not as definitely false.

Here an anti-Millian might raise a number of questions. She might start by querying the claim that the Jones case, like the case of Pierre, places our

normal practices of belief attribution under strain. As Kripke brings out, there
do not seem to be any unproblematic answer to the questions 'Does Pierre
believe that London is pretty?', 'Does Pierre believe that London is not pretty?'
By contrast, the anti-Millian might say, there are perfectly unproblematic
answers to the questions 'Does Jones believe that Cicero was bald?', 'Does
Jones believe that Tully was bald?' Or maybe there aren't; but this would need
to be shown. And, the anti-Millian might say, it cannot be shown simply by
noting that Jones's situation is strikingly like Pierre's (in involving two names
that actually refer to the same thing but are taken to refer to different things).

Secondly, the argument against intersubstitutability that Kripke considers
presupposes DQ. But it looks as though the opponent of intersubstitutability
can argue against it without invoking DQ. For example, she could say:

> Suppose Jones discovers to his surprise that Cicero and Tully are the same
> person. If asked, 'Did you use to believe that Cicero and Tully were differ-
> ent persons?', he will answer yes. If asked, 'Did you use to believe that Cicero
> and Cicero were different persons?', he will answer no. Since Jones's answers
> are correct, proper names are not always interchangeable in contexts involv-
> ing knowledge and belief.

Whatever its merits, this argument does not presuppose DQ.[43]

Furthermore, if the opponent of intersubstitutability is a descriptivist, she
may well have good reason not to invoke DQ. As Kripke has emphasized, a
(classical) descriptivist may allow that the sense I assign to a name may not
be the sense you assign to it (or else, that the sense I assign to a name may
not be the sense you assign to a homophonic and co-referential name).
Suppose that S associates 'Cicero' with sense σ, and S' associates 'Cicero' with
a different sense σ^*. Then S cannot infer that S' believes that Cicero is F from
the fact that S' sincerely and reflectively assents to 'Cicero is F'.

For much the same reason, Kripke notes, a descriptivist may well reject C.
Pierre, she may say, assigns one sense to 'London' and another sense to
'Londres'. So 'London is pretty' (in Pierre's English idiolect) means something
different from—is not (strictly speaking) a translation of—'Londres est jolie'
(in Pierre's French idiolect).

So, the descriptivist might say, DQ, C, and the intersubstitutability thesis are
all false: Pierre's case is sufficient to establish the falsity of the conjunction of C
and DQ, and Jones's is sufficient to establish the falsity of the intersubstitutability thesis. It may initially appear difficult to gainsay the conjunction of DQ and
C—but only until we see the 'straightforward absurdities' acceptance of that
conjunction commits us to. (In this sense, the descriptivist might say, Pierre's

---

[43] That the denier of intersubstitutability need not invoke DQ is pointed out by D. Sosa in 'The
Import of "A Puzzle About Belief"', *Philosophical Review*, 105 (1996), 381–2. If a denier of intersubstitutability did appeal to DQ, I imagine that she would do so when asked to justify her assumption that,
say, Jones believes that Cicero was bald and that Tully was not. Even then, she might leave DQ out of it,
and simply appeal to the ways we ordinarily report and do not report our own and each other's beliefs.

case is grist for her mill.) As for the intersubstitutability thesis, it doesn't even have the sort of prima facie unassailability that DQ and C have.

So, the descriptivist may continue: I can, but the Millian cannot, avoid saying paradoxical things about Pierre's or Jones's doxastic situation, because I can, and the Millian cannot, treat the pairs of terms ⟨'London', 'Londres'⟩, and ⟨'Cicero', 'Tully'⟩ as non-synonymous: both Pierre's case and Jones's favour a descriptivist account of names over a purely referential one.

Kripke objects that the descriptivist cannot in fact avoid the paradoxes of Pierre. Suppose that in his monolingual French period Pierre learns the name 'Londres', and associates it with a certain set of uniquely identifying propert-ies—where these do *not* include being pretty. Suppose also that, after coming to England, Pierre learns the name 'London' and associates it with the very same set of uniquely identifying properties. Then, given that the uniquely identifying properties associated with a name fix its sense, the descriptivist must say that 'London' (in Pierre's English idiolect) and 'Londres' (in Pierre's French idiolect) have the same sense.

Even if Pierre associates 'Londres' and 'London' with the same set of uniquely identifying properties, Kripke insists, he may not be aware of this fact. When asked (in French) which uniquely identifying property he associates with 'Londres', he replies (say), 'l'être la ville où la reine d'Angleterre vive'. When asked (in English) which uniquely identifying property he associates with London, he replies 'being the city where the queen of England lives'.[44] Still, he need not identify the person he calls 'la reine d'Angleterre' (in French) with the person he calls 'the queen of England' (in English), or identify the property he calls 'l'être la ville où la reine d'Angleterre vive' (in French) with the property he calls 'being the city where the queen of England lives' (in English). If he doesn't, then he may give sincere reflective assent to both 'Londres est jolie' and 'London is not pretty', in spite of associating 'Londres' and 'London' with the same sense.

If this is right, then the descriptivist can't after all avoid the problems Pierre raises for the Millian. On her account, there should be no problem about dis-quotation within an idiolect. So if Pierre sincerely and reflectively assents to 'Londres est jolie', 'Pierre croit que Londres est jolie' is true in Pierre's French idiolect. And if Pierre sincerely and reflectively assents to 'London is not pretty', 'Pierre believes that London is not pretty' is true in Pierre's English idi-olect. Moreover, if 'London' and 'Londres' have the same sense, then 'Pierre croit que Londres est jolie' (in Pierre's French idiolect) can be translated into 'Pierre believes that London is pretty' (in Pierre's English idiolect). So, if we build into the Pierre story that he assigns the same sense to 'London' and 'Londres', then (it is true in Pierre's English idiolect that) Pierre believes that London is pretty and that London is not pretty, and we are back to Kripke's 'straightforward absurdities'.

[44] Note that, just as 'London' and 'Londres' intuitively mean exactly the same thing, the sense-providing descriptions 'the queen of England' and 'la reine d'Angleterre' intuitively mean exactly the same thing.

But could a sufficiently reflective Pierre who assigns the same sense to 'London' and 'Londres' be unaware that he does? I take it that the pure descriptivist would say not. On her view, each name is a priori equivalent to and synonymous with a purely qualitative definite description. So, the descriptivist would say, if a speaker is asked whether two of the names in her idiolect (or her idiolects) n and n' have the same sense, she can determine a priori which purely qualitative definite description n is synonymous with, which purely qualitative definite description n' is synonymous with, and whether or not the two definite descriptions make reference to the same purely qualitative condition. (The assumption is that if P and P$'$ are purely qualitative properties, it can be determined a priori whether P = P$'$.) In that case, if Pierre associates 'London' and 'Londres' with the same sense, sufficient reflection will reveal this to him. As a corollary, if Pierre associates both 'London' and 'Londres' with the same sense, and Pierre (sincerely) assents to both 'Londres est jolie' and 'London is not pretty', his assents are not (all) sufficiently reflective to betoken belief. As Kripke seems to grant, the pure descriptivist need not lose sleep over Pierre (see 'APAB', sect. III).

We have already seen in Section 1, though, that purely descriptivist accounts of proper names are not viable. So the question becomes whether an impure descriptivist can exclude the possibility that Pierre sincerely and reflectively assents to both 'Londres est jolie' and 'London is not pretty', even though he assigns the very same sense to 'Londres' and 'London'.

Here the impure descriptivist might say: It is only because 'Cicero' and 'Tully' have different senses (in Jones's idiolect) that Jones can sincerely and reflectively assent to 'Cicero was bald' and 'Tully was not bald', in spite of the co-referentiality of 'Cicero' and 'Tully'. *Pari ratione*, it is only because 'London' and 'Londres' have different senses (in Pierre's two idiolects) that Pierre can sincerely and reflectively assent to 'Londres est jolie' and 'London is not pretty', in spite of the co-referentiality of 'Londres' and 'London'.

The impure descriptivist is within her rights to use 'sense' in such a way that anyone in a situation like Jones's or Pierre's must assign different senses to the relevant names. Jones and Pierre are in the situation they are in only because they (in some sense) conceive of Cicero and Tully, or London and Londres, in different ways. If the descriptivist wants to call Pierre's conception-of-Londres his 'Londres'-sense, and Pierre's (different) conception-of-London his 'London'-sense, she is perfectly free to do so. But then she will face a dilemma. Are senses so conceived meanings, or not? If they are not, then the claim that 'London' and 'Londres', or 'Cicero' and 'Tully' have different senses is perfectly compatible with the Millian insistence that 'the first and only meaning' of a name is its referent.[45] Suppose, on the other hand, that senses so conceived are meanings. It seems possible that, when Pierre originally learns the name

---

[45] The phrase is taken from David Kaplan, who says that a variable's first and only meaning is its value ('Demonstratives', 2).

'London', he associates the same sense (the same uniquely identifying properties) with 'London' that he already associated with 'Londres'—although he is unaware of that fact, for the reasons adduced above. And it seems perfectly compossible with this last supposition that Pierre subsequently comes to assent sincerely and reflectively to 'London is F', in spite of his continuing indisposition to sincerely and reflectively assent to 'Londres est F' (where 'F' is the translation into French of the English predicate 'F'). For example, it could be that the sense that Pierre first associates with 'Londres' and subsequently with 'London' is neutral with respect to London's pulchritude, and that Pierre (after having assigned a sense to 'London') comes to assent sincerely and reflectively to 'London is not pretty' while remaining indisposed to assent sincerely and reflectively to 'Londres est jolie'. If senses are meanings, as well as whatever makes it possible for Pierre to assent sincerely and reflectively to 'London is F', but not 'Londres est F' (in spite of the co-referentiality of 'London' and 'Londres'), it follows that what Pierre meant by 'London' before he came to assent sincerely and reflectively to 'London is not pretty' is different from what Pierre meant by 'London' afterwards. Surely, though, Pierre doesn't start meaning something different by 'London' just because he comes to believe (as he would put it) 'that London is not pretty'!

So, the impure descriptivist must say, there are two possibilities. It may be that Pierre initially assigned to 'London' a sense that had, as it were, already been taken by 'Londres'. In that case, Pierre is in a position to ascertain that fact a priori, and his subsequent disposition to assent sincerely to 'London is pretty' but not 'Londres est jolie' is unreflective. Or it may be that Pierre's subsequent dispositions are (fully) reflective, in which case the sense Pierre initially assigned to 'Londres' (and still assigns to it) was different from the sense he initially assigned to 'London' (and still assigns to it). What is impossible is that Pierre (unwittingly) initially assigned the same sense to 'London' and 'Londres' but can only ascertain that fact a posteriori.

But why is this impossible? Hasn't Kripke shown that the descriptivist ought to regard it as possible? The pure descriptivist has an answer to this question, but it is unavailable to the impure descriptivist. So she might try a different tack. Suppose that the sense Pierre assigns to 'Londres' somehow has the *name* 'Londres'—but not the name 'London'—'built into it'; and that the sense Pierre assigns to 'London' has the name 'London'—but not the name 'Londres'—built into it.[46] Then, the thought would be, the sense of 'London' and the sense of 'Londres' will be distinct, and their distinctness will be ascertainable a priori. All Pierre needs to do is reflect, to see that his 'London'-sense and his 'Londres'-sense 'contain' different names and are accordingly different senses.

---

[46] For example, the uniquely identifying property associated with 'London' might be *being a city named 'London'*, and the uniquely identifying property associated with 'Londres' might be *being a city called 'Londres'*.

This way of avoiding the paradoxes of Pierre has some very unattractive consequences. To start with, as Kripke emphasizes, it is difficult to believe that 'London' does not mean 'Londres' (and that there could not be a word of French that was phonetically discernible from 'London' and nevertheless meant 'London').[47] Also, it seems that Pierre-type paradoxes can arise involving natural kind terms. If the descriptivist does not use the same (metalinguistic) strategy to avoid them, she won't avoid (all) the paradoxes of Pierre. If she does use that strategy, she will have to say that 'lapin' does not mean 'rabbit', and that there could not have been a French word that was phonetically discernible from 'rabbit' and nevertheless meant 'rabbit'. She may even have to say that there could not have been a French word that was phonetically discernible from 'hot' but meant 'hot'. In short, if the metalinguistic strategy provides a general solution to Pierre-type problems, it does so at the (prohibitive) cost of making much or most of what we say in English untranslatable into French ('APAB', sect. III and n. 36).

We can also ask whether the descriptivist is thinking of the name 'London' as something that is a name of just one thing (the capital of England), or thinking of it as something that is a name of many things (one in England, another in Ontario, etc.). Suppose she is thinking of it in the first way. Then it is by no means clear that whether two senses have the same names or different names 'built into them' will always be ascertainable a priori. Suppose we modify the Pierre story as follows: when Pierre was very young, his English-speaking aunt taught him English and sometimes talked to him about a city called 'London'. Much later in life Pierre moves to England and learns that the city he now lives in is called 'London'. He remembers (dimly) that his great-aunt used to tell him about a city called 'London'. But he remembers almost nothing about what his great-aunt told him about London, and he knows that there are 'lots of Londons'. So he wonders whether London (the city his great-aunt used to talk to him about) is London (the city he now lives in). Suppose that the sense Pierre assigns to 'London' (the city his great-aunt used to talk to him about) contains a name, and that the sense Pierre assigns to 'London' (the city he now lives in) contains a name. Do they contain the same name or different names? Presuming that names are individuated in such a way that one name can only have one bearer, it depends. If the city that Pierre's great-aunt told him about was London, England, they contain the same name. If it was London, Ontario, they contain different names. The crucial point is that there is no reason to think that Pierre will be able to

---

[47] We unhesitatingly translate (a) 'She knows that Florence is in Tuscany' into Italian as (b) 'Sa che Firenze è in Toscana'; we do not translate (c) 'She knows that the city called "Florence" is in Tuscany' into Italian as (d) 'Sa che la città' che si chiama "Firenze" è in Toscana' (for she might know that a city called 'Florence' is in Tuscany without knowing that a city called 'Firenze' is in Tuscany; she might believe that the city called 'Firenze' is in the Veneto). If 'Florence' means 'the city called "Florence"', and 'Firenze' means 'la città' che si chiama "Firenze"' (so that (a) is synonymous with (c), and (b) with (d)) how is this difference to be explained?

determine a priori (by 'sense-inspection') whether or not he is using two 'London'-names or just one.[48]

Nor would it help the descriptivist to suppose that 'London' is one name with many bearers. Why did the descriptivist build names into the senses of 'London' and 'Londres' in the first place? Because it seemed that if the 'London'-sense and the 'Londres'-sense did not have qualitatively different 'components', Pierre might not be able to tell a priori whether the 'London'-sense was the 'Londres'-sense. Suppose the descriptivist thinks of senses as having not just qualitative components but also names built into them— where names are thought of as multireferential. If she says that senses are identical if they have all the same qualitative components and all the same names built into them (in the same way), then it will turn out that sense no longer determines reference. (Think of a symmetrical universe in which everything has a qualitatively indiscernible duplicate on the other side of the universe and a language in which any name of any thing on one side of the universe is also a name of that thing's duplicate on the other side of the universe.) Since the descriptivist holds that sense determines reference, she will have to say that senses can be distinct, even if they have all the same qualitative components and all the same (multireferential) names built into them (in the same way). Whether or not we have two senses or one can depend on facts concerning the non-qualitative and 'non-nominal' components of senses (whatever exactly those might be). And in certain Pierre-type cases these last facts will not be ascertainable a priori. So once again it could turn out that (i) the sense σ Pierre associates with name n is in the fact sense σ^* he associates with the name n', and (ii) Pierre, in spite of being a competent user of n and n', has no a priori access to the identity of σ with σ^*. Once again, Pierre spells trouble for the descriptivist as well as for the Millian.[49]

---

[48] Someone may protest that in the case under discussion Pierre can tell a priori that he has two 'London' names, because he can tell a priori that he associates just one of those names with a sense that includes *being the city my great-aunt told me about*. Again, though, it seems very implausible that 'being the city my great-aunt told me about' is part of what one of Pierre's 'London'-names *means*. Consider a simpler case. When Pierre is a child, his great-aunt often talks to him about a city called Bruges. For a while, Pierre thinks of Bruges as the city that has such-and-such properties (the ones his great-aunt has told him about). Later on he forgets what those properties were, and thinks of Bruges as the city his great-aunt talked to him about as a child. As an adult discussing his childhood with his sister, he learns that his great-aunt took his sister on holiday to Bruges. He eventually forgets that his great-aunt used to tell him about Bruges, and thinks of Bruges as the beautiful city that his great-aunt took his sister on holiday to. Some years later he finds out that Bruges is in Belgium (and not Holland, as he had thought) and spends a weekend there ... It is very implausible to suppose that, in the story just told, Pierre has a sequence of (homophonic but non-synonymous) 'Bruges'-names, one of which means 'the city my great-aunt talked to me about as a child'.

[49] Kripke discusses a related 'one-name' version of the Pierre case involving Peter, who sincerely assents to 'Paderewski had musical talent' but not to 'Paderewski had musical talent', because he does not realize that Paderewski (the statesman) is Paderewski (the musician). He concludes that, in order to avoid Paderewski as well as Pierre paradoxes, the descriptivist would have to reject DQ, or at least reject a homophonic translation of the name 'Paderewski' from Peter's idiolect into ours ('APAB', n. 37).

Kripke's defence of Millianism against what might be called 'the argument from non-intersubstitutability' is a defence in a rather qualified sense. He admits that Millianism apparently commits us to a thesis of intersubstitutability that (together with apparently self-evident ancillary premisses) has strongly counter-intuitive (for Kripke, 'straightforwardly absurd') consequences. So reflection on intersubstitutability should certainly—to some extent—shake whatever faith we (antecedently) had in Millianism. But this should not lead us to under-appreciate the subtlety and interest of that 'defence'. Kripke succeeds, I think, in showing that it is by no means obvious, and by no means as clear as one might have thought, that the (apparent) failure of substitutivity in belief contexts decisively favours Fregean theories of names over Millian ones.

3, REFERENCE AND CAUSALITY

For a (pure) descriptivist, when someone uses a (non-empty) name, that name has as its referent (on that occasion) the thing that uniquely satisfies (or comes sufficiently close, and closer than anything else to satisfying) a (purely qualitative) condition associated with that name.[50] If, as Kripke has argued, this is a fundamentally wrong-headed picture of how names refer, what should we put in its place?

Well, says Kripke, a name is typically introduced via a baptism. This may be a baptism by ostension, or by a reference-fixing description ('N & N', 302). In the case of baptism by ostension, the baptizer is typically in more or less direct perceptual contact with the *nominandum*, and may introduce the name by means of a formula involving an impure demonstrative (a demonstrative plus a sortal) such as (say) 'I name this ship *Titanic*'. In the case of baptism via reference-fixing description, there may be no perceptual contact with the *nominandum*, and the baptizer may introduce the name by means of a formula such as (say), 'I shall call the planet causing such-and-such perturbations in the orbit of Uranus *Neptune*' or 'I shall call the set of natural numbers ω'.[51]

---

[50] I say 'when someone uses a name, that name has as its referent', rather than 'when someone uses a name, she refers' to make clear that what is at issue here is what Kripke calls *semantic reference* rather than *speaker's reference*. Suppose you see Smith raking leaves and mistake him for Jones. When you say, 'I see Jones is raking leaves', the name you use on that occasion refers to Jones. But there is a sense in which you are referring to Smith (though you mistook him for Jones). The speaker's referent is given by a specific intention to refer to a certain individual (in this case, the man raking the leaves); it need not coincide with (in this case, does not coincide with) the semantic referent of the name the speaker uses to refer to the individual she intends to refer to. See Kripke's 'Speaker's Reference and Semantic Reference', in P. French, T. Uehling, and H. Wettstein (eds.), *Contemporary Perspectives in the Philosophy of Language* (Minneapolis: University of Minnesota Press, 1979).

[51] Kripke notes that it may be possible to subsume baptism by ostension under the category of baptism by reference-fixing description ('N & N', 303). It may not be necessary, but it won't hurt, to reiterate

So, initially at least, a name has as its referent the thing the name's introducer baptized with that name.[52] We would have a pleasingly simple account
of the reference of proper names—call it *the baptismal account*—if we said
that, whenever someone uses a (non-empty) name (whether the user is the
name's introducer or someone the introducer passed on the name to), that
name has as its referent (on that occasion) the thing the name's introducer
baptized with that name. But this account will work only if no one ever uses
a name that refers (on that occasion) to something other than that name's
original (baptismal) referent. And Kripke appears to think that sometimes
people do just that. In discussing the role of intentions to co-refer in the
preservation of reference, he writes:

[For reference to be preserved] when the name is 'passed from link to link', the receiver
of the name must, I think, intend when he learns it to use it with the same reference as
the man from whom he heard it. If I hear the name 'Napoleon', and decide it would be
a nice name for my pet aardvark, I do not satisfy this condition. ('N & N', 302)

In the footnote that accompanies the last sentence of this passage he adds:

I can transmit the name of the aardvark to other people. For each of these people, as
for me, there will be a certain sort of causal or historical connection between my use of
the name and the Emperor of the French, but not of the required type.

When the person—call him S—who has decided to call his aardvark
'Napoleon' says 'I'm taking Napoleon to the beach', it is true, not just that S
refers to his aardvark, but that the name he uses refers (on that occasion) to
his pet aardvark (cf. note 49 above). This is because he has (consciously)
'opted out' of the referential practice that begins with the baptism of
Napoleon, and includes the use of 'Napoleon' by the person who passed it on
to S; he's decided to put an old name to new referential use. As a result, the
name he uses does not (on that occasion) refer to its original (baptismal) referent. And if you get the name 'Napoleon' (solely) from S on an occasion when
he is using it as a name of his pet aardvark, then, when you use the name
'Napoleon', that name will refer (on those occasions) to S's aardvark—unless,
of course, you decide to opt out of the referential practice initiated by S.

---

why this admission doesn't raise the prospect of a descriptivist account of reference by the back door,
with the reference-fixing description providing the associated description for a name. For starters:
according to the descriptivist (as construed by Kripke), the associated description ('the F', 'the thing
with a weighted most of the properties in F') abbreviates or is synonymous with the name with which
it is associated. Thus any fully competent user of the name must be in a position to know a priori that
if anything is *n*, it satisfies the associated description, and that if anything uniquely satisfies the associated description, *n* does. But, on Kripke's account, there is no requirement that a competent user of
the name know, or be in a position to know a priori, that either of these conditionals hold. The reference-fixing description does not provide the meaning of the proper name; it is to the name as scaf-
folding is to a building under construction.

[52] If the introducer did not succeed in baptizing anything—say, because the (purportedly)
reference-fixing description is not uniquely satisfied, or because the demonstrative used failed to
refer—then the 'empty name' introduced has no referent.

Moreover, Kripke thinks, it can happen that someone uses a name without that name's referring (on that occasion) to its baptismal referent, even though that person has not made any conscious decision to opt out of an established referential practice:

Gareth Evans has pointed out that...cases of reference shifts arise...from one real entity to another of the same kind. According to Evans, 'Madagascar' was a native name for a part of Africa; Marco Polo, erroneously thinking that he was following native usage, applied the name to an island. Today the usage of the name as a name for an island has become so widespread that it surely overrides any historical connection with the native name...So real reference can shift to another real reference. ('N & N', addenda, 768)

Neither Marco Polo, nor the Europeans to whom Marco Polo passed on the name 'Madagascar', saw themselves as opting out of an existing referential practice. Still, Kripke says, when we use the name 'Madagascar', the name we use refers to an island off Africa and not the portion of mainland that is the original baptismal referent of 'Madagascar'. Why?

A present intention to refer to a given entity...overrides the original intention to preserve reference in the historical chain of transmission. The matter deserves extended discussion. But the phenomenon is perhaps roughly explicable in terms of the predominantly social character of the use of proper names...we use names to communicate with other speakers in a common language. This character dictates ordinarily that a speaker intend to use a name the same way as it was transmitted to him, but in the 'Madagascar' case this social character dictates that the present intention to refer to an island overrides the distant link to native usage. (ibid. 768–9)

There are two points here. First, Kripke is agreeing with Evans that it would be silly to maintain that, say, Madagascar is actually a portion of the African mainland, and ordinary people, geography teachers, atlases, and so on, are all in error. Secondly, Kripke is suggesting that this fact is to be accounted for in terms of the fact that there is a widespread (and entrenched) intention on the part of current speakers to refer to an island off Africa when they use the name 'Madagascar'. This seems right. Suppose that Marco Polo had been apprised of his mistake immediately after making it (hence at a time when intentions to refer to an island when using the name 'Madagascar' were neither widespread nor entrenched). Then he would presumably have deferred to the usage of his interlocutor, and said, 'All right, the island isn't (called) Madagascar. What is the island called?'[53] A (general, social) practice of applying 'Madagascar' to an island would never have got off the ground, and 'Madagascar' wouldn't now be the name of an island off Africa.

---

[53] If, upon hearing that the native name 'Madagascar' referred to a portion of the African mainland, Marco Polo had insisted on going on calling the island 'Madagascar', and had induced the rest of us to participate in this referential practice, then we'd have another case of reference shift by conscious opt-out, and not a case different in kind from the Napoleon case.

Kripke emphasizes that he is not attempting to offer a full-blown theory of the reference of proper names ('N & N', 300). He is not saying:

> A name *n* refers to a thing *t* if and only if *t* is the thing that the introducer of *n* baptized with *n*

or even

> A name *n* refers (on an occasion of its use) to a thing *t* if and only *t* is the thing that the introducer of *n* baptized with *n*, and there is an unbroken chain of intentions to co-refer linking the introduction of *n* with the current use of *n*.

He is instead saying something more like:

> A name *n* typically refers (on an occasion of its use) to the thing *n*'s introducer baptized with *n*—at least as long as there is no conscious opting out in the chain between the introduction of the name and the occasion of use (as in the Napoleon case), and there isn't a widespread and entrenched intention, at the time of use, in the community of the user, to refer to something different from the baptismal referent of *n*.

A name might on a given occasion of use refer to something that is not its original baptismal referent, because a referential shift has occurred, through conscious opting out (the Napoleon case) or misunderstanding (the Madagascar case). Alternatively, a name might on a given occasion of use refer to something that is not its original baptismal referent, because that name was not introduced by a baptism (on any reasonably narrow construal of baptism).

Suppose that the inhabitants of a small town are facing a very hard winter: torrential autumn rains have ruined crops and caused severe flooding, so food and housing are at a premium. One day a horse-drawn coach stops in front of the town hall. A stranger climbs out of the coach, goes up to the mayor, and hands him a staggering amount of money, saying, 'This is to help the village get through the winter.' Before the dumbstruck mayor can find his voice, the stranger has climbed back into his coach and the coach has started to pull away. As it does, the mayor asks, 'Who was that man?' The coach-driver answers 'Dunno', in his Québécois-accented English. The mayor and the others present think that the coach-driver has called the man 'Donneau'. The mysterious benefactor is never seen or heard from again. But the villagers, many of whom would never have got through the winter without 'Donneau', keep his memory green: the village commissions a statue of him, has a 'Donneau day' every year on the anniversary of his appearance, and renames the town hall 'Donneau Hall' in his honour. In the story 'Donneau' is clearly the villagers' name for the mysterious benefactor, although it was not conferred on him by baptism. (None of the villagers intended to fix the reference of 'Donneau' (either by dubbing or by description): they intended to apply an antecedently referring name to the benefactor.)

Here is another way in which a name can get a referent, without a (genuine) baptism ever taking place. Suppose that a physicist tries to introduce a name for a theoretical entity—say, a certain kind of force—by stipulating that F shall be the force that has property P. In fact, though, no force has that property. At this point, either of two things might happen. The would-be baptizer, or some of her colleagues, might quickly discover that no force has the property P, and conclude that 'F' never referred to anything. Alternatively, they might do a lot of theorizing about F, and decide that F not only had the property used to fix the reference of 'F', but also had a family of other properties **P**. Suppose that, over time, physicists come to regard the properties in **P** (but not property P)as the theoretically and explanatorily important properties of F. When asked what F is, physicists say 'the force with the properties in **P**', and only those well versed in the history of physics know that F was ever linked with P. Suppose further that there really is a (unique) force that has the properties in **P**, and that that force is the one physicists detect when they take themselves to be detecting F. Then it seems reasonable to conclude that at some time after the introduction of the name 'F' a referential shift occurred, so that 'F' now refers to a certain force (the **P**-force) even though it did not do so when it was introduced. If so, we have another way in which reference can be fixed non-baptismally.

Again, though, none of this is an objection to Kripke, since he recognizes that names need not always be introduced by an identifiable initial baptism, although they typically are ('N & N', 768).[54]

If we wanted a picture of the reference of proper names with a somewhat broader application than Kripke's, rather than simply appealing to baptisms we might appeal to a broader notion of 'reference-fixing circumstances'. The idea would be, roughly, that, although a name typically has its reference fixed by a baptism, it may also have its reference fixed by circumstances that are a baptismal surrogate. In the Donneau case the baptismal surrogate might be the villagers' applying 'Donneau' to the mysterious benefactor.

In both the Donneau case and the Madagascar case we have someone (initially) applying a name to something in the mistaken belief that he is applying the name to the same thing the person from whom he got the name applies it to. (In the Donneau case the villagers are wrong about there being a name 'Donneau' that the coach-driver applies to the mysterious benefactor; in the Madagascar case Marco Polo is wrong about there being a name ('Madagascar') that the natives apply to the island.) So we might say that in both the Donneau case and the Madagascar case we have reference tracing back to an (error-involving) baptismal surrogate.

---

[54] Perhaps because Kripke recognizes the non-necessity of an initial baptism in the addenda to *Naming and Necessity*, rather than in the main text, his recognition has not always been noted. Jaegwon Kim, for example, writes that 'according to Kripke's causal account of naming, for a person to refer to an object by the use of a name is for this use to be connected by an appropriate causal chain with the original baptismal act whereby the object was so named' ('Perception and Reference Without Causality', *Journal of Philosophy*, 74 (1977), 613).

Incidentally, if we did go this route, and if we adopted a suggestion for which Kripke shows some sympathy, perhaps we could after all endorse something rather like what I called earlier the baptismal account of reference.

In the preface to *Naming and Necessity* Kripke notes that some have argued that proper names are not rigid, on the grounds that typically the same name actually refers to more than one individual. He then says:

It is true that in the present monograph I spoke for simplicity as if each name had a unique bearer. I do not in fact think, as far as the issue of rigidity is concerned, that this is a major oversimplification. I believe that many important theoretical issues about the semantics of names (probably not all) would be largely unaffected had our conventions required that no two things shall be given the same name . . . For language as we have it, we could speak of names as having a unique referent if we adopted a terminology, analogous to the practice of calling homonyms distinct 'words', according to which uses of phonetically the same sounds to name distinct objects count as distinct names. This terminology certainly does not agree with the most common usage, but I think it may have a great deal to recommend it for theoretical purposes. (*N & N*, 7–8)

If I understand Kripke here, he is suggesting that it might be no bad thing, when doing the theory of reference, to adopt a terminology according to which, if I say (truly) 'Windsor is in Berkshire', and you say (truly) 'Windsor is in Ontario', then we are using different (though phonetically and orthographically indiscernible) names. (Compare: It is standard, and arguably no bad thing, in doing mereology (the theory of parts and wholes), to adopt a terminology according to which everything is a part of itself, in spite of the fact that we don't ordinarily speak of things as parts of themselves.)

If, however, the person who says 'Windsor is in Berkshire' is using a different name than the person who says 'Windsor is in Ontario', inasmuch as the name the first person used refers to a town in Berkshire and the name the second person uses refers to a (different) town in Ontario, then it would seem that, if the man who passed 'Napoleon' on to S says 'Napoleon was Corsican' and S says 'Napoleon eats termites', they are using different (though phonetically and orthographically indiscernible) names. And if the native whom Marco Polo misunderstood says 'Madagascar is a portion of the African mainland' and the European who learned 'Madagascar' from Marco Polo says 'Madagascar is an island off Africa', they are using different (though phonetically and orthographically indiscernible) names.

But if the natives applied one name to a portion of the African mainland, and Marco Polo and other Europeans applied another to an island—if the man from whom S got the name 'Napoleon' applies one name to the emperor, and S applies another to an aardvark—then those cases don't involve one and the same name starting out with one referent and ending up with another; they don't involve passing on a name without passing on its reference.

Of course, even if we think that names are 'unireferential', we shall have to say that, in the Madagascar case, *something* is passed on from the natives to

Marco Polo to the Europeans—to wit, 'Madagascar'.[55] We could call that something—which is presumably individuated by its phonetic or orthographic properties rather than its referential ones—the *expression* of a name. (The idea would be that in something like the same way the vocable 'bank' can express two concepts, or the sentence 'Someone went to the bank' can express two propositions, 'Madagascar' or 'Napoleon' can express more than one name.) A certain nominal expression is passed on from the natives to Marco Polo to present-day Europeans. But present-day Europeans aren't in possession of any name for the portion of African mainland that the natives had a name for. The Napoleon case is different. At least as long as the details of the story are filled in in the right way, someone passes on to S not just the expression of the name 'Napoleon', but the name itself. When S decides to call his aardvark 'Napoleon' (and carries out his intention), S produces a new name (whose reference he fixes by something like ostensive baptism); the new name and the name S acquired from the person who told him who Napoleon was have the same expression. (Thus S can say truly both 'Napoleon was Corsican' and 'Napoleon eats termites', using the same expression of a proper name, but different proper names, on the two occasions.[56])

It might be thought, and Kripke suggests, that a simple baptismal account of the reference of proper names gives the wrong results in cases where users of a proper name have wittingly or unwittingly lost an overriding intention to conform to the referential practice initiated by the introducer of that name, and acquired an overriding intention to use that name to refer to something different from that name's baptismal referent. But this seems to presuppose a 'multireferential' conception of names. If we adopt a unireferential conception, we can say that present-day Europeans use a different name from the one the natives used, and S uses a different name from the one his 'Napoleon'-source used (at least on those occasions when S intends to speak of an aardvark rather than of an emperor). In the Napoleon case nothing prevents us from saying of both of S's 'Napoleon' names that they refer to their baptismal referent. In the Madagascar case we cannot say that the name present-day Europeans use for the island off Africa refers to its baptismal referent, because that name was not introduced by baptism. But, as I have suggested, one might try to articulate a conception of reference-fixing circumstances that covers both baptisms proper and baptismal surrogates. Given such a conception, and

---

[55] For simplicity, I ignore the fact that the natives' name for the bit of African mainland was presumably not phonetically or orthographically, indiscernible from the name that Europeans use, and that the name Italians apply to the island off Africa is phonetically, though not orthographically, discernible from the name the English apply to the island.

[56] Did the natives pass on the name 'Madagascar' to Marco Polo, as well as the expression? Perhaps. As I have suggested, if Marco Polo's misunderstanding had been cleared up immediately, he presumably would have deferred to established usage. So he presumably originally had an overriding intention to co-refer with the person he got 'Madagascar' from—an overriding intention that present-day Europeans do not share.

a unireferential understanding of names, one might advance a quasi-baptismal account of the reference of proper names, according to which, roughly,

> A name *n* refers to a thing *t* just in case *t* is the thing that is the object of the baptism, or baptismal surrogate, with which the name *n* was introduced.[57]

When a speaker uses a name to refer to a thing, it will typically be true that (*a*) the person's use of the name on that occasion has among its causes the baptismal event (or surrogate thereof, to cover cases like Donneau's) in which the name was introduced. It will also typically be true that (*b*) the speaker has been (directly or indirectly) causally affected by the thing named.

As Kim notes, Kripke has often been interpreted as holding that a speaker cannot use a name that refers (on that occasion of use) to a thing unless both these conditions are satisfied, so that there is a 'causal' or 'historical' connection running from the thing named to the speaker's use of that name.[58] This causal connection is thought of as involving the referent's figuring in the causal history of the introduction of the name, which in turn is among the remote causes of the speaker's use of the name. Thus James Maxwell figures in the causal history of the (ostensive) introduction of the name 'James Maxwell', which in turn is among the remote causes of my current use of that name. Or Neptune figures in the causal history of the perturbations in the orbit of Uranus that are among the causes of Leverrier's fixing the reference of 'Neptune' via the description 'the cause of the perturbations in the orbit of Uranus', which in turn is among the remote causes of your using the name 'Neptune'.

Some passages might be thought to favour this interpretation. In the addenda to *Naming and Necessity* Kripke is discussing whether or not we should say that a story describing a substance with the appearance of gold is describing gold. Kripke writes: 'What substance is being discussed must be determined as in the case of proper names: by the historical connection of the story with a certain substance' ('N & N', 764). This suggests that a story cannot be about gold, and a proper name cannot refer to gold, unless the story or the name is historically connected (in the right way) with the substance gold. Similarly, in the aardvark passage cited earlier Kripke says that those who get the name 'Napoleon' from the person who uses it as a name of aardvark do

---

[57] Of course, whether we think of names as unireferential or multireferential, it will still be true that there are mechanisms that originally fix reference, and mechanisms that transmit the reference thus fixed; and, as Kripke says, just as intentions to refer will play a crucial role in the fixing of reference, intentions to co-refer will play a crucial role in the transmission of reference.

[58] Kim, 'Perception and Reference Without Causality', 613. See also G. Evans, 'The Causal Theory of Names', *Proceedings of the Aristotelian Society*, 80 (1976), 197, where Evans writes, 'There is something absurd in supposing that the intended referent of some perfectly ordinary use of a name by a speaker could be some item utterly isolated (causally) from the user's community and culture...I would agree with Kripke in thinking that the absurdity resides in the absence of any causal relation between the item concerned and the speaker.'

not, when they use it, use a name that refers to an emperor, because 'for each of those people... there will be a certain sort of causal or historical connection between my use of the name and the Emperor of the French, but not one of the required type'. ('N & N', 96). Kripke seems to be saying here that, in order for someone's name to refer (on a given occasion of use) to (the emperor) Napoleon, there must be the right sort of causal connection between the person's use of that name and the emperor (presumably, one running from the emperor to the person's use of the name, via the introduction of that name).

In fact, though, it is very doubtful that a speaker can use a name that (on that occasion) refers to a thing only if that thing figures in the causal history of the speaker's use of that name on that occasion. To start with, as Tyler Burge has noted, it seems as though someone might be able to come up with a concept of gold in the absence of any causal contact with gold—if, say, she had enough knowledge of chemistry.[59] So—if Kripke's story had been written by a fantastically gifted amateur chemist on a desert island—it seems it might be about gold, even in the absence of any causal-historical connection between the story and that substance. Once we have said this, there seems little reason to deny that the writer of the story could have a name for gold, even if her uses of that name were not causal-historically connected to gold.

Also, if the referent must figure in the causal history of the name that refers to it, then one cannot name a thing that does not yet exist. But, to update an example of David Kaplan's, one could at least attempt to introduce the name 'Newman 2' by saying something like: ' "Newman 2" shall designate (fully) rigidly whatever individual "the first child born in the twenty-second century" actually (and flexibly) designates.' If I performed this reference-fixing ceremony (and if the reference-fixing description turned out not to be improper), wouldn't I thereby introduce a name of an individual that does not yet exist?

This is admittedly an odd case. But there seem to be far more ordinary cases in which we actually introduce names for things before they exist. Didn't the people who built the *Titanic* have a name for her even before she came into existence? It is unclear exactly when the *Titanic* began to exist. Still, at some stage in her production her builders would have been happy to say that they were building the *Titanic* but not happy to say that the *Titanic* already existed. To take another example, it seems that my publisher already has a name for this book (it appears in my contract). As I write these words, though, the book does not yet exist (although parts of it do). Again, don't we have names for stretches of the future? Isn't 'Saturday' a name of the part of the future that starts at midnight tonight? (I write these words on Friday.) Well, perhaps 'Saturday' isn't happily thought of as a proper name, any more than 'tomorrow' is. But what would stop me from introducing a proper name for Saturday–tomorrow? If I introduced and used a

---

[59] T. Burge, 'Other Bodies', in A. Woodfield (ed.), *Thought and Object: Essays on Intentionality* (Oxford: Clarendon Press, 1982), n. 18.

name for tomorrow (today), I would be using a name for a thing that doesn't figure in the causal history of its introduction (or subsequent use).

There are also cases in which, it seems, we could introduce, or do introduce, and use names for things that never have and never will figure in any causal histories. To vary an example discussed by Nathan Salmon, it seems that I could introduce a name, 'Lauranda', for the human being who would have existed if the sperm cell from which my first daughter (Laura) came had fertilized the egg cell from which my second daughter (Amanda) came. (How might I introduce such a name? Not by ostensive baptism, obviously. But why couldn't I do it by reference-fixing description, as follows: 'Lauranda' shall refer to the being who would have existed if the sperm cell Laura came from had fertilized the egg cell Amanda came from? (I assume that there is a unique (possible) individual who would have existed had those particular gametes got together.[60]))

Even if it is impossible to have names for *inactualia*, we could and do have names for things that (it is almost universally supposed) don't appear in any causal histories, such as the number π.

Because Kripke thinks of 'π' as a name, and shows no sign of holding the (eccentric) view that π is causally 'ert', it is natural to conclude that he does not after all suppose that we can name only those things that have left causal traces on us. Moreover, Nathan Salmon reports that, in lectures and in conversation, Kripke has maintained that nothing would prevent us from introducing names for *inactualia* such as Lauranda.[61] A name like 'Sherlock Holmes' is not, for Kripke, a name of a merely possible individual, because there is no unique *possibile* that name could sensibly be taken to refer to;[62] but the same doesn't hold for names like 'Lauranda'.

To recap: A referential use of a name might be thought to require a causal connection between that use and some event that introduced the name used. A referential use of a name might also be thought to require a causal connection between that use and the thing named. While Kripke appears to endorse the first claim, he does not endorse the second. Kripke's 'causal account of reference' is often thought to have important affinities with causal theories of perception and causal theories of knowledge. But it is important not to overlook the differences between the former and the latter. Champions of a causal theory of perception hold that to perceive a thing or an event is to be causally related to that thing or that event in the right sort of way. Champions of a

---

[60] It might be objected here that it makes no sense to say there is a unique non-actual individual who would have existed in such-and-such circumstances. For a defence of the idea that it does make sense, see K. Fine, 'Prior on the Construction of Possible Worlds and Instants', in A. N. Prior and K. Fine, *Worlds, Times, and Selves* (Amherst: University of Massachusetts Press, 1977).

[61] See Salmon, *Reference and Essence*, 39 and n. 41.

[62] 'Several distinct possible people, and even actual ones such as Darwin or Jack the Ripper, might have performed the exploits of Holmes, but there is none of whom we can say that he would have *been* Holmes. For if so, which one?' ('N & N', 764).

causal theory of knowledge hold that to know a fact is to be causally related to that fact in the right sort of way. But Kripke does *not* hold that to refer to a thing is to be causally related to it in the right sort of way. For Kripke, the relational predicates 'refers to' and 'names' are quite unlike relational predicates such as 'perceives' or 'detects' (or 'is affected by') and like predicates such as 'fears', 'seeks', and 'describes', in being intentional (in Brentano's sense of the term). You cannot perceive *a* or detect *a* if *a* does not exist.[63] But you can still fear, or seek, or describe *a*; and (Kripke allows) you can still name and refer to *a*.

4, NATURAL KIND TERMS

For a pure descriptivist about proper names, if someone is a fully competent user of the name 'Aristotle', then for some purely qualitative definite description *the F*, it is necessary that, and that user could know a priori that, someone is Aristotle if and only if he is the F. (Again, to accommodate a Searlean version of pure descriptivism, we allow that *the F* may be either a straightforward definite description, or a definite description of the type *the thing having a weighted most of qualitative properties $Q_1 \ldots Q_n$*.)

For a pure descriptivist about geometrical sortals, if someone is a fully competent user of a sortal such as *square*, then, for some purely qualitative indefinite description *an F*, it is necessary that, and the user could know a priori that, something is a square if and only if it is an F. (That is, the user could know a priori that something is a square if and only it is F simply by reflecting on her concept *square*: in what follows, for brevity, I shall often leave this inexplicit.)

Pure descriptivists about proper names have usually also been pure descriptivists about not just geometrical sortals, but also biological sortals (*human*, *mammal*) and chemico-compositional predicates (*made of water, made of gold*).[64] They have accordingly supposed that if someone is a fully competent user of, say, the sortal *human*, then, for some purely qualitative indefinite description *an F*, it is necessary that, and that user could know a priori that, someone is human if he or she is an F.[65] For the pure descriptivist about biological sortals, *a thing which is human* is a trivial example of such an indefinite

[63] At least, you cannot perceive or detect what never has and never will exist. Perhaps you can perceive individuals that no longer exist (faraway stars); perhaps you can only perceive events that are no longer occurring.

[64] For some references, see Salmon, *Reference and Essence*, ch. 4, sect. 6, *et passim*. Note that one couldn't be a pure descriptivist about, say, the name 'Aristotle', and suppose that 'analysis' for 'Aristotle' is given by 'the man (male human) who ...', without also being a pure descriptivist about the sortal *human*.

[65] Again, to accommodate Searlean accounts of biological sortals, the indefinite description might be either straightforward, or an indefinite description of the form *a thing with a weighted most of qualitative properties $Q_1 \ldots Q_n$*.

description. But pure descriptivists about biological sortals and chemico-compositional predicates have supposed that, if someone is a competent user of the sortal *human*, then, for some purely qualitative indefinite description *an F*, it is non-trivial that, and necessarily true that, and that user could know a priori that, something is a human if and only if it is an F. *Mutatis mutandis*, they have said the same thing about chemico-compositional predicates such as *made of gold*. In other words, they have supposed that, whenever someone competently uses a biological sortal or a chemico-compositional predicate, there is a purely qualitative 'analysis' or 'analytic definition' of that sortal or predicate that its user could know a priori (by dint of reflection on the concept she associates with that sortal or predicate).

Someone might reject a pure descriptivist account of ordinary proper names, and accept a pure descriptivist account of geometrical sortals and biological sortals and chemico-compositional predicates.[66] Such a person might naturally hold that, although there are no (purely qualitative) definite descriptions that provide the 'analysis' of ordinary proper names, there are (purely qualitative) definite descriptions (either of a straightforward sort, or of the sort, *the thing with a weighted most of properties $Q_1 \ldots Q_n$*) that provide the 'analysis' of names like 'gold', '*Homo sapiens*', 'electricity', *et similia*. It might be, for example, that 'gold', '*Homo sapiens*', and so on are names of conjunctive qualitative properties,[67] and the analysis for such names is given by specifying the (qualitative) 'conjunct properties' that, as it were, 'add up to' the conjunctive qualitative property.

Indeed, if Kripke is right, Mill rejected a pure descriptivist account of ordinary proper names, but accepted a pure descriptivist account of terms like 'cow' and 'human' (and *Homo sapiens*):

Mill counts both predicates like 'cow', definite descriptions, and proper names as names. He says of 'singular' names that they are connotative if they are definite descriptions but non-connotative if they are proper names. On the other hand, Mill says that *all* 'general' names are connotative; such a predicate as 'human being' is defined as the conjunction of certain properties which give necessary and sufficient conditions for humanity—rationality, animality, and certain physical features. ('N & N', 322).[68]

---

[66] Suppose that, for the reasons adduced at the end of the first section of this chapter, someone thinks that there is no purely qualitative definite description such that, necessarily, a thing satisfies that description if and only if it is Mount Faverghera. That person may think that there is a purely qualitative indefinite description such that, necessarily, a thing satisfies that indefinite description if and only if it is a square. And she may think any competent user of the sortal *square* knows a priori, or could after sufficient reflection come to know a priori, that a thing is a square if and only if it is an F, where it is necessary that a thing is a square if and only if it is an F.

[67] As I mentioned earlier, it is not uncommonly supposed that mass terms such as 'gold' or 'water' denote properties: Montague supposes that 'gold' denotes the property of being a piece of gold, and 'water' the property of being a body of water. See his 'The Proper Treatment of Mass Terms in English', in F. J. Pelletier (ed.), *Mass Terms: Some Philosophical Problems* (Dordrecht: D. Reidel, 1979), 173–9.

[68] Kripke's footnote 66 of 'Naming and Necessity' suggests that in this context he is thinking of the properties given by the connotation of a general name as purely qualitative—as 'pure properties' without a referential element of their own.

One can see how someone might find this view at least initially attractive. After all, there is something very odd about the idea that 'Bessie' has a definition; it does not seem nearly so odd to suppose that 'cow' (or '*Bos taurus*') has a definition (even if it is not immediately obvious that it will have a purely qualitative *definiens*).

In the first two lectures of *Naming and Necessity* Kripke targets (what he takes to be) the Russellian–Fregean account of proper names. In the third he targets (what he takes to be) the Millian (as well as Russellian and Fregean) account of natural kind terms.[69] There is an asymmetry here that may be worth underscoring. In Lectures I and II Kripke tries to show that a pure descriptivist account of proper names is untenable. In Lecture III he tries to show that a pure descriptivist account of sortals *of a certain kind*—natural kind sortals—is untenable. He nowhere states or even suggests that a pure descriptivist account of geometrical sortals such as *square*, or sortals such as *mountain*, are untenable.[70]

According to the Millian view, if someone is a fully competent user of the sortal *tiger*, there is a purely qualitative (reductive) analytic definition of *tiger* (a non-trivial necessary truth of the form *Something is a tiger if and only if it is an F*, where F is purely qualitative) that that user could know a priori. We might hope to find such a (reductive) analytic definition in a dictionary. Kripke notes that, for the *Shorter Oxford English Dictionary*, a tiger is 'a large carnivorous quadrupedal feline, tawny yellow in colour with blackish transverse stripes and white belly'. But, he points out, this entry does not provide us with an analytic definition of 'tiger'. For the existence of three-legged tigers or stripeless tigers cannot be ruled out a priori ('by definition'), on the grounds that the concept *tiger* subsumes the concept *four-legged* or the concept *striped* ('N & N', 317).[71] A defender of a more or less Millian view of natural kind terms might say here that being a tiger is a matter of having most of, or a weighted most of, the set of qualitative properties appearing in the *Shorter Oxford English Dictionary*.[72] But, Kripke objects, we might find out that tigers were neither quadrupedal, nor tawny yellow, nor striped, nor white-bellied,

---

[69] Where 'natural kind term' covers names like '*Cavia cavia*' and 'water', sortals like *guinea pig*, and predicates like 'is electrically charged'.

[70] See 'N & N', 327: 'my argument implicitly concludes that *certain* general terms, those for natural kinds, have a greater kinship with proper names than is generally realized' (my emphasis). See also 'N & N', 323 and n. 66, where Kripke says that the sort of pure descriptivist account that Mill wants to give for all 'general names' may be right for some general names. Kripke never suggests in the first two lectures that the sort of descriptivist account that Russell or Searle would want to give for all proper names may be right for some proper names.

[71] Indeed, something stronger can be said. The existence of three-legged or stripeless (or non-white-bellied) tigers cannot be ruled out at all. For there are tigers that are not (and never were, and never will be) four-legged, striped, or white-bellied: tigers that die in the womb, before they develop sufficiently to have any of the properties in question. So it is not true, much less necessary, or a priori, that all tigers have four legs, or stripes, or white bellies.

[72] Although there is a question about whether *being feline* and *being carnivorous* are qualitative properties.

nor ... That tigers don't have any of the (alleged) identifying marks attributed to them by the *Shorter Oxford English Dictionary* is very unlikely, but not to be excluded on a priori or conceptual grounds ('N & N', 318).

Just as there might turn out to be tigers with none of the properties that appear in the 'definition' of 'tiger' above, Kripke holds, there might turn out to be non-tigers with all of those properties. Suppose we discovered some large, quadrupedal, tawny yellow, blackish-transverse-striped, white-bellied, carnivorous animals that had an internal structure very different from the internal structure of mammals and were in fact reptiles. We would conclude, not that some tigers are reptiles, but that some animals satisfying the 'dictionary description' of *tiger* were not tigers at all, but animals of a different species altogether.[73] So it is no more a priori that only tigers satisfy the dictionary description of *tiger* than it is that all tigers satisfy that description ('N & N', 317).

Here some might object: All this shows that you won't find a purely qualitative (reductive) analytic definition of *tiger* in the *Shorter Oxford English Dictionary*. But, they might add, this is perfectly compatible with the claim that, if someone is a competent user of the sortal *tiger*, there is some purely qualitative (reductive) analytic definition (non-trivial necessary truth of the form, *Something is a tiger if and only if it is an F*, where F is purely qualitative) which she could know a priori, simply by reflecting on her tiger-concept.

This objection underestimates the scope of the considerations Kripke brings to bear against pure descriptivist accounts of sortals like *tiger*. Kripke argues that

(1) whether or not something is a tiger depends on whether or not it has the right sort of internal structure to be a tiger

and

(2) what the right sort of internal structure to be a tiger actually is, is an empirical matter, and cannot be ascertained a priori.

Suppose the pure descriptivist about biological sortals concedes (1) to Kripke. Then she must face the following question: How could the competent user of 'tiger' know a priori that all and only the things that satisfy an (alleged) purely qualitative (reductive) analytic *definiens* of 'tiger' have the right sort of internal structure to be tigers? An obvious answer would be: The right sort of internal structure to be a tiger is simply the sort of internal structure built into the concept *tiger*, and the purely qualitative (reductive) analytic *definiens* of 'tiger'. (Suppose someone were to ask how the competent user of *square* could know a priori that all and only the things that satisfy the (alleged) purely

---

[73] 'But these animals wouldn't satisfy the description of tigers in the dictionary, because they wouldn't be feline.' This may be true, if felinity is a matter of belonging to a particular biological family. But if it is, Kripke thinks, then it is an empirical discovery, rather than an a priori truth, that tigers are feline. So, Kripke concludes, this response is of no use to the defender of a Millian account of natural kind terms ('N & N', 317).

qualitative (reductive) analytic *definiens* of 'square' have the right sort of 'geometrical structure' to be squares. The pure descriptivist about geometrical sortals would respond that the right sort of geometrical structure is just the sort of geometrical structure built into the concept *square* and the purely qualitative (reductive) analytic *definiens* of 'square'.[74]) But if the pure descriptivist gives this answer, then she is committed to supposing that empirical inquiry is not needed to determine what the right sort of internal structure to be a tiger actually is, any more than it is needed to determine what the right sort of 'geometrical structure' to be a square is. So, if Kripke is right to suppose both that being a tiger depends on having the right sort of internal structure, and that it is an empirical and not an a priori matter what that sort of internal structure is, then the pure descriptivist can't give the obvious answer to the question 'How could you know a priori that all the things that satisfy the (alleged) purely qualitative (reductive) analytic *definiens* of 'tiger' have the right sort of internal structure to be tigers?' And it is hard to see what other answer she could give. Thus her troubles do not depend (solely) on her particular choice of a purely qualitative (reductive) analytic definition of 'tiger'.

Note that the pure descriptivist has these worries because she wants to say that, if someone is a competent user of *tiger*, there is a purely qualitative *reductive* (non-trivial) analytic *definiens* which necessarily applies to all and only tigers, and which that user could know a priori applies to all and only tigers. For suppose that the pure descriptivist allowed the analytic *definiens* for 'tiger' to be a 'tiger'. Then there would be no problem about how it could be true that (*a*) it is necessarily true that, and the competent user of the name knows a priori that, all and only things satisfying the *definiens* are tigers; (*b*) whether or not something is a tiger depends on whether or not it has the right sort of internal structure to be a tiger; and (*c*) what the right sort of internal structure to be a tiger actually is is not something that can be ascertained a priori.

If, however, the pure descriptivist about sortals like *tiger* does not require that there be a purely qualitative reductive analytic definition of 'tiger', then pure descriptivism about biological sortals does not go beyond the (metaphysical) claim that *being a tiger* and the like are purely qualitative properties. (Compare: If the pure descriptivist about proper names allows the analysis-providing definite description associated with, say, the name 'Socrates' to be 'the Socratizer' (insisting that Socratizing is a purely qualitative property that resists specification in terms of more basic qualitative properties), then pure descriptivism about proper names reduces to the (metaphysical) claim that *being Socrates* is a qualitative property.) Kripke's arguments don't tell against a pure descriptivist account of biological sortals if that account does not go beyond the thesis that biological kinds have purely qualitative essences, just as Kripke's arguments don't tell against a pure descriptivist account of proper

---

[74] Which might be, say, *a plane rectilinear and rectangular figure with four equal sides.*

names if that account does not go beyond the thesis that individuals have purely qualitative essences. But Kripke never intended his arguments to tell against descriptivism about proper names construed in that (weak) way,[75] and I take it he never intended his arguments to tell against descriptivism about biological sortals construed in that (correspondingly weak) way.

Kripke's arguments against a pure descriptivist account of chemico-compositional predicates (such as *made of gold*) and names for chemical substances (such as *gold*) follow the lines of his arguments against a pure descriptivist account of biological sortals. What, Kripke asks, might the (necessarily true, a priori knowable) purely qualitative reductive analytic definition of 'gold' be? Not, as Kant sometimes seems to suggest, *a yellow metal*. We could find out that we have all been the victims of an illusion, and gold is actually blue rather than yellow: this (epistemic) possibility cannot be ruled out by reflecting on the concept of gold. Or we could find out that something that is not gold is nevertheless a yellow metal ('N & N', 315–16). Of course, yellowness is only one of the marks we use to identify gold (and, anyway, is all gold yellow?). So, the descriptivist might say, the purely qualitative reductive analytic *definiens* for 'gold' is not *the yellow metal*, but *the metal that is yellow(?), malleable, ductile...*—where the description makes reference to all the properties we originally used to identify gold—or perhaps *the thing with a weighted most of the properties {being metallic, being yellow, being malleable, being ductile...}*. But, Kripke replies, we could find out that gold doesn't actually have any of the properties we originally 'identified' it by. Moreover, Kripke notes, there might be something that has all the marks we originally used to identify gold, but is not gold, because it has the wrong sort of composition to be gold. Indeed there is such a thing: iron pyrites, or 'fool's gold'.

As in the case of biological sortals, the pure descriptivist could say that there is a purely qualitative reductive analytic definition of gold, even if neither Kant nor anybody has put his finger on it. Or she could say that, unlike Kant, we have such a definition: *the element with atomic number 79*.

For reasons of a sort already discussed, Kripke would find both these responses ineffectual. He would say that, inasmuch as nothing can be gold without having the right sort of composition, the pure descriptivist about gold will have to say that any competent user of the name 'gold' could know a priori that anything satisfying the purely qualitative (reductive) analytic *definiens* of gold has the right sort of composition to be gold. If asked how the competent user could have such knowledge, her answer would have to be that the right sort of composition is built into the concept of gold and the purely qualitative reductive analytic definition of gold; and this cannot be reconciled with the fact that it is an empirical, and not an a priori, matter what sort of composition something has to have to be gold.

---

[75] As we have seen, Kripke does not pronounce on whether individuals have purely qualitative essences.

As for the second response, Kripke agrees that it is necessary that gold is the element with atomic number 79 ('N & N', 319–21). But, as we have seen, he would say that it was nevertheless an empirical discovery that gold has atomic number 79.[76]

A defender of a pure descriptivist account of chemico-compositional predicates and names of chemical substances might base her defence thereof on a thesis endorsed by a number of philosophers, and recently championed by Gabriel Segal.[77] As Segal sees it, before the discovery of the chemical composition of water, 'water' expressed a concept whose 'extension conditions' (as Segal puts it) leave open the possibility that something is water, even though it is not H_2O. If there had been any of Putnam's XYZ anywhere in the universe (XYZ being a substance that has the same 'superficial' properties as water, but a quite different chemical composition),[78] the term 'water' then in use would have been true of it. As a result of certain empirical discoveries, the term 'water' (for most speakers) has ceased to express what Segal calls a *motley* concept—applying to whatever stuff has the right sort of superficial properties—and has come to express a natural kind concept, applying only to stuff with the right internal constitution. The extension conditions of the term 'water' have accordingly changed, so that we can now say truly (using a term that expresses the 'scientific' concept of water) that if there is (or if there were) such a thing as XYZ, it isn't (it wouldn't be) water. If the term 'water' expresses the pre-1750, pre-scientific water-concept, it is just not true that something must be H_2O to be in the extension of 'water'; if the term 'water' expresses the current water-concept, it is. Similarly, if 'gold' express the pre-scientific gold-concept, it is just not true that something must have atomic number 79 to be in the extension of 'gold'; if 'gold' expresses the scientific gold-concept, it is. And, assuming that 'tiger' in its current scientific use is applied only to members of *Felis tigris*, if 'tiger' expresses the pre-scientific tiger-concept, it is not true that something must be a member of *Felis tigris* to be in the extension of 'tiger'; if 'tiger' expresses the scientific tiger-concept, it is.[79]

So, a defender of pure descriptivism might say, before we made certain empirical discoveries about the internal structure of tigers, we had a pre-scientific

[76] On the other hand, Kripke would grant that a chemist could introduce a name for a substance none of whose samples had been observed, fixing its reference via a description of the form 'the substance with atomic number n'. In such a case, Kripke would allow, the reference-fixer would know a priori that, if that substance exists, it has atomic number n (cf. 'N & N', 328). Still, Kripke would say, it is not a precondition of being a competent user of the term 'gold' that one be able to know (by reflection on one's gold-concept) that gold has atomic number 79.

[77] See Segal, *A Slim Book About Narrow Content* (Cambridge, Mass: MIT Press, 2000), ch. 4.

[78] See Putnam, 'The Meaning of Meaning', in Putnam, *Mind, Language, and Reality* (Cambridge: Cambridge University Press, 1975).

[79] See Segal, *A Slim Book About Narrow Content*, 82: 'Prior to the speaker's finding about the underlying nature of the examples they are talking about, the term they use is not yet a natural kind term. At that point, the term may naturally be applied to Twin Earth counterparts of the terrestrial natural kind.'

concept of tiger—one that was less 'structurally demanding' than our current scientific tiger-concept. Both the pre-scientific and the scientific concept are purely qualitative, and whichever one is at issue, what the right sort of structure is for being a tiger is a question that can be settled a priori. What is an empirical discovery, and could not be ascertained a priori, is that all and only tigers (in the pre-scientific, less structurally demanding sense of 'tiger') have the particular kind of internal structure that tigers (in the scientific, more structurally demanding sense of 'tiger') necessarily have, and may be ascertained a priori to have by anyone who has a full grasp of the scientific tiger-concept. Similarly, there is a (less compositionally demanding) pre-scientific gold-concept and a (more compositionally demanding) scientific gold-concept, and—whichever of the two concepts 'gold' expresses—what the right sort of composition is for being gold is a question that can be settled a priori, though it is an empirical fact that the substance which is gold (in the pre-scientific, less compositionally demanding sense) has the sort of composition that gold (in the scientific, more compositionally demanding sense) necessarily has, and may be ascertained a priori to have by anyone with a full grasp of the scientific gold-concept.

But, Kripke would respond, we would (now) classify animals with the wrong sort of internal structure to be tigers—say, reptilian internal structure—as non-tigers, however much they had the usual external appearances and identifying marks of tigers, 'not because, as some people would say, the old concept of tiger has been replaced by a new scientific definition. I think this is true of the concept of tiger *before* the internal structure of tigers had been investigated' ('N & N', 318). Similarly, Kripke holds, there aren't two different concepts of *gold* or *metal*—one phenomenological and pre-scientific, the other 'compositional' and scientific ('N & N', 315 and 316).

Pre-scientifically, we apply the term 'tiger' to certain animals and the term 'gold' to a certain kind of stuff. For Kripke, we already surmise at this stage that all the things to which we apply the term 'tiger' have the same internal structure, and that all the stuff to which we apply the term 'gold' has the same chemical composition. Our surmise is defeasible: we may find out that some of the things to which we apply the term 'tiger' have one sort of internal structure, and the rest have a completely different sort. In such a case, we may decide, there have (surprisingly) turned out to be two kinds of tigers ('N & N', 318). Similarly, we may find out that there are two kinds of gold, with different compositions ('N & N', 328), in much the way that there are two kinds of jade (jadeite and nephrite). But *if* we are right in supposing that the things to which we apply the term 'tiger' all have the same internal structure, and that all the stuff to which we apply the term 'gold' has the same chemical composition, then 'tiger'—right from the start—applies only to animals with that internal structure, and 'gold' applies only to stuff with that chemical composition. *Pace* Segal, Kripke would say, it is not true that 'The [pre-scientific] term "water" ... was a term the extension conditions of which did not confine it to any specific natural kind, but left open the possibility of its being true of many natural kinds' (*A Slim Book About Narrow Content*, 73). The term 'water' had the same 'extension conditions' in 1750 as it has now: it was true then, as it is now, that a substance in the actual world or some alternative possible world is water (in that world) just in case it is H_2O (in that world). What has changed is that *what we know about the world* no longer leaves open the possibility that 'water' is true of any natural kind but H_2O. In the same way, the term 'gold' had the same 'extension conditions' before we knew gold was the element with number 79 as it does now, and expressed the same concept as it does now; it is just that we now know, rather than defeasibly surmise, that gold has a certain composition (to wit, the one that goes along with having atomic number 79).

Or, at least, we know, relative to ordinary, Moorean standards for knowledge. A sceptic might say that, for all we know, gold does not have atomic number 79: the entire theory of the atomic and molecular structure of matter to which *Gold has atomic number 79* belongs might turn out to be a mistake. We could not refute such a sceptic by pointing out that, simply by reflecting on our current (scientific) gold-concept, we may come to know a priori that something is gold if and only if it has atomic number 79. For there is a sense in which *gold has atomic number 79*, along with all the rest of our theory of the atomic and molecular structure of matter, might turn out to be mistaken— the sense in which *P might turn out to be mistaken* is tantamount to *It cannot be excluded a priori that P* (cf. 'N & N', 319).

So the pure descriptivist about *gold* (or *tiger*) cannot after all resist Kripke's argument by positing pre-scientific and scientific senses of 'gold' (or 'tiger'), and saying that there is a contingent and a posteriori link between the pre-scientific concept and a certain structure (or composition) and a necessary and a priori link between the scientific concept and that structure (or composition). Pure descriptivist accounts of natural kind terms, like pure descriptivist accounts of proper names, misclassify a posteriori knowledge (our knowledge that water is made of hydrogen and oxygen, say, or our knowledge that Gödel discovered the incompleteness of arithmetic) as a priori knowledge.

As we have seen, Kripke does not attempt to replace pure descriptivist theories of the reference of proper names with a better theory. Instead, he offers a picture of how proper names typically acquire and conserve reference. In the same way, rather than offering a new and improved theory of the reference of names of natural kinds and natural kind sortals, Kripke offers us a picture of how such terms acquire and conserve reference. Here's how it works:

In the case of proper names, the reference can be fixed in various ways. In an initial baptism it is typically fixed by an ostension or a description. Otherwise, the reference is usually determined by a chain, passing the name from link to link. The same observations hold for such a general term as 'gold'. If we imagine a hypothetical (admittedly somewhat artificial) baptism of the substance, we must imagine it picked out as by some such 'definition' as, 'Gold is the substance instantiated by the items over there, or at any rate, by almost all of them'. . . . I believe that in general, terms for natural kinds (e.g. animal, vegetable, and chemical kinds) get their reference fixed in this way; the substance is defined as the kind instantiated by (almost all of) a given sample. The 'almost all' qualification allows that some fools' gold may be present in the sample. If the original sample has a small number of deviant items, they will be rejected as not really gold. If, on the other hand, the supposition that there is one uniform substance or kind in the initial sample proves more radically in error, reactions can vary: sometimes we may declare that there are two kinds of gold, sometimes we may drop the term 'gold'. (These possibilities are not supposed to be exhaustive.)…the original sample gets augmented by the discovery of new items… More important, the species name may be passed from link to link, exactly as in the case of proper names, so that many who have seen little or no gold can still use the term. Their reference is determined by a causal (historical) chain…('N & N', 328, 330–1)

On the Kripkean picture the way that terms like 'gold' or 'tiger' acquire their reference is like the way that proper names acquire their reference, and especially like the way that proper names like 'Jack the Ripper' acquire their reference. In both the case of 'Jack the Ripper' and 'gold' (or 'tiger') the intentions of the name's introducer and the world each play a part in fixing reference. Suppose that the baptismal name of 'Jack the Ripper' was John Doe. Why did 'Jack the Ripper' (when introduced) refer to John Doe, and not, say, to Winston Churchill? Because the person who introduced the name 'Jack the Ripper' intended it to refer to the person who committed such-and-such murders (whoever that turned out to be), and the person who committed those murders was John Doe (and not Winston Churchill). Why did 'gold' (when introduced) refer to the element with atomic number 79 rather than, say, iron pyrites? Because the person who introduced the term 'gold' intended it to refer to the substance (all or at least almost all) these samples were samples of (whatever that turned out to be), and the substance that (all or at least almost all) these samples were made of is the element with atomic number 79 (and not iron pyrites). Why did the sortal 'tiger' (when introduced) have in its extension members of the species *Felis tigris* rather than, say, members of the species *Mesocricetus auratus*? Because the introducer of the sortal 'tiger' intended to pick out things of the same kind as those things (whatever they turned out to be), and those things were members of the species *Felis tigris* (and not of the species *Mesocricetus auratus*).[80]

---

[80] Because the world has a part to play in (originally) fixing the reference of a natural kind term— a part we needn't know everything about—Kripke's picture allows that what it takes to belong to a natural kind, and indeed, what a natural kind *is*, can be (and often is) a matter for empirical discovery. Scientists may wonder what gold is, and find out that it is the element with atomic number 79. (See 'N & N', 330: 'In general, science attempts…to find the nature, and thus the essence (in the philosophical sense) of the kind.') The pure descriptivist account of natural kinds makes the reference-fixing of natural kind terms a unilateral affair, and leaves no room for finding out empirically what it takes to belong to a natural kind, or what a natural kind is. On that account, prompted perhaps by the 'clustering' of certain co-instantiated qualitative properties, we 'draw up the blueprint' for a natural kind— a concept of a natural kind, formulable in a purely qualitative (reductive) analytic *definiens*. After we've done that, we are in a position to know all there is to know about what it takes to belong to that kind, and about what that kind is, by looking at the blueprint, although empirical inquiry is typically needed to determine how natural kinds are instantiated and co-instantiated. There is a certain irony in the fact that traditionally empiricists have embraced the sort of pure descriptivist account of natural kind terms that makes knowledge of (scientific) kinds a non-empirical, a priori matter.

Someone might be puzzled—at any rate, I was puzzled—about how to reconcile the account of reference-fixing for natural kind terms in the passage cited above with what Kripke says elsewhere about the aposteriority of the hypothesis that tigers form a kind. At 'N & N', 318, he writes:

Since we have found out that tigers do indeed, as we suspected, form a single kind, then something not of this kind is not a tiger. Of course, we may be mistaken, in supposing that there is such a kind. In advance, we suppose that they probably do form a kind. Past experience has shown that usually things like this, living together, looking alike, mating together, do form a kind. If there are two kinds of tigers that have something to do with each other but not as much as we thought, then maybe they form a larger biological family. If they have absolutely nothing to do with each other, sometimes then there are really two kinds of tigers. This all depends on the history and what we actually find out.

Judging from the 'N & N', 328, passage cited earlier, the introducer of 'tiger' fixes its reference via a formula along the lines of:

> *Tiger* is the kind of thing that all or almost all of these things (the items in the baptismal sample) are.

Now, as Kripke indicates in a part of the 'N & N', 328, passage that I have summarized, he holds that someone who fixes the reference of a term via a reference-fixing description is in a position to know a priori that anything that term applies to satisfies that reference-fixing description (for an extended discussion of this view, see Chapter 2, section 2). So, if the reference-fixing description for 'tiger' is *the kind of thing that almost all of these things (the things in the baptismal sample) are*, then the introducer of the term 'tiger' is in a position to know a priori that, if there are tigers, they all are or form or belong to the same kind (namely, the kind that all or almost all of the items in the baptismal sample for 'tiger' belong to).

In the 'N & N', 318, passage, though, Kripke seems to say that the introducer of the term 'tiger' may find out that tigers after all do not form a kind; it may turn out that, even if each tiger belongs to one or other of the two kinds of tiger, there is no single kind to which all tigers belong.

So the puzzle is: if the introducer of the term 'tiger' is in a position to know a priori that, if there are tigers, they all belong to the kind to which all or almost all of the items in the baptismal sample belong (whatever that kind might turn out to be), how can the question of whether tigers form a kind be one that must be settled empirically?

The answer, I take it, is that, when Kripke says that it is an empirical question whether or not tigers form a kind (or form a single kind), what he has in mind is that it is an empirical question whether or not tigers form a *uniform* kind with a uniform structure. (See again 'N & N', 328, where Kripke says that, when we fix the reference of a term for a chemical substance or an animal or vegetable kind, we (defeasibly) suppose that all (or at any rate almost all) of the items in the baptismal sample belong to the same *uniform* substance or kind.) So, when we fix the reference of the term 'gold' via a formula along the lines of

> *Gold* is the substance that all or almost all of these items (the ones in the baptismal sample) instantiate,

we know a priori that, if there is such a thing as gold, then there is some substance that all the things that are gold instantiate. But it is an empirical question whether that substance will be uniform (will have a uniform composition). Gold might, for all the introducer of the term 'gold' knows, turn out to be like jade, which is not a uniform substance with a uniform composition, inasmuch as there are two different kinds of jade, with different compositions. In the same way, the introducer of the term 'tiger' knows a priori that, if there are tigers, they all belong to the same kind, in an undemanding sense of 'kind' that embraces both uniform and 'multiform' kinds. But it is an empirical question whether there is a uniform kind (with a uniform internal structure) to which all tigers belong.

Kripke avers that his picture of how natural kind terms acquire and conserve their reference is even further from a full-blown theory than his picture of how proper names acquire and conserve their reference (see 'N & N', 331). Two respects in which Kripke's account of the reference of natural kind terms is incomplete are worthy of note.

As we have seen, Kripke holds that animal, vegetable, and chemical kind terms get their reference fixed via a formula of the type *K is the kind instantiated by all (or almost all) of the items in the (baptismal) sample*. One might think that in this formula 'the kind instantiated by all or almost all of the items in the sample' meant 'the one and only kind instantiated by all or almost all of the items in the sample'. But, upon reflection, it seems that this cannot be right.

Kripke thinks that the baptismal sample for a natural kind term K may, and sometimes does, turn out to be (more or less evenly) split between two different kinds of K. To (re)use Putnam's example, the baptismal sample for 'jade' may turn out to be (more or less evenly) split between jadeite and nephrite. In this sort of case we can say that jade is the 'multiform' or 'disjunctive' kind that all or almost all of the items in the baptismal sample instantiate. But if we countenance the multiform kind *jade*, then we shall have to say that there are (at least) two different kinds to which all or almost all of the items in the baptismal sample for 'jadeite' belong: the more uniform kind, jadeite, and the less uniform kind, jade. The only way to avoid the conclusion that the items in the baptismal sample for 'jadeite' all or almost all belong to each of two different kinds is to deny the existence of the multiform kind, jade. But if we go that route, we shall have to say there is no kind to which all or almost all of the items in the (jadeite and nephrite) baptismal sample for 'jade' belong. In that case, given the Kripkean account of how a term like 'jade' has its reference fixed, there won't be two kinds of jade; instead, there will be no such thing as jade.

If I am right, Kripke cannot consistently maintain that there is at most one kind that all or almost all of the items in the baptismal sample for a given kind instantiate. It is in any case clear that the items in a baptismal sample that instantiate one biological or chemical kind will typically instantiate other biological or chemical kinds, given that such kinds come in hierarchies (hamsters are mammals, gold is a metal, and so on).

In light of these considerations, charity dictates that in the Kripkean reference-fixing formula

> K is the kind instantiated by all or almost all of the items in the baptismal sample

we take the reference-fixing description to be incomplete or inexplicit. K is the *right* kind instantiated by all or almost all of the items in the baptismal sample—the kind that is instantiated by all or almost all of those items, and satisfies some additional conditions (not specified by Kripke).

But what might a more explicit reference-fixing formula (with a more explicit reference-fixing description) be? We might try:

> K is the kind that is instantiated by all or almost all of the items in the baptismal sample, and is uniform, if some uniform kind is instantiated by all or almost all of the items in the sample.

As long as the items in a baptismal sample never all or almost all instantiate more than one uniform kind, or more than one multiform kind, the revised reference-fixing description will be satisfied by at most one kind.

Uniformity is, however, a matter of degree (*mammal* is a less uniform kind than *guinea pig*, but more uniform than *animal*). This suggests that a more explicit reference-fixing formula should make reference to maximal uniformity rather than uniformity. An attempt along these lines would be:

> K is the most uniform kind instantiated by all or almost all of the items in the baptismal sample.

This proposal has the merit of explaining why, when the reference of 'jadeite' is fixed with respect to a sample of jadeite items, the term 'jadeite' does not pick out the (multiform) kind jadeite-or-nephrite, even if all the items in the sample instantiate that kind. It also explains why, when the reference of 'jade' is fixed with respect to a sample of jadeite and nephrite items, 'jade' doesn't refer to a kind even less uniform than jadeite-or-nephrite.

It is open to doubt, though, whether the proposed reference-fixing formula will give intuitively the right results. It seems that someone could introduce the term 'gold', and only later find out that gold comes, or at any rate could come, in slightly different forms, corresponding to different isotopes of the element with atomic number 79.[81] It seems that this could happen even if the baptismal sample for 'gold' consisted entirely or almost entirely of just one form of gold. It is hard to see how the reference-fixing formula proposed above can accommodate this possibility: if the baptismal sample consists entirely or almost entirely of one form of gold, then the most uniform kind instantiated by all or most of the sample will presumably be gold-in-that-form rather than gold.[82]

Also, a biologist might discover some (conspecific) animals that did not belong to any known genus or even any known family. If she had a penchant for classificatory completeness, she might introduce one term for the species of those animals, a second for the genus, and a third for the family. (She would be open to the possibility of finding animals that belonged to the newly dis-covered genus, but not the newly discovered species.) Again, it is difficult to see how the proposed reference-fixing description could accommodate this possibility: the kind terms for the genus or family of the newly discovered ani-mal will not refer to the most uniform kind instantiated by all or almost all of the baptismal sample.

To recapitulate: we can (uncharitably) take Kripke to be offering a fully explicit account of how natural kind terms (typically) acquire reference—one that presupposes that, for each baptismal sample, there is at most one kind instantiated by all or almost all of the items in that sample.[83] Alternatively, we can (more charitably) take Kripke to be offering a partial, less than explicit, account of how natural kind terms (typically) acquire reference. But if we think of Kripke's account in this way, I don't see any obvious way of complet-ing that account without drawing on elements that do not appear in Kripke's discussion of how natural kind terms get their reference fixed.[84]

There is another respect in which Kripke's account of how natural kind terms acquire reference appears incomplete. As we have seen, Kripke sug-gests that we think of the introducer of the term 'gold' as fixing its reference via some such formula as 'Gold is the substance instantiated by the items over there, or at any rate, by almost all of them' ('N & N', 328). And he suggests that this gives us a good picture of how natural kind terms in general acquire their reference.

---

[81] In 'Natural Kind Terms and Recognitional Capacities', *Mind*, 107 (1998), 275–303, Jessica Brown points out that gold is one of a number of elements (including arsenic, aluminium, sodium, fluorine, cobalt, and manganese) that naturally occurs in just one isotope.

[82] Someone might object here that if the original sample for 'gold' contains only $_{79}Au^{197}$, then that (and no other isotope of Au) is what gold is. Similarly, they might say, if the original sample for 'water' does not contain heavy water, then heavy water is not water. This does not mesh very well with the way we actually use words like 'gold' and 'water'; it seems at most indeterminate, rather than false, that heavy water is (a kind of) water.

[83] That Kripke's account of how natural kind terms acquire reference, so construed, is inadequate has been pointed out by Segal and many others.

[84] For a promising account of how natural kind terms acquire reference that relies crucially on notions extraneous to Kripke's account, see Brown, 'Natural Kind Terms and Recognitional Capacities'.

Taken literally, though, a reference-fixing formula of the sort Kripke suggests for 'gold' will not yield the intuitively right extensions for natural kind terms. This is because in all or almost all cases, whenever an item falling under a natural kind has a certain location, another item not falling under that natural kind will have the same location.

Suppose, for example, that there is a tiger here now. Then there is also an aggregate of tiger cells here now, and the body of a tiger. But, it seems, the last two items are each different from the tiger, since the tiger will (probably) outlast the aggregate of cells, and the tiger's body will (probably) outlast the tiger.[85] Also, neither the aggregate of tiger cells, nor the tiger's body, is a tiger. If the aggregate of tiger cells and the tiger's body were tigers, there would be three tigers in the same place at the same time: the tiger we originally said was there, the aggregate of tiger cells (which, we have seen, is not the (original) tiger), and the tiger's body (which again is not the (original) tiger). But there aren't three tigers in the same place at the same time; so the tiger currently shares a location with at least two non-tigers (an aggregate of tiger cells and a tiger's body).

Similarly, suppose there is a portion of water here now. Then there is also a portion of hydrogen-and-oxygen here now. But the portion of hydrogen-and-oxygen is different from the portion of water, since the former, but not the latter, existed before its constituents were (as Peter van Inwagen would put it) 'arranged H_2O-wise'. And the portion of hydrogen-and-oxygen is not a portion of water different from the portion of water we originally said was there, since there is just one portion of water in this (exact) place at this time. So the portion of water shares a location with a non-portion-of-water.

Suppose, then, that the introducer of the term 'tiger' fixes its reference via the formula '*Tiger* is the (right) kind instantiated by all or almost all of the items over there'. Given how many and varied the items over there are, 'tiger' will either fail to refer (if the kind's being right requires that it is non-heterogeneous in certain ways), or pick out a highly disjunctive kind (*animal of the species Felis tigris or aggregate of tiger cells or tiger's body or ...*). Similarly, if someone introduces the term 'water', or 'gold', via the formula '*Water* (or *gold*) is the (right) kind instantiated by all or almost all of the items over there', the term introduced will either fail to refer, or refer to a kind less uniform than, and different from, water (or gold).

Moral: if we think of the introducer of the term 'tiger' (or 'gold') as fixing that term's reference via the formula '*Tiger* (or *gold*) is the kind instantiated by all or almost all of the items over there', we must understand 'the items over there' as 'the (right) items over there', just as we must understand 'the kind' as 'the (right) kind'. The term 'tiger' could not have had its reference fixed by the formula

> *Tiger* is the (right) kind instantiated by all or almost all of the items over there

---

[85] Though we shall see that this last claim is contestable; Ch. 4, sect. 1.

because 'the items over there' does not pick out an appropriate sample for fixing the reference of 'tiger'.[86] If we want to provide a Kripkean account of how 'tiger' has its reference fixed, we must accordingly replace the description 'the items over there' in the above formula. I am unsure, though, what Kripke would want to replace it with.

It is easy enough to find a description that picks out tigers, but not tiger bodies, tiger-cell-aggregates, or other non-tigers: 'the tigers over there' is one such. I take it, though, that Kripke would not want to say that 'tiger' has its reference fixed via the formula

> *Tiger* is the (right) kind instantiated by all or almost all of the tigers over there.

Alternatively, we might try replacing 'the items over there' by 'the animals over there', so that the reference-fixing formula would be:

> *Tiger* is the (right) kind instantiated by all or almost all of the animals over there.

Since tigers' bodies, tiger-cell-aggregates, and the like are not animals, there is no problem about the appropriateness of the sample picked out by the description. But if we think of the reference of 'tiger' as fixed by the above formula (and we have Kripkean views on the a priori) then we shall conclude that it is a priori, for the reference-fixer, that tigers are animals; and Kripke would not accept this last claim (see 'N & N', 319).

Going in a different direction, we might think of the introducer of 'tiger' as fixing its reference via the formula

> *Tiger* is the (right) kind instantiated by all or almost all of *these things*,

where 'these things' picks out what Kripke calls the 'paradigmatic instances' of the kind.

But suppose we ask the introducer of the term 'tiger', 'What kind of thing is a tiger?', and she answers, 'The same kind of thing as *these things*' (pointing at some tigers). Why, when she points and uses the term 'these things,' does she pick out some tigers, rather than, say, some tiger bodies, or some aggregates of tiger cells? A natural answer is that she intends, when using the demonstrative phrase 'these things', to refer to the things that are F, and the tigers over there, rather than the tiger bodies, or the aggregates of tiger cells, are the things that are F. But what exactly is F here? This is more or less the question we started with when we asked: Given that the introducer of the term 'tiger' cannot have intended that the term 'tiger' apply to all or almost all of the things over there, which of the things over there did he intend the term 'tiger' to apply to?

---

[86] Unless, of course, 'the items over there' is taken as elliptical, or having a contextually restricted extension, in a way that would need to be spelled out.

Having considered two ways in which Kripke's account of how natural kind terms acquire reference appears incomplete, we may consider two ways in which it might be thought to be mistaken.

We have already seen that, for Kripke, given the way that 'gold' and 'water' had their reference fixed, it is necessary, albeit a posteriori, that gold = the element with atomic number 79, and that water = H_2O. Not a few philosophers have had doubts about these claims. Some of these philosophers have held that it is simply true that water could have had a different atomic structure.[87] Others, such as David Lewis, have maintained only that it is un-false (outside a particular context):

> Like any up-to-date philosopher of 1955, I think 'water' is a cluster concept. Among the conditions in the cluster are: it is liquid, it is colourless, it is odourless, it supports life. But, *pace* the philosophers of 1955, there is a lot more to the cluster than that. Another condition in the cluster is: it is a natural kind. Another condition is indexical: it is abundant hereabouts. Another is metalinguistic: many call it 'water'. Another is both metalinguistic and indexical: *I* have heard of it under the name 'water.' When we hear that XYZ off on Twin Earth fits many of the conditions but not all, we are in a state of semantic indecision about whether it deserves the name 'water'... When in a state of semantic indecision, we are often glad to go either way and accommodate our own usage to the whims of our conversational partners...So if some philosopher, call him Schmutnam, invites us to join him in saying that the water on Twin Earth differs in chemical composition from the water here, we will happily follow his lead. And if another philosopher, Putnam...invites us to say that the stuff on Twin Earth is not water...we will just as happily follow his lead. We should have followed Putnam's lead only for the duration of that conversation, then lapsed back into our accommodating state of indecision. But, sad to say, we thought that instead of playing along with a whim, we were settling a question once and for all. And so we came away lastingly misled.[88]

Notice, though, that when Kripke and Putnam argued that nothing but H_2O could have been water, they did not see themselves as inviting us to accommodate them in their (possibly idiosyncratic) inclination to classify only H_2O as water. It is not as though they were saying: 'Why not suppose that XYZ (the superficially water-like substance with the different microstructure from H_2O) wouldn't be—couldn't be—water? What bad things would happen if we did that?' Instead, they were trying to persuade us that—even before they had brought the matter to our attention—we had already been (on balance) inclined to classify XYZ as non-water.[89] And they did persuade us—or at least

---

[87] See e.g. D. H. Mellor, 'Natural Kinds', *British Journal for the Philosophy of Science*, 28 (1977), 302–3, and D. Ackermann, 'Natural Kinds, Concepts, and Propositional Attitudes', in P. French, T. Uehling, and H. Wettstein (eds.), *Contemporary Perspectives in the Philosophy of Language* (Minneapolis: University of Minnesota Press, 1979), 480.

[88] D. Lewis, 'Reduction of Mind', in S. Guttenplan (ed.), *The Blackwell Companion to the Philosophy of Mind* (Oxford: Basil Blackwell, 1998), 424.

[89] 'If there were a substance...which had a completely different atomic structure from that of water, but resembled water in these [qualitative] respects, would we say that some water wasn't H_2O? I think not. I think we would say that just as there is a fool's gold, there could be a fool's water' ('N & N', 323).

most of us—of that. Their success is unsurprising, if we really did have that inclination *ex ante*. It seems considerably more surprising if, as Lewis seems to suppose, we were actually in a state of semantic indecision *ex ante*.

Also, suppose Lewis is right. Then, when a representative sample of (theoretically unbiased) people are asked (in a non-tendentious way) for a ruling on whether XYZ is or is not water, we would not expect a negative answer to predominate. For myself, I would be surprised if a negative answer did not predominate, in much the way I would be surprised if philosophically uncorrupted people didn't for the most part rule that we don't have knowledge in Gettier cases. Of course, this may simply reflect my prior theoretical commitments. But there is apparently at least some psychological research suggesting that (presumably theoretically unbiased) 10-year-olds tend to say that XYZ is not water.[90] The issue is an empirical one. If it should turn out that only philosophers baulk at classifying XYZ as water, I am ready to defer in my usage to the non-philosophical majority, and say that 'water', like 'glue', is not the name of a kind with a chemical essence.[91]

A different sort of objection to Kripke's identification of water with H_2O rests on the idea that, unlike H_2O, water can (indeed, almost invariably does) contain impurities—bits that are not hydrogen or oxygen (or parts of hydrogen or oxygen).[92] Noam Chomsky has argued that issues involving impurities show not just that water is not H_2O, but also that physico-chemically indiscernible things may differ with respect to being water, so that being water cannot simply be a matter of having the right atomic structure (whatever that structure is). As Chomsky sees it, whether something is water depends on human interests and concerns. If I fill a cup from an ordinary kitchen tap, what I get is a cup of water. If I dip a teabag into it, I now have a cup of tea, and not water, in spite of the fact that, from a chemical point of view, both before and after, what I had in my cup was H_2O, together with a negligible amount of impurities. Moreover, Chomsky argues, suppose that the tap I filled my cup from had been connected to a reservoir in which tea had been dumped as a new kind of purifier. In that case, my cup would be filled with water—not tea—even if the water that had come out of my tap had been chemically indiscernible from the tea that (in the original story) I made by dipping a teabag into (ordinary) tapwater.[93]

Someone might respond to these objections by saying that 'water' has a variety of senses, and only in the strictest sense of 'water' is it true that water

---

[90] See F. C. Keil, *Concepts, Kinds, and Cognitive Development* (Cambridge, Mass.: MIT Press, 1989).

[91] Imagine a possible world in which all matter has a homoeomerous structure. As long as there is some appropriately sticky substance in that world, there is glue in that world (and not simply 'fool's glue'). Hence 'glue'—unlike, say, 'hydrogen'—is not a name for a kind of substance that essentially has a certain sort of chemical composition.

[92] See e.g. B. Aune, 'Determinate Meaning and Analytic Truth', in G. Debrock and M. Hulswit (eds.), *Living Doubt* (Dodrecht: Kluwer Academic Publishers, 1994).

[93] N. Chomsky, 'Language and Nature', *Mind*, 104 (1995), 22.

is H_2O.[94] Alternatively, one might say that, although the word 'water' is unambiguous, it can be applied in accordance with stricter or laxer standards, and only when the standards are maximally strict is it true that water is H_2O.[95] On either view, it might be conceded that Aune speaks the truth when he says, 'Water as it occurs in nature is mostly but not entirely H_2O,' or that Chomsky speaks the truth when he says, 'Chemically indiscernible substances may differ with respect to being water'; even though Kripke is right in maintaining that water just is H_2O. What Aune and Chomsky are saying would not, on the view at issue, contradict what Kripke is saying, since the relevant senses or standards of application are different.

In fact, though, I am doubtful about the idea that there is a sense of 'water' which means 'H_2O give or take a few impurities'. Surely the following statement is simply false (not false relative to some sense of 'water', or some standard of application for the term 'water'):

> (T) It could be that: there is nothing in the Thames except water (and its constituents), and there is arsenic in the Thames.

If 'water' unambiguously, standard-independently rigidly designates H_2O, there is no mystery about why T is (unambiguously, standard-independently) false. 'There is nothing in the Thames except water (and its constituents)' is true only if everything in the Thames is (a portion of) water or a constituent thereof. If 'water' rigidly designates H_2O, then this last statement is true only if everything in the Thames is (a portion of) H_2O or a constituent thereof. Now 'There is arsenic in the Thames' is true only if something in the Thames is (a portion of) arsenic. But, necessarily, no portion of arsenic is a portion of H_2O or a constituent thereof. So if 'There is nothing in the Thames except water (and its constituents)' is true, then 'There is arsenic in the Thames' is false; whence T is false.

Suppose, though, that there were a sense of 'water' which meant 'H_2O give or take a few impurities'. Then there would be a sense of 'water' according to which a bit of arsenic could be a (trace) constituent of a portion of water, and T should have a true reading.

Still, someone may insist:

> Water is the stuff that is in the Thames and comes out of the taps. The stuff that is in the Thames and comes out of the taps undeniably contains impurities (bits that are neither hydrogen nor oxygen nor constituents thereof). So how can water be H_2O?

The portions of water that are in the Thames or come out of the taps do contain impurities. But, I am inclined to think, they are not partly composed of (partly constituted by) impurities, any more than a desk drawer is partly

---

[94] See H. Putnam, *Representation and Reality* (Cambridge, Mass.: MIT Press, 1988), 293.
[95] See B. Abbott, 'A Note on the Nature of "Water"', *Mind*, 106 (1997), 311–19.

composed of its contents, or a region of space is partly composed of the objects it contains. So nothing prevents us from saying both that water is H_2O and that the water that comes out of the taps, or is in the Thames, contains impurities.

'But (portions of) salt water or rose-water surely are partly constituted by things that aren't (portions of) hydrogen or oxygen, or the constituents thereof. Moreover, non-potable water wouldn't be non-potable, and tapwater wouldn't taste the way it does, if non-potable water and tapwater were made of nothing but hydrogen and oxygen.'

On the most natural way of understanding the term, it does seem as though salt water is constituted not just of water, but also of salt. *Mutatis mutandis*, the same may hold for rose-water, non-potable water, and tapwater. But if it does, then salt water isn't water, and neither is rose-water, non-potable water, or tapwater. To be sure, salt water, or rose-water, is *mostly* water. But neither one is water, any more than enriched flour is flour. Enriched flour is not (just) flour; it is flour + something else. Salt water and rose-water are not (just) water; they are water + something else.[96]

In sum, the fact that water can contain impurities does not seem to show either that water isn't H_2O, or that whether or not something is water can depend on something other than that thing's atomic structure. A strong form of Kripke's thesis—that '"water" rigidly designates H_2O is unambiguously and standard-independently true'—seems to me both initially plausible, and defensible, in the face of the arguments offered against it by Aune and Chomsky (and the quite different arguments of Lewis).

In *Naming and Necessity*, in opposition to the (then) prevalent view, Kripke argued that *gold = the element with atomic number 79* and *water = H₂O* are necessary, even though they are a posteriori. In that same work, again in opposition to the (then) prevalent view, he argued that *cats are animals* and *tigers are mammals* are a posteriori, even though they are necessary. True, Kripke maintained, cats are animals, and anything (actual or possible) that looked as much as you please like a cat but was an automaton, or a demon, couldn't be a genuine cat. Still, he argued, that cats are animals rather than automata or demons is an (unsurprising) empirical truth rather than a truth we have access to simply in virtue of grasping the concept of cat:

Cats might turn out to be automata, or strange demons... planted by a magician. Suppose they turned out to be a species of demons. Then on [Putnam's] view, and I think also my view, the inclination is to say, not that there turned out to be no cats, but that cats have turned out not to be animals as we originally supposed. The original

---

[96] See my 'Matter and Actuality in Aquinas', *Revue Internationale de Philosophie*, 204 (1998), sect. III. The temptation to think that statements such as 'Rose-water is water' and 'Tapwater is water' must be true is, I think, reduced when we consider that we are not inclined to say that rose-water is tapwater. Of course, someone *can* mean by 'tapwater' the water (H_2O) that comes out of the taps. But when they say things like 'Tapwater owes much of its taste to chlorine', I doubt they do mean that.

concept of cat is: *that kind of thing*, where the kind is identified by the paradigmatic instances. ('N & N', 319)

We could have discovered that the actual cats we *have* are demons. Once we have discovered, however, that they are *not*, it is part of their very nature that, when we describe a counterfactual world in which there were such demons around, we must say that the demons would not be cats. It would be a world containing demons masquerading as cats. Although we could say cats *might turn out to be demons*, of a certain species, given that cats are in fact animals, any cat-like being which is not an animal, in the actual world or in a counterfactual one, is not a cat. ('N & N', 321)

There are two points in the first passage. First, the introducer of the term 'cat' fixed its reference by applying it to some paradigmatic instances. (We can imagine a palaeontologist introducing an animal kind term K in the absence of any paradigmatic instances, via a formula of the sort 'K is the kind of animal whose species is the transitional species between this species and that species'. In such a case, Kripke would say, it would be a priori (for the introducer of the term, at least), as well as necessary, that Ks are animals. But 'cat' did have its reference fixed in this sort of instance-independent way.)

Secondly, as Kripke sees it, the introducer of the term 'cat' did not build animality, or any condition that entails animality, into the description that fixed the reference of 'cat'. The introducer no doubt surmised that cats are animals; for this reason she bore a quite different epistemic relation to *cats are animals* than, say, the introducer of the term 'gold' bore to *gold is the element with atomic number 79*. For all that, Kripke is inclined to say, animality was no more built into the reference-fixing description for 'cat' than elementhood was built into the reference-fixing description for 'gold'. More generally, the reference-fixing description for 'cat' and the original concept of cat are 'permissive' or 'non-committal' inasmuch as they leave it an open question not just, say, whether cats are mammals or reptiles, but also whether cats are animals, or automata, or demons, or…

True, 'cat' might have been introduced via a reference-fixing description that made no reference to animality, and left open the (epistemic) possibility that cats are automata or demons. Whether it actually was so introduced I don't know. After all, consider a term like 'grapefruit' or 'sunflower'. Isn't it possible that the description with which the introducer of the term 'grapefruit' fixed its reference involved being a fruit, and that the description with which the introducer of the term 'sunflower' fixed its reference involved being a flower? To take another example, the word 'deer' seems to have come from a word which in Old English meant 'animal' and in Middle English had its extension narrowed to members of the family *Cervidae*. Isn't it possible that the reference-fixing description for 'deer' (in the narrower sense) involved being an animal?

If, however, it is possible that the reference-fixing description for 'sunflower' involved being a flower, and that the reference-fixing description for 'deer' involved being an animal, then it also seems possible that the reference-fixing description for 'rose' involved being a flower, and that the reference-fixing description for 'cat' involved being an animal.

Even granting that the description with which the introducer of the term 'cat' fixed its reference made no reference to animality, it is unclear whether the reference-fixing description for 'cat' and the original concept of cat are as permissive as Kripke suggests.

Let us suppose that the person who introduced the term 'cat' was in fact an English explorer who came across cats in Persia. The explorer then came back to England and passed the word 'cat' on to us, who use it to this day.

Now consider an alternative possible world whose history diverges from the history of the actual world shortly after the explorer has seen and baptized cats. In the alternative world someone in Persia goes to elaborate lengths to convince the explorer that he is the victim of a hoax, and that the things he saw, and took to be animals, and called cats, were not genuine animals at all, but very lifelike cloth stuffed animals—cloth stuffed animals that no more represent any real species of animal than, say, a cloth stuffed unicorn does. The explorer—who only saw one or two (motionless, soundly sleeping) cats, at twilight and not very close up—is persuaded that the things he saw were not real animals at all, but cloth stuffed animals. What will he conclude? It seems to me that he might naturally conclude that he was a victim of a hoax, and there aren't actually any cats. He might naturally regard the cloth stuffed animals not as cats, but as props used by the hoaxers to get him to believe in cats.

Suppose that, upon coming to believe that the things he saw and called cats were cloth stuffed animals, the explorer would conclude that there aren't any cats. That would suggest that the introducer of the term (from the start) had an intention not to apply the term 'cat' to cloth stuffed animals (in other words, to apply the term 'cat' only to non-cloth-stuffed-animals). Given that original referential intention, it would be a priori (for him, at least) that cats are not cloth stuffed animals.

Suppose now that, instead of coming to believe that the things he saw and called cats were ordinary cloth stuffed animals, the explorer had come to believe that the things he saw and called 'cats' were mobile, radio-controlled cloth stuffed animals, or cleverly constructed automata. It still seems to me that the explorer might naturally have concluded that there were no cats, and the mobile, radio-controlled cloth stuffed animals, or the cleverly constructed automata, were not cats, but (very sophisticated) props used by the hoaxers to get him to believe in cats. For this reason, it is not clear to me that what Kripke calls 'the original concept of cat' really is permissive enough to leave it an open question whether or not cats are automata.

Here is a slightly different case that suggests that the original concept of *cat* may not be permissive in the way Kripke seems to suggest. Suppose that this time the Persians convince the explorer that the things he saw, and thought were animals, and called cats, were actually taxidermized animals rather than living ones. As I shall argue in Chapter 4, section 1, taxidermized animals (and, more generally, dead animals) are not animals, any more than fossilized leaves are leaves. Suppose that our explorer accepts this (Aristotelian) claim. Upon coming to believe that the things he saw were taxidermized animals, which of his previous beliefs would he hold onto, and which of his beliefs would he give up? If the original concept of *cat* is permissive in the way Kripke appears to think, it seems that he would hold onto the belief that the things he saw, and called cats, were cats, and give up the belief that cats are animals. (Cats, he would conclude, are not animals, but taxidermized animals.) But I find it intuitively more plausible that he would instead hold onto the belief that cats are animals, and give up the belief that the things he saw, and called cats, were cats; he would conclude that the things he thought were paradigmatic instances of the cat kind were not after all cats, but (taxidermized) corpses of cats. This suggests that it was part of the original referential intention of the explorer to apply the term 'cat' to an organism, rather than to the (dead, or for that matter, the living) body of an organism.

We can bring out the same point with a sortal-involving variant of the Madagascar case discussed by Evans and Kripke. Suppose that the explorer sees what he takes to be an animal, but is in fact a taxidermized (feline) corpse. Rather than coining the name 'cat' for the thing he sees, the explorer asks his native guide what *that* is called (pointing to the thing he thinks is an animal, but is in fact a taxidermized corpse). The guide, who is aware that what the explorer is pointing at is a corpse, responds, 'barandi', where 'baran' is the native term for cats, and 'barandi' is the native term for 'corpse-of-a-cat'. The explorer doesn't realize that the guide thinks (indeed, knows) that the explorer and the guide are looking at a taxidermized corpse (just as the guide doesn't realize that the explorer *isn't* aware that the explorer and the guide are looking at a taxidermized corpse). The misunderstanding is never cleared up, and the explorer returns to his own country, and passes on the name 'barandi' to his countrymen, who come to share his belief that a barandi is a kind of animal, rather than a corpse of a certain kind of animal. (If you show them something that they can see is a taxidermized cat, and ask them whether it's a barandi, they answer, 'Well, not really . . . it's a dead, stuffed barandi.')

When the explorer and his countrymen use the term 'barandi', I take it, it has as its extension members of the species *Felis catus*; the same cannot be said for the guide or the other natives. It's like the case brought to light to Evans: when Europeans use the name 'Madagascar', it has as its referent an island off the coast of Africa, though the same cannot be said for those from whom Marco Polo got the name 'Madagascar'. In the case of Madagascar, Kripke plausibly suggests that the shift in reference occurs because Europeans have an 'overriding intention' to use the name 'Madagascar' to refer to an island off Africa, rather than to a bit of the African mainland ('N & N', 768–9). In the 'barandi' case one could similarly plausibly suggest that the shift in reference occurs because the explorer and his countrymen have an overriding intention

to use the sortal *barandi* to refer to living beings of a certain kind rather than
to corpses of living beings of a certain kinds. If, however, we attribute such an
intention to the explorer and his countrymen, we cannot also say that the
explorer uses the term 'barandi' to refer to things that he surmises are living
beings, but might for all that turn out to be corpses (or stuffed cloth animals,
or automata). In other words, we cannot think of the (explorer's) 'original
concept of' barandi as permissive in the way Kripke appears to think it is.

In the barandi case the explorer has a concept before he has a sortal predicate expressing that concept. Rather than, say, coining the word 'cat' to express
that concept, he appropriates what he (mistakenly) takes to be the native sortal predicate that already expresses that concept. But the concept the explorer
has is the very same concept he would have had if he had coined the word 'cat'
to express it; the concept he has, it seems, is just the concept of *cat*. So if the
'original concept' of *barandi* is less permissive than Kripke suggests the original concept of *cat* is, then the original concept of *cat* is less permissive than
Kripke suggests the original concept of *cat* is.

I do not mean to argue here that Kripke is wrong in (tentatively) asserting
that cats might have turned out not to be animals.[97] Perhaps the original concept of *cat* allowed for our finding out that cats are actually a weird sort of
plant. Nor am I arguing that Kripke is wrong in (tentatively) asserting that
cats might have turned out not to be organisms: perhaps the original concept
of *cat* allowed for our finding out that cats are actually clusters of different
organisms (the way sponges are, if I am not mistaken). I mean only to suggest
that it is an open and difficult question just how permissive the original concept of *cat* actually is.

It may also be worth emphasizing that we cannot assume at the outset that
there will always be a fact of the matter about whether the original concept of
cat is permissive in a certain respect. Perhaps the original referential intentions
of the person who introduced the term 'cat' were vague, so that the original
concept of *cat* neither definitely allowed nor definitely precluded the (epistemic) possibility of finding out that cats don't have a certain feature we all
assume they have. Perhaps *being an organism*, or *being organic*, or *being a non-
automaton* is such a feature.

After all, it does not seem obviously implausible that, if we found out that
all the things we had thought were cats were actually, say, automata, we'd have
to *decide* whether to say that cats have turned out not to exist, or to say that
cats have turned out to be automata. The former sounds somehow more
natural to me, but I don't know that the latter would be an outright mistake;
perhaps it would simply be a less natural decision, in the circumstances.

---

[97] I say 'tentatively' in light of the already cited passage in which Kripke says that if the things we
take to be cats turned out to be demons, then, on Putnam's view, 'and I think also my view, the inclination is to say, not that there turned out to be no cats, but that cats have turned out not to be animals
as we originally supposed' ('N & N', 319).

2

Necessity

Modal logic (the logic of possibility and necessity) goes back at least as far as Aristotle's *Prior Analytics*, and flourished in the high Middle Ages.[98] After the demise of scholasticism the subject received comparatively little attention, especially from empiricist philosophers. But the first half of the twentieth century saw a renewed interest in modal logic, arising at least in part from interest in the nature of implication.

In their *Principia Mathematica* Russell and Whitehead employed a two-place sentence connective called 'the horseshoe' ($\supset$). They stipulated that formulae of the form $A \supset B$ were true *unless* A was true and B was false. (To put this another way, they stipulated that $A \supset B$ had the same truth-conditions as Not: both A and not-B.) When it was true that $A \supset B$, Russell and Whitehead said that *A materially implied B*.

For any formula A, either A materially implies ~A (that is, the negation of A), or ~A materially implies A. Moreover, a false A materially implies B, for any B, and a true A is materially implied by any B whatever. All of this makes it look as though 'material implication' has little to do with implication as usually understood: we don't ordinarily suppose that falsehoods imply anything you please, or that truths are implied by anything you please. Nor do we ordinarily suppose that any statement either implies or is implied by its negation.

For these reasons, C. I. Lewis maintained that there is a sense of 'implication' not captured by Russell and Whitehead's (truth-functional) notion of material implication.[99] In this other sense of 'implies', A's materially implying B is a necessary but insufficient condition for A's implying B; Lewis accordingly called this sort of (not merely material) implication *strict implication*. What is needed for A to strictly, and not merely materially, imply B? A natural thought is that the truth of A must guarantee the truth of B. And a natural way of understanding the notion of guarantee is as follows: A guarantees the truth of B just in case it is logically impossible that: A and ~B. In fact, Lewis held that A strictly implies B just in case it is logically impossible that: A and ~B. He then

---

[98] Cf. S. Knuuttila, 'Modal Logic', in N. Kretzmann, A. Kenny, and J. Pinborg (eds.), *The Cambridge History of Later Medieval Philosophy* (Cambridge: Cambridge University Press, 1982).

[99] Cf. C. I. Lewis, 'Implication and the Algebra of Logic', *Mind*, 21 (1912), 522–31.

explicated the notion of logical impossibility (and the related notions of logical possibility and necessity) axiomatically. The idea was to set out a system of axioms for a logic of modalities and strict implication akin to the sorts of axiomatic systems for propositional (non-modal) logic found in Russell and Whitehead and others.[100]

Lewis provided not one but five axiomatic systems for propositional modal logic—the so-called S1–S5. The systems are numbered according to their increasing strength: all the theses of S1 are theses of each of S2–S5, but each system contains theses not contained by any of its predecessors in the sequence. Thus, where 'A ⇒ B' means 'A strictly implies B', '◇A' means 'It is (logically) possible that A' and '□A' means 'It is (logically) necessary that A', all of the following are theses of S5:

(a) $\Diamond A \Rightarrow \Box \Diamond A$
(b) $\Box A \Rightarrow \Box \Box A$
(c) $A \Rightarrow \Box \Diamond A$
(d) $\Box A \supset A$
(e) $\Diamond (A \, \& \, B) \Rightarrow \Diamond A$.

(a) and (c) are not theses of any of S1–S4; (b) is not a thesis of any of S1–S3; (e) is a thesis of each of S1–S5 except S1, and so on.

Lewis was not the only logician working in the first half of the twentieth century to provide axiomatic systems of modal logic. Drawing on some ideas of Gödel, Robert Feys propounded system T, characterized by the pair of axioms:

$$\Box p \supset p \text{ (sometimes called 'the axiom of necessity')}$$

and

$$\Box (p \supset q) \supset (\Box p \supset \Box q).\text{[101]}$$

System T is stronger than S1 and S2, weaker than S4 and S5, and independent of S3. Becker discusses a different system, often called the Brouwerian (or Brouwersche) system, whose distinctive modal axiom is:

$$p \supset \Box \Diamond q.\text{[102]}$$

The Brouwerian system of modal logic is stronger than S1–S3 (and T), weaker than S5, and independent of S4.

All the modal logics discussed so far have been propositional. That is, while they contain all the theses of first-order propositional logic, they do not

---

[100] See C. I. Lewis and C. H. Langford, *Symbolic Logic* (New York: Dover, 1932). A detailed and very clear discussion of Lewis's axiomatic systems is found in G. E. Hughes and M. J. Cresswell, *An Introduction to Modal Logic* (London: Methuen, 1968), 213–55.

[101] Cf. R. Feys, 'Les Logiques nouvelles des modalités', *Revue Néoscholastique de Philosophie*, 40 (1937), 517–53.

[102] Cf. O. Becker, 'Zur Logik der Modalitäten', *Jahrbuch für Philosophie und Phänomenologische Forschung*, 11 (1930), 497–548.

contain the theses of first-order logic with quantifiers. Before 1950 some—though not a great deal—of work was done on axiomatic systems of quanti-fied modal logic. Thus Ruth Barcan Marcus set out systems of quantified modal logic based on Lewis's S2 and S4.[103] As in the case of propositional modal logic, logicians ended up formulating a family of axiomatic systems of varying strength. Where '$(\forall x)Fx$' means 'Everything is F', some but not all of the systems contained the following pair of theses:

(BF) $(\forall x)\Box Fx \supset \Box(\forall x)Fx$

and

(CBF) $\Box(\forall x)Fx \supset (\forall x)Fx$.

'BF' stands for 'Barcan formula', and 'CBF' for 'converse Barcan formula'.[104] BF says that if everything is necessarily F, then necessarily everything is F; CBF says that if necessarily everything is F, then everything is necessarily F.

While the wealth of axiomatizations of modal logic proposed in the first half of the twentieth century shed light on the nature of possibility and neces-sity, they also raised a number of questions. Given that some of the systems of propositional modal logic were stronger than others, were the weaker ones too weak to capture all the truths of propositional modal logic? Or were the stronger ones too strong to capture only truths of propositional modal logic? Some of Lewis's systems—e.g. S1—seem clearly too weak to capture all the truths of propositional modal logic. (As we have seen, in S1 one cannot prove that 'Possibly A and B' strictly implies 'Possibly A'.) But are all of the theses of S5 truths of modal logic? Is it in fact the case that whatever is actually true, or possibly true, is necessarily possibly true? Similar questions arise for the weaker and stronger systems of quantified modal logic: should a quantified modal logic contain among its theses the Barcan formula or its converse?

It may be that the possible and the necessary, as Aristotle would put it, are spoken in many ways. And it may be that whether or not a given axiomatic system of propositional or modal logic is 'too weak' or 'too strong' depends on the kind of possibility and necessity (or the sense of 'possibility' and 'necessity') at issue. Thus it has been suggested that S5 is the appropriate logic for metaphysical possibility and necessity, but not for physical possibility and necessity. Even a system as weak as T has been thought to be too strong to be an appropriate logic for deontic necessity, in virtue of containing the axiom $\Box p \supset p$.[105]

---

[103] Cf. R. C. Barcan, 'A Functional Calculus of First Order Based on Strict Implication', *Journal of Symbolic Logic*, 11 (1946), 1–16.

[104] The Barcan formula is so-called in honour of Ruth Barcan Marcus, who discussed the related formula $\Diamond(\exists x)Fx \Rightarrow (\exists x)\Diamond Fx$.

[105] Deontic necessity is the kind of necessity alluded to in such statements as 'Promises must be kept'. That promises must be kept does not entail that they are in fact kept; hence the so-called 'axiom of necessity' fails.

Even if we have in mind a particular sort of possibility and necessity, there are still questions about whether, say, the characteristic axiom of S5, or the Barcan formula, is a logical truth about that sort of possibility and necessity. Moreover, there are questions about what kind of assumptions about the nature of possibility and necessity are, so to speak, behind the choice of a particular axiomatic system of propositional or quantified modal logic. Kripke's semantics for modal logic shed light on both sorts of questions.[106]

In setting out that semantics, Kripke begins by introducing the idea of a *model structure*. A model structure is a triple (**G**, **K**, **R**) where **G** is an element of **K**, **K** is a set, and **R** is a reflexive relation on **K**. Intuitively, he says, we may think of things this way: **G** is the actual world, **K** is a set of possible worlds, and **R** is a relation of relative possibility.

In 'Semantical Considerations on Modal Logic' Kripke does not say much about what sort of thing a possible world is, and different philosophers have conceived of them in rather different ways. According to some philosophers, possible worlds are structural universals; according to others, they are sets of sentences or propositions; according to at least one, they are concrete universes. But whatever exactly possible worlds are, they are in some ways like, and in some ways unlike, ordinary possibilities. Possible worlds are like ordinary possibilities in being the sort of things that may or may not be 'actualized' ('actual') or not. Some possibilities (e.g. the possibility that gravity exists) are actualized or realized, and others (e.g. the possibility that gravity screens exist) are not. Similarly, among the set of possible worlds **K**, some worlds—actually, some one world—is actual (**G**), and (all the) others are not.[107] On the other hand, ordinary possibilities may be either compossible or incompossible with each other. (Possibilities are compossible if they can be jointly actualized, and incompossible otherwise.) By contrast, no two possible worlds are compossible (in other words, the only possible world that a given possible world is compossible with is itself). So there is only one actual world, although there are many actual (or actualized) possibilities.

In order for possible worlds to do the sort of work Kripke wants, they must be parameters of truth and falsity. In other words, they must be the sorts of things that formulae can be true or false with respect to (or 'at').

As for the relation **R**, it should be understood in this way: to say that world H_1 stands in **R** to world H_2 is to say that whatever is true at H_2 is possible at H_1. Given that **R** is reflexive, we know that each world is possible relative to itself: this captures the intuitive idea that whatever is actually true at a world is a fortiori possibly true at that world.

---

[106] In the exposition that follows, I draw on Kripke's 'Semantical Considerations on Modal Logic', *Acta Philosophica Fennica*, 16 (1963), 83–94; repr. in L. Linsky (ed.), *Reference and Modality* (Oxford: Oxford University Press, 1977).

[107] As we shall see, though, although Kripke's semantics requires that there be at least (and at most) one actual world, it does not require (though it allows) that there be at least one 'alternative' (or 'merely possible') world.

Kripke says that φ is a *model* for a model structure (G, K, R) if φ is a function which assigns to each pair consisting of an atomic formula P and a possible world H exactly one of {truth, falsity}. A model fixes the truth-values of all the atomic formulae at all the possible worlds in the model structure. Once this is done, the truth-values of the non-atomic formulae are fixed inductively as follows: assuming φ (A, H) and φ(B, H) have been defined for all H in K, φ (A & B, H) = truth if φ(A, H) = truth = φ(B, H); φ(A & B, H) is falsity otherwise. φ(~A, H) = truth if φ(A, H) = falsity; φ(~A, H) = falsity otherwise. Finally, φ($\Box$A, H) = truth if φ(A, H') = truth at every H' in K such that H stands in R to H'; φ($\Box$A, H) = falsity otherwise. This last clause tells us that $\Box$A is true at a possible world H if and only if A is true at every world that is possible relative to H.

The reader may find it helpful to think of model structures and models as the joint determinants of the truth or falsity of a formula at a possible world. Sometimes the truth or falsity of a formula at a world will be left open by the choice of a model structure, and fixed only by the choice of a model. Suppose, for example, that our model structure is (G, K, R), where H $\in$ K. Whether or not the atomic formula P comes out true at H will depend on whether we choose a model φ that assigns truth to the pair <P, H>, or a different model φ' that assigns falsity to it. Other times, the truth or falsity of a formula at a world will be determined by the choice of a model structure, independently of the choice of a model. Suppose, for example, we choose a model structure in which G is the only element of K. Then the formula ~(P & ~$\Box$P) will come out true at G, whichever model we choose for that model structure. ($\Box$P is true at a world G (on a model) just in case P is true at every world that is possible relative to G (on that model). If the only world possible relative to G is G itself, then whenever P is true at G (on a model), so is $\Box$P.)

Sometimes the truth or falsity of a formula at a world will be independent, not just of the choice of a model, but also of the choice of a model structure. The reader who works through the definitions will be able to see that ~(P& ~P) and ~($\Box$P & ~P) are true at G on any model associated with any model structure.

Consider the set of formulae F that are true at G on any model associated with any of Kripke's model structures. A natural question is whether F contains all or only the formulae that are theses of the various systems of propositional modal logic discussed above. As Kripke showed, it turns out that F contains all but not only the theses of S1 and S2, and only but not all the theses of the Brouwersche system, S4, and S5. In fact, F contains all and only the theses of Feys's system T. Thus Kripke's model structures and models provide us with a semantics for Feys's system T.

If we want a semantics for a system weaker than T—say, S2—we need to allow K to contain 'non-normal' possible worlds. A possible world H is non-normal if no possible world is possible relative to H (not even H itself). If non-normal worlds are allowed in K, we must replace the requirement that R be reflexive with the (weaker) requirement that **R** be *quasi-reflexive*. (A relation is reflexive just in case, it always holds between a thing and itself; a relation is quasi-reflexive just in case, whenever it holds between a thing and something or other, it holds between that thing and itself.[108])

If, on the other hand, we want a semantics for a system stronger than T, we need to require more of the relation **R** than that it be reflexive. If we require that **R** also be symmetric, we get a semantics for the Brouwersche system; if we require that **R** also be transitive, we get a semantics for S4; if we require that **R** also be symmetric and transitive, we get a semantics for S5.[109] To see why this works, it may help to look at a couple of examples. (For ease of exposition, I shall suppress reference to models and model structures.)

Suppose that at a possible world **H** it is true that A, and false that □◇A. If □◇A is false at **H**, then ◇A must be false at some world **H**′ that is possible relative to **H**. If ◇A is false at **H**′, then A must be false at every world that is possible relative to **H**′. But A couldn't be false at every world possible relative to **H**′, if **H** were possible relative to **H**′ (since, we are supposing, A is true at **H**). So **H** must not be possible relative to **H**′, even thought **H**′ is possible relative to **H**. This shows that, if the characteristic axiom of the Brouwersche system (A ⊃ □◇A) is false at a world (on a model, in a model structure), then the relative possibility relation for that model structure is non-symmetric. So, by requiring that the relative possibility relation be symmetric, we can ensure that the characteristic axiom of the Brouwersche system comes out true at **G** on any model associated with any model structure.

Again, suppose that at some world **H** it is true that □A and false that □□A. Then A is true at every world that is possible relative to **H**, but □A is *not* true at every world that is possible relative to **H**. There must accordingly be a world **H**′ such that (i) **H**′ is possible relative to **H**, and (ii) A is true but □A is false at **H**′. Since □A is false at **H**′, there is a world **H**″ such that (i) **H**″ is possible relative to **H**′, and (ii) A is false at **H**″. So we know that A is false at some world (namely, **H**″) that is possible relative to some world (namely, **H**′) that is possible with respect to **H**, even though A is true in every world that is possible relative to **H** (since, we are supposing, □A is true at **H**). The only way this can happen is if relative possibility is a non-transitive relation. So, by requiring that the relative possibility relation be transitive, we can ensure that, whenever □A is true, so is □□A. In other words, by requiring that **R** be transitive, we can ensure that the characteristic axiom of S4 comes out true at **G** on any model associated with any model structure.

The semantics Kripke provides for the systems of propositional modal logic under discussion does not by itself answer the questions alluded to earlier.

---

[108] For details of the semantics for S2 and S3, see Kripke's 'Semantical Analysis of Modal Logic II: Non-normal Modal Propositional Calculi', in J. W. Addison, L. Henkin, and A. Tarski (eds.), *The Theory of Models* (Amsterdam: North Holland Publishing Company, 1965).

[109] A relation R is symmetric if, whenever *x* stands in R to *y*, *y* stands in R to *x*, and transitive if, whenever *x* stands in R to *y*, and *y* stands in R to *z*, *x* stands in R to *z*.

We can still ask whether T is too weak, or S5 is too strong, or T and S5 are both too weak, to capture all and only the (propositional) logical truths about possibility and necessity (or about a certain kind of possibility and necessity, or a certain sense of 'possibility' and 'necessity'). The question of whether, say, S5 is too strong can be recast as the question of whether it is a truth of (propositional modal) logic that relative possibility (or the sort of relative possibility associated with a certain kind of possibility and necessity) is an equivalence relation;[110] but recasting the question in those terms doesn't in any obvious way suggest how it should be answered.[111] Still, Kripke's semantics provides insight into a family of axiomatic systems of propositional modal logic. We understand those systems better when we see that they in effect represent different choices about how 'accessible' we require possible worlds to be to each other (and to themselves).

In 'Semantical Considerations on Modal Logic' Kripke also offers a semantics for quantified modal logic. In a quantified first-order language we don't simply have propositional variables (P, Q, and so on) and truth-functional operators and connectives ($\sim$, $\supset$, &, and so on). We also have names (a, b, and so on), variables ($x_1 \ldots x_n, y_1 \ldots y_n$, and so on), predicates ($P_1, P_2, \ldots, P_n, Q_1, Q_2 \ldots, Q_n$, and so on), and quantifiers ($\forall$, $\exists$). When we put together names and predicates in the right way, we get a sentence; e.g. P_1a, which says that a has the (one-place) property P_1, and R_2ab, which says that a stands in the (two-place) relation R_2 to b. When we put together variables, predicates, and quantifiers in the right way, we get a (closed) sentence; e.g. '$(\exists x)P_1x$', which says that there is an x such that x has property P_1 (or, more idiomatically, that something is P_1), or '$(\forall x)(\exists y)R_2xy$', which says that every x stands in the (two-place) relation R_2 to some y or other.

In giving a semantics for a quantified (first-order) language, it is standard to introduce the idea of a model. In Kripke's semantics for propositional modal logic, (what he calls) a model is a function that assigns a truth-value to a pair consisting of a world and an atomic formula. In the present context a model is an ordered pair $<\mathbf{D}, \mathbf{I}>$, where $\mathbf{D}$ is a domain or set of individuals, and $\mathbf{I}$ is an interpretation. An interpretation is a function that assigns semantic values to certain expressions of the language. A name such as a is assigned an individual in the domain $\mathbf{D}$. A one-place predicate letter P_1 is assigned a subset of $\mathbf{D}$ (intuitively, the set of individuals in the domain that have the property P_1). A two-place predicate R_2 is assigned a subset of the Cartesian product of $\mathbf{D}$ with itself—that is, a set of ordered pairs both members of which belong to $\mathbf{D}$. (Intuitively, the predicate R_2 is assigned the set of ordered pairs of individuals i and i' in $\mathbf{D}$ such that i stands in the two-place relation R_2 to i'.) An n-place relational predicate R_n is assigned a subset of the nth Cartesian product of $\mathbf{D}$ with itself.

With this in place, it is straightforward to say under what conditions a sentence built up from names and predicates is true in a model $\mathbf{M}$. 'P_1a' is true

---

[110]   An equivalence relation is a reflexive, symmetric, and transitive relation.
[111]   Some of the considerations relevant to an answer will be discussed in Ch. 3, sect. 2.

in a model **M** just in case the individual in the domain that (the associated interpretation) **I** assigns as the semantic value of *a* is an element of the subset of the domain that **I** assigns as the semantic value of P_1. 'R_2ab' is true in **M** just in case a certain ordered pair (consisting of the individual that **I** assigns as the semantic value of *a*, and the individual that **I** assigns as the semantic value of *b*) is an element of the subset of the Cartesian product of **D** with itself that **I** assigns as the semantic value of R_2.

Under what conditions should we say that a quantified sentence such as '$(\exists x)P_1x$' or '$(\forall x)(\exists y)R_2xy$' is true in a model **M**? Here the answer is not straightforward, since the interpretation associated with a model does not assign values to variables such as *x* and *y*. The usual strategy is to appeal to *assignments of values to variables* (or *value assignments*, for short) as well as to an interpretation. Formally, an assignment of values to variables is a function that assigns to each variable in the language an element of **D**. Intuitively, we may think of an assignment of values to variables as a kind of temporary pretence that the variable *x* is a name of this element in the domain, that the variable *y* is a name of that element in the domain, and so on. Using the notion of a value assignment, we get a general and systematic method for evaluating the truth of sentences like '$(\exists x)P_1x$' and '$(\forall x)(\exists y)R_2xy$' with respect to a model. If the reader is not familiar with how this works, a worked example may help. (If the reader is familiar with how this works, she may skip the next two paragraphs.)

Suppose our domain **D** contains just two individuals: *i* and *i'*. And suppose that our interpretation **I** assigns to the two-place predicate R_2 the set consisting of the ordered pair $<i, i'>$ and the ordered pair $<i', i>$. If we want to evaluate the truth or falsity of '$(\forall x)(\exists y)R_2xy$' on **M**, we start with the subformula 'R_2xy'. Since **I** does not assign any semantic value to the variables *x* and *y*, we cannot evaluate the truth or falsity of 'R_2xy' with respect to **M**. But there is no problem about evaluating the truth of 'R_2xy' with respect to a particular value assignment **v** and a particular model **M**: we say that 'R_2xy' is true with respect to **v** and **M** just in case the ordered pair consisting of what **v** assigns to *x* and what **v** assigns to *y* is an element of the subset of the Cartesian product of **D** with itself that **I** assigns to R_2. For our purposes, we may pretend that there are only four value assignments for the language: $\mathbf{v}_1$, which assigns *i* to the variable *x*, and *i'* to the variable *y*; $\mathbf{v}_2$, which assigns *i* to the variable *x*, and *i* to the variable *y*; $\mathbf{v}_3$, which assigns *i'* to the variable *x* and *i* to the variable *y*, and $\mathbf{v}_4$, which assigns *i'* to the variable *x* and *i'* to the variable *y*. (Actually, there will be many value assignments that start out the way, say, $\mathbf{v}_1$ starts out, and assign different values to variables other than *x* and *y*. But since the formula whose truth-value we want to evaluate only contains no other variables except *x* and *y*, we can ignore this for the sake of simplicity.)

The reader can verify that 'R_2xy' comes out true with respect to $\mathbf{v}_1$ and **M** and $\mathbf{v}_3$, and **M**, and false with respect to $\mathbf{v}_2$ and **M** and $\mathbf{v}_4$ and **M**. But should we say that '$(\exists y)R_2xy$' is true or false with respect to, say, $\mathbf{v}_2$ and **M**? We need a rule for evaluating the truth or falsity of the existentially quantified formula

'$(\exists y)R_2xy$' with respect to a value assignment $\mathbf{v}$ and a model $\mathbf{M}$. If the reader reflects, she will see that we get intuitively the right results if we say that '$(\exists y)R_2xy$' is true for a given value assignment $\mathbf{v}$ and model $\mathbf{M}$ just in case for some value assignment $\mathbf{v}'$, $\mathbf{v}'$ differs from $\mathbf{v}$ at most with respect to what it assigns to the variable y, and 'R_2xy' is true with respect to $\mathbf{v}'$ and $\mathbf{M}$. Applying this rule, we see that '$(\exists y)R_2xy$' is true with respect to $\mathbf{v}_1$ and $\mathbf{v}_3$. (After all, for some value assignment $\mathbf{v}'$ differing from $\mathbf{v}_1$ at most with respect to what it assigns to y, 'R_2xy' is true with respect to $\mathbf{v}'$ and $\mathbf{M}$ (let $\mathbf{v}' = \mathbf{v}_1$.) Similarly, for some value assignment $\mathbf{v}'$ differing from $\mathbf{v}_3$ at most with respect to what it assigns to y, 'R_2xy' is true with respect to $\mathbf{v}'$ and $\mathbf{M}$ (let $\mathbf{v}' = \mathbf{v}_3$.) But '$(\exists y)R_2xy$' is also true with respect to $\mathbf{v}_2$ and $\mathbf{M}$, and $\mathbf{v}_4$ and $\mathbf{M}$. 'R_2xy' is false with respect to $\mathbf{v}_2$ and $\mathbf{M}$. But for some $\mathbf{v}'$ differing from $\mathbf{v}_2$ at most with respect to what it assigns to y, 'R_2xy' is true with respect to $\mathbf{v}'$ and $\mathbf{M}$. (Let $\mathbf{v}' = \mathbf{v}_1$.) Again, although '$R_2xy$' is false with respect to $\mathbf{v}_4$ and $\mathbf{M}$, for some $\mathbf{v}'$ differing from $\mathbf{v}_4$ at most with respect to what it assigns to y, 'R_2xy' is true with respect to $\mathbf{v}'$ and $\mathbf{M}$. (Let $\mathbf{v}' = \mathbf{v}_3$.) So '$(\exists y)R_2xy$' is true with respect to any value assignment $\mathbf{v}$ and $\mathbf{M}$. What about the sentence we started with: '$(\forall x)(\exists y)R_2xy$'? We need a rule for evaluating the universally quantified formula '$(\forall x)(\exists y)R_2xy$' with respect to a value assignment $\mathbf{v}$ and model $\mathbf{M}$. Again, the reader should be able to see that we get intuitively the right results if we say that the formula '$(\forall x)(\exists y)R_2xy$' is true with respect to, say, $\mathbf{v}_1$ just in case, for every value assignment $\mathbf{v}$, if $\mathbf{v}$ differs from $\mathbf{v}_1$ at most with respect to what it assigns to x, '$(\exists y)R_2xy$' is true with respect to $\mathbf{v}$ and $\mathbf{M}$. Given that '$(\exists y)R_2xy$' is true on each of $\mathbf{v}_1$–$\mathbf{v}_4$, this condition will obviously be met. So '$(\forall x)(\exists y)R_2xy$' is true with respect to $\mathbf{v}_1$ and $\mathbf{M}$. By the same reasoning, it is true with respect to $\mathbf{v}_2$ and $\mathbf{M}$, $\mathbf{v}_3$ and $\mathbf{M}$, and $\mathbf{v}_4$ and $\mathbf{M}$. So, just as '$(\exists y)R_2xy$' came out true with respect to $\mathbf{v}$ and $\mathbf{M}$ whichever $\mathbf{v}$ we choose, '$(\forall x)(\exists y)R_2xy$' comes out true with respect to $\mathbf{v}$ and $\mathbf{M}$ whichever $\mathbf{v}$ we choose. Now, if a sentence is true on a model cum value assignment whichever value assignment we choose, we may say that it is true on that model (just as, if a sentence comes out false on a model cum value assignment whichever value assignment we choose, we may say that it is false on that interpretation). So in the case under discussion we can say, not just that '$(\forall x)(\exists y)R_2xy$' is true with respect to (say) $\mathbf{v}_1$ and $\mathbf{M}$, but also that it is true with respect to $\mathbf{M}$. Since '$(\forall x)(\exists y)R_2xy$' says that everything stands in relation R_2 to something or other, and since, given our assumptions about $\mathbf{M}$, everything does stand in relation R_2 to something or other, this is just the result we want. The interested reader may verify for herself that, if we use the procedure just sketched to determine the truth-value of '$(\exists y)(\forall x)R_2xy$' with respect to $\mathbf{M}$, it comes out false. This is again what we want, since, on our assumptions about $\mathbf{M}$, there isn't any one thing that stands in R_2 to everything (including itself).[112]

---

[112] My (rapid) exposition leaves out many of the subtleties of the semantics for quantification: the interested reader is referred to any good introductory logic text, e.g. G. J. Massey, *Understanding Symbolic Logic* (New York: Harper & Row, 1970), pt. IV and app. F. The approach to the semantics of quantification expounded here is not quite the same as Kripke's in that the value assignments assign values to all the variables of the language and not just to all the free variables in a given formula.

So much for (non-modal) first-order logic with quantification. In offering a semantics for quantified modal logic, Kripke starts with a language that contains, in addition to the operators $\Box$ and $\sim$, and the connective &, the quantifiers $\exists$ and $\forall$, an infinite supply of variables x, y, z, and an infinite supply of predicates P_0, P_1, ..., P_n, Q_0, Q_1 ..., Q_n. Sentences of the language include: P_0 (which is, in effect, a notational variant of P in a propositional logic), P_2ab, $(\exists y)(\forall x)R_2xy$, P_1a & $\sim\Box(\exists y)Q_1y$, $(\forall x)(\forall y)\Box R_2xy$, and so on. In providing a semantics for the language, Kripke begins by introducing the idea of a quantificational model structure. A quantificational model structure is a model structure (G, K, R), together with a function ψ which assigns to each H in K a set of individuals $\psi(H)$ called the domain of H. Intuitively, $\psi(H)$ is the set of individuals existing at the possible world H; it need not be the same as $\psi(G)$ (the set of individuals existing at the actual world) or $\psi(H')$ (the set of individuals existing at a non-actual world different from H). Kripke then introduces the idea of a quantificational model for a quantificational model structure. A quantificational model is a function $\varphi(P_n, H)$, where P_n ranges over predicate letters of arbitrary n-adicity, and H ranges over the elements of K. If $n = 0$, φ assigns to (P_n, H) either truth or falsity. If $n \geqslant 1$, φ assigns to (P_n, H) a subset of the nth Cartesian product of U with itself, where U is what we might call 'the superdomain', that is, the union of all the domains associated with any of the elements of K. Intuitively, φ determines the truth-value of a propositional variable at any given world and the extension of a one-place or many-place predicate at any given world.

Kripke then provides an inductive definition for every formula A and H in K of the truth-value of $\varphi(A, H)$ relative to an assignment of values to the free variables of A. Where P_0 is a propositional variable, P_0 is true at H if $\varphi(A, H) = $ truth, and false if $\varphi(A, H) = $ falsity. Where $P_n(x_1 \ldots x_n)$ is an atomic formula (for n greater than 0), and $\mathbf{v}$ is an assignment of elements $a_1 \ldots a_n$ of the superdomain U to the variables $x_1 \ldots x_n$, $\varphi(P_n(x_1 \ldots x_n), H)$ is truth relative to $\mathbf{v}$ if the ordered n-tuple $<a_1 \ldots a_n> \in \varphi(P_n, H)$; otherwise $\varphi(P_n(x_1 \ldots x_n), H)$ is falsity relative to $\mathbf{v}$. With this basis in place, truth-conditions for complex formulae can be given inductively. We have already seen the inductive steps for $\sim$, &, and $\Box$ in Kripke's semantics for propositional modal logic. The inductive step for universally quantified formulae works as follows: assume that in the formula in $A(x, y_1, \ldots, y_n)$ the only variables occurring freely are x, $y_1 \ldots y_n$.[113] Assume also that for each value assignment to the free variables in $A(x, y_1 \ldots y_n)$, x, the truth-value of $\varphi(P_n(x_1 \ldots x_n), H)$ has been defined.

[113] An occurrence of a variable in a formula is *free* if it is not *bound* by an existential or universal quantifier. Thus in the formula '$(\exists y)R_2xy$' the occurrence of x is free, but the occurrence of y is bound. A sentence containing free variables is sometimes called an *open* sentence; a sentence containing at most bound variables is sometimes called a *closed* sentence. If a sentence is open, it may come out true with respect to a model on some value assignments and false with respect to that model on other value assignments. (In the worked example this happened with 'R_2xy.') If it is closed, it will be true with respect to a model and any one assignment just in case it is true with respect to that model and any other assignment (as we saw in the case of '$(\forall x)(\exists y)R_2xy$').

Then say that $\varphi((\forall x)A(x, y_1 \ldots y_n), H)$ is truth relative to an assignment of elements of U $b_1 \ldots b_n$ to $y_1 \ldots y_n$ if $\varphi(A(x, y_1 \ldots y_n), H)$ is truth relative to every assignment of $a, b_1 \ldots b_n$ to $x, y_1 \ldots y_n$, respectively, where $a \in \psi(H)$; otherwise $\varphi((\forall x)A(x, y_1 \ldots y_n), H)$ is falsity relative to that assignment.

Notice that, on Kripke's semantics, $\varphi(P_2xy, H)$ may be truth relative to a value assignment **v**, even though the elements of the superdomain that **v** assigns to x and y are not in $\psi(H)$. That is because Kripke does not require that $\varphi(P_n, H)$ be a subset of $(\psi(H))_n$. (So, for example, the extension of the one-place predicate P_1 at world H may be a set of individuals none of which belong to the domain of H.) It should also be emphasized that in giving the truth-conditions for $\varphi((\forall x)A(x, y_1 \ldots y_n), H)$ relative to an assignment of $b_1 \ldots b_n$ to $y_1 \ldots y_n$, Kripke does not require that $\varphi(A(x, y_1 \ldots y_n))$ come out true at H relative to every assignment of $a, b_1 \ldots b_n$ to $x, y_1 \ldots y_n$. He requires only that $\varphi(A(x, y_1 \ldots y_n))$ come out true at H for every such assignment where x is assigned an individual a in $\psi(H)$. This has the effect that '$(\forall x)P_1x$' comes out true at a world H (relative to any assignment) as long as everything in the domain of H is P_1—even if not everything in the superdomain U is P_1. As Kripke puts it, in H we quantify only over the individuals existing in H.

After setting out his semantics for quantified modal logic, Kripke goes on to show that on it neither the Barcan formula nor its converse comes out valid. We consider a model structure (G, K, R), where $K = \{G, H\}$ (with G distinct from H), and R is K_2 (so that R holds between any world in K and any other, as well as between any world and itself). We suppose that $\psi(G) = \{a\}$, and that $\psi(H) = \{a, b\}$ (where a and b are distinct). Lastly, we define, for a one-place predicate P_1, a model φ in which $\varphi(P_1, G) = \{a\}\varphi(P_1, H)$. $\Box P_1x$ is true at G when x is assigned a as its value. Since there isn't anything in G's domain except a, there won't be an assignment relative to which '$(\forall x)\Box P_1x$' comes out false at G. But 'P_1x' is false at H relative to an assignment that assigns b to x. Hence '$(\forall x)P_1x$' is false at H. Since H is possible relative to G, '$\Box(\forall x)P_1x$' is false at G. We have a counter-model to the Barcan formula (which says that $(\forall x)\Box Fx \supset \Box(\forall x)Fx$).

So, if we make some natural assumptions—that 'everything is F' is true at a world just in case everything in the domain of that world is F, and that the domain of the actual world may be a proper subset of the domain of an alternative possible world—we should conclude that the Barcan formula is not a truth of (quantified) modal logic. The reader who has found this overly technical and abstract may find it helpful to think of the predicate P_1 in Kripke's counter-model as *being identical to* a. Intuitively, as long as the quantifier ranges (only) over things in the domain of G, it is true at G that everything is necessarily identical to a. (The only thing in G is a, and a couldn't very well have been different from itself.) On the other hand, it isn't true at G that necessarily everything is identical to a, since there is a world that is possible relative to G in which it is false that everything is identical to a (namely, H).

To get a counter-model to the converse of the Barcan formula, Kripke starts with the same model structure that figured in the counter-example to the

Barcan formula. But this time he supposes that $\psi(G) = \{a, b\}$ and $\psi(H) = \{a\}$, where a and b are distinct, and that $\varphi(P_1, G) = \{a, b\}$, and $\varphi(P_1, H) = \{a\}$. Since '$(\forall x)P_1 x$' is true at both G and H, $\varphi(\Box(\forall x)P_1 x, G) = $ truth. But $\varphi(P_1 x, H) = $ falsity, when x is assigned b as its value (since it is not the case that $b \in \varphi(P_1, H)$). So, relative to that assignment, $\varphi(\Box P_1 x, G) = $ falsity. Hence $\varphi((\forall x)\Box P_1 x, G) = $ falsity, and we have a counter-example to the validity of the converse Barcan formula. As Kripke notes, we may think of the predicate P_1 as expressing existence. (Necessarily, everything exists; but not everything necessarily exists.) Again, someone might insist that, whenever $\Box(\forall x)Fx$ is true, so is $(\forall x)\Box Fx$, but only by taking issue with the (natural) assumptions about quantification and domain variation built into Kripke's semantics.[114]

Because of the work of Kripke and others,[115] the decade between 1955 and 1965 was an exciting time for modal logic—especially the semantics of modal logic. Nevertheless, before the appearance of *Naming and Necessity* quite a few philosophers had doubts about the philosophical respectability of modal logic, and indeed about the philosophical respectability of the concepts of possibility and necessity. The philosopher most responsible for this state of affairs was Willard Van Orman Quine.

For Quine, modal logic is philosophically disreputable because it commits us to the following (untenable) disjunction: either there is a way to sort truths into analytic truths and synthetic truths, or there is a way to sort the properties of an object into those which it has necessarily or essentially and those which it has contingently or accidentally.

Suppose, Quine says, we ask a proponent of a modal logic of 'strict' necessity what it is for a statement to be necessary. She may follow Lewis and Carnap in saying that 'It is (strictly) necessary that A' is true if and only if A is an analytic truth ('true by virtue of meaning and independently of fact'). But, Quine thinks, there is no good reason to suppose there are any analytic truths.[116] Moreover, suppose necessity is explicated in terms of analyticity in

---

[114] To get a counter-example to the Barcan formula, we suppose that H's domain contains individuals not in G's domain; to get a counter-example to its converse, we suppose that G's domain contains individuals not in H's. I do not mean to suggest here that the natural assumption that different possible worlds should be allowed different domains is the only defensible one. Timothy Williamson offers interesting arguments against that assumption (and in favour of the validity of the Barcan formula and its converse) in 'Bare Possibilia', *Erkenntnis*, 48 (1998), 257–73.

[115] Cf. J. Hintikka, 'Modality and Quantification', *Theoria*, 27 (1961), 119–28; S. Kanger, *Provability and Logic* (Stockholm: Almquist & Wiksell, 1957); R. Montague, 'Logical Necessity, Physical Necessity, Ethics, and Quantifiers', *Inquiry*, 3 (1960), 159–69.

[116] Cf. Quine, 'Two Dogmas of Empiricism', in Quine, *From a Logical Point of View* (New York: Harper Torchbooks, 1961), 36: 'It is obvious that truth in general depends on both language and extralinguistic fact ... Thus one is tempted to suppose in general that the truth of a statement is somehow analyzable into a linguistic component and a factual component. Given this supposition, it next seems reasonable that in some statements the factual component should be null; and these are the analytic statements. But, for all its a priori reasonableness, a boundary between analytic and synthetic statements simply has not been drawn. That there is such a distinction to be drawn at all is an unempirical dogma of empiricists, a metaphysical article of faith.'

the way just suggested. Then it will make no sense to ask whether or not an individual has a certain property necessarily. After all, it doesn't make sense to ask whether an individual is such that it is an analytic truth that *it* has a certain property. Assuming for the sake of argument that there are analytic truths, '*t* is F' may be an analytic truth, and '*t′* is F' a synthetic truth, even though *t* = *t′*. For example, '9 equals 9 if anything does' is analytic, but 'The number of the planets equals 9 if anything does' is synthetic. So we cannot ask of the individual that is both the number 9 and the number of the planets whether it is analytic that *it* equals 9 if anything does, any more than we can ask of the individual who is both Giorgione and Barbarelli whether *he* has the property of being so-called because of his size. So, on the understanding of necessity under discussion, we cannot ask of the individual that is both the number 9 and the number of the planets whether necessarily *it* equals 9 if anything does.

If that question makes no sense, then neither do open sentences of the form '□F*x*'. As we have seen, an open sentence of the form '□F*x*' is true (relative to a value assignment) just in case a certain individual (the one that the value assignment assigns to *x*) is such that *it* is necessarily F. And if open sentences like '□F*x*' don't make sense, neither do closed sentences such as '(∃*x*)□F*x*' and '(∀*x*)□F*x*'.

The moral is that, if we explicate necessity in terms of analyticity in the way that Lewis and Carnap did, we shall have to say that the necessity operator (or the possibility operator) can only meaningfully prefix closed sentences. But, Quine thinks, this trivializes modal logic. If we cannot, as Quine puts it, 'quantify across a modal operator', we might as well just quote sentences, and say that they are analytic—leaving modal operators out of it.[117]

So why not abjure the explication of necessity in terms of analyticity? Why not say that '□F*x*' is true relative to a value assignment just in case *being F* is a necessary or essential property of the individual that the assignment assigns to *x*? Because, Quine says, it makes no more sense to say of a particular individual that it is necessarily or essentially F than it does to say of a particular individual that it is or is not so-called because of its size. It is true (let us suppose) that 9 is necessarily odd; it is false (let us suppose) that the number of the planets is necessarily odd; but we cannot say of the individual that is both the number 9 and the number of the planets either that it has the property of being odd necessarily, or that it has the property of being odd contingently.[118] The idea that a thing, independently of the way it is described, might have some of its properties necessarily or essentially, and others contingently or accidentally, is one Quine finds incomprehensible and indefensible:

Mathematicians may conceivably be said to be necessarily rational and not necessarily two-legged; and cyclists necessarily two-legged and not necessarily rational. But what of

---

[117] See Quine, 'Reference and Modality', in Quine, *From a Logical Point of View*, 156.
[118] Ibid. 147–50.

an individual who counts among his eccentricities both mathematics and cycling? Is this concrete individual necessarily rational and contingently two-legged or vice versa? Just insofar as we are talking referentially of the object, with no special bias toward a background grouping of mathematicians as against cyclists or vice versa, there is no semblance of sense in rating some of his attributes as necessary and others as contingent. Some of his attributes count as important, and others as unimportant, yes; some as enduring and others as fleeting; but none as necessary or contingent... Curiously, a philosophical tradition does exist for just such a distinction between necessary and contingent attributes... But however venerable the distinction, it is surely indefensible.[119]

There are actually two targets here. The first is (what Quine calls) Aristotelian essentialism—the view that some of the properties of a thing are essential to it and some accidental. The second is the view that the properties of a thing are either essential to it or accidental to it. (The second view, but not the first, is consistent with the idea that all the properties (or none of the properties) of an individual are essential to it.)[120]

Actually, Quine is less sweeping in his dismissal of modal logic than our discussion so far might suggest. Although he thinks that any logic of what he calls 'strict' or 'extreme' necessity will be indefensible (if it is essentialist) or not worth defending (if it explicates necessity in terms of analyticity), he does not make the same claim for other sorts of necessity, such as physical necessity. Still, Quine has often been taken to be hostile towards modality in any of its guises.

Kripke responds to Quine's dismissal of essentialism as follows:

Some philosophers have distinguished between essentialism, the belief in modality *de re*, and a mere advocacy of necessity, the belief in modality *de dicto*. Now some people say: Let's give you the concept of necessity. A much worse thing, something creating great additional problems is whether we can say of any particular that it has necessary or contingent properties, even make the distinction between necessary and contingent properties.... Whether a *particular* necessarily or contingently has a certain property depends on the way it's described... So it's thought: was it necessary or contingent that Nixon won the election? It might seem contingent, unless one has some view of some inexorable processes... But this is a contingent property of Nixon only relative to our referring to him as 'Nixon' (assuming 'Nixon' doesn't mean 'the man who won the

---

[119] W. V. O. Quine, *Word and Object* (Cambridge, Mass.: MIT Press, 1960), 199.

[120] In a similar vein, Quine opposes not just the idea that some truths are analytic and some synthetic, but also the idea that truths are either analytic or synthetic: he would be no happier with the (apparently Leibnizian) view that all truths are analytic than he would with the (apparently Leibnizian) view that all of the properties of a thing are essential to it. Robert Stalnaker notes that someone might respond to Quine's attack on Aristotelian essentialism by maintaining either that all of a thing's properties are essential to it, or that none are, and sketches a view on which none (or almost none) are (see his 'Anti-Essentialism', in P. French, T. Uehling, and H. Wettstein (eds.), *Midwest Studies in Philosophy* (Minneapolis: University of Minnesota Press, 1979). In spite of the interest of that view, I doubt that Quine would find it significantly less uncongenial than Aristotelian essentialism, just because, on it, a thing 'of itself, and by any name or none' is the subject of modal properties. (Quine says in the passage just cited that *none* of the attributes of our mathematical cyclist count as *either* necessary or contingent.)

election at such and such a time'). But if we designate Nixon as 'the man who won the election in 1968', then it will be a necessary truth, of course, that the man who won the election in 1968, won the election in 1968 … It has even been suggested in the literature, that though a notion of necessity may have some sort of intuition behind it (we do think some things could have been otherwise; other things we don't think could have been otherwise), this notion [of a distinction between necessary and contingent properties] is just a doctrine made up by some bad philosopher, who (I guess) didn't realize that there are several ways of referring to the same thing … it is very far from being true that this idea [that a property can be held to be essential or accidental to an object independently of its description] is a notion which has no intuitive content, which means nothing to the ordinary man. Suppose that someone said, pointing to Nixon, 'That's the guy who might have lost.' Someone else says, 'Oh no, if you describe him as "Nixon", then he might have lost; but of course, describing him as the winner, then it is not true that he might have lost.' Now which one is being the philosopher here, the unintuitive man? It seems to me obviously to be the second. ('N & N', 265)

The view that things have at least some of their properties accidentally (and some of their properties essentially) is not just the view of Aristotle, or Aquinas, but also the view of the man on the Clapham omnibus—or so Kripke argued. And he persuaded, if not everyone, nearly everyone.

For much of the first half of the twentieth century modality had a somewhat marginal place in analytic philosophy. Kripke contributed more to its 'demarginalization' than any other analytic philosopher. He did this by providing a semantics for modal logic (or rather, for a family of (propositional and quantified) modal logics); by vigorously and effectively addressing Quinean worries about whether quantification into modal contexts made sense; and by bringing modal issues into various central debates in philosophy. We have already seen an instance of this in Kripke's account of names; we shall see another in his treatment of the mind–body problem. The 'remodalization' of metaphysics and the philosophy of language may retrospectively come to be thought of as Kripke's most important contribution to twentieth-century philosophy. Those of us who, as undergraduates, learned philosophy from Quineans think of Kripke as a philosopher who (almost single-handedly) transformed the philosophical landscape.

2, THE NECESSARY A POSTERIORI AND THE CONTINGENT A PRIORI

In discussions of modal logic in the 1950s and 1960s the necessity or otherwise of identity statements involving (only) proper names was a *vexata quaestio*. There was an argument that seemed to show that if it is true that, $a = b$, then it is necessarily true—to wit:

(1) $(\forall x)(\forall y)((x = y) \supset (Fx \supset Fy))$ the indiscernibility of identicals

(2) $(\forall x)\Box(x = x)$ the necessity of self-identity

(3) $(\forall x)(\forall y)((x = y) \supset [\Box(x = x) \supset \Box(x = y)])$ by instantiation from (1), with $\Box(x = \text{—})$ playing the role of F.

(4) $(\forall x)(\forall y)((x = y) \supset \Box(x = y))$ from (2) and (3).

(5) $a = b \supset \Box(a = b)$ by instantiation from (4).[121]

On the other hand (it was generally appreciated) whether Hesperus is identical to Phosphorus was once an open empirical question: though Hesperus and Phosphorus turned out to be the same planet, they might have turned out to be different planets. Even now it is conceivable that astronomers have got things very badly wrong: Hesperus and Phosphorus might turn out to be different planets. In that case, it would appear, that Hesperus is different from Phosphorus is extremely unlikely, but not *impossible*. Now 'Hesperus' and 'Phosphorus' are names. So, it appears, sometimes it is true that a and b are identical but not impossible that a and b are distinct—in which case a statement like (5) can be false.

One might attempt to dissolve this puzzle by denying that (3) really is an instance of (1), on the grounds that there is no such property as *being necessarily identical to x*. (We may call this the Quinean response, since, as we have seen, Quine would deny that the idea of such a property is coherent, and would accordingly deem both (2) and (3) nonsense.) Alternatively, one might accept the soundness of the formal argument above, and still maintain that 'Hesperus = Phosphorus' is contingently true, on the grounds that 'Hesperus' and 'Phosphorus' (unlike a and b in the argument) are not really names. (We may call this the Russellian response, since Kripke attributes it to Russell.)

Like Russell, Kripke holds that if a and b are names, and '$a = b$' is true, then '$a = b$' could not have been false.[122] For, he supposes, if a and b are names, then they are rigid designators. If they are rigid designators, and they designate one and the same thing with respect to the actual world, they cannot pick out different things with respect to an alternative possible world. Likewise, if 'Hesperus' and 'Phosphorus' are rigid, and designate the same thing with respect to the actual world, they cannot designate different things with respect to any possible world. But, Kripke thinks, 'Hesperus' and 'Phosphorus' are rigid. So if *Hesperus = Phosphorus* is true, it is necessarily true (or at least weakly necessarily true).[123]

But can't we imagine a possible world in which Hesperus and Phosphorus are different planets? If so, won't there be a possible world in which Hesperus

---

[121] See 'Identity and Necessity', in M. Munitz (ed.), *Identity and Individuation* (New York: New York University Press, 1971), 136; hereafter 'I & N'. Kripke says the above argument 'has been stated many times in recent philosophy', though he does not provide references.

[122] At least, '$a = b$' could not have been false if either a or b had existed. Kripke is neutral about whether '$a = b$' is true, false, or truth-valueless with respect to a possible world in which neither a nor b exists ('I & N', 137; and 'N & N', 311).

[123] A statement is *weakly necessary* just in case it is true with respect to all the worlds in which all of the things mentioned therein exist ('I & N', 137).

and Phosphorus are different planets? Not if Hesperus and Phosphorus are actually the same planet, Kripke answers. Suppose for *reductio* that at the actual world **G** Hesperus and Phosphorus are the same planet, and at an alternative world **H** Hesperus and Phosphorus are different planets. At **G** there is one thing that is both Hesperus and Phosphorus. At **H** there are two things, different from each other, one of which is Hesperus and the other of which is Phosphorus. Now the two (different) things at **H** cannot both be the one thing at **G** that is both Hesperus and Phosphorus. At least one of them must be something else. But to suppose that 'the Hesperus of **G**' is one thing and 'the Hesperus of **H**' is something else is tantamount to supposing that Hesperus might have been something else—which is false. Likewise, to suppose that 'the Phosphorus of **G**' is one thing and 'the Phosphorus of **H**' is something else is tantamount to supposing that Phosphorus might have been something else—which is false.

There is no problem about imagining a possible world in which there are two different planets, one of which appears here in the evening (where Hesperus actually appears), and the other of which appears there in the morning (where Phosphorus actually appears)—one of which is called (at that world) 'Hesperus', and the other of which is called (at that world) 'Phosphorus'. But if what we are imagining is a genuinely possible world, it is not a world in which *Hesperus* is different from *Phosphorus*; it is a world in which something that is *called* '*Hesperus*' and resembles Hesperus in certain respects is different from something that is *called* '*Phosphorus*' and resembles Phosphorus in certain respects (cf. 'I & N', 155–6; 'N & N', 308).

Again, though, mightn't Hesperus and Phosphorus have turned out to be (or turn out to be) different planets? Doesn't this show that it is, after all, possible that Hesperus and Phosphorus are different planets?

Here Kripke draws a distinction between what might have turned out to be the case and what might turn out to be the case. Whatever might have turned out to be the case might have been the case, and is the case at some possible world. So, given that it is actually and necessarily true that Hesperus and Phosphorus are not different planets, it could not have turned out that Hesperus and Phosphorus are different planets.[124] But what might turn out to be the case need not obtain at any possible world. So it might turn out that Hesperus and Phosphorus are different planets. Kripke explicates this distinction at issue in note 72 of *Naming and Necessity*:

Some of the statements I myself make above may be loose and inaccurate in this sense. If I say, 'Gold *might* turn out not to be an element', I speak correctly; 'might' here is *epistemic* and expresses the fact that the evidence does not justify *a priori* (Cartesian) certainty

---

[124] See 'N & N', 333: 'The inaccurate statement that Hesperus might have turned out not to be Phosphorus should be replaced by the true contingency mentioned earlier in these lectures: two distinct bodies might have occupied in the morning and the evening, respectively, the very positions actually occupied by Hesperus–Phosphorus–Venus.'

that gold is an element...If I say, 'Gold *might have* turned out not to be an element', I seem to mean this metaphysically, and my statement is subject to the correction noted in the text.

In this note, and the part of the text it accompanies, Kripke appears to be making two points. First, he is distinguishing metaphysical possibility—which requires being the case at some possible world—from epistemic possibility, which does not. (On a natural way of understanding epistemic possibility, something is epistemically possible (for a person, at a time) just in case, for all that person knows then, it is true. But in note 72 Kripke understands epistemic possibility somewhat differently, in such a way that, if anything is epistemically impossible for a person at a time, its falsity is knowable a priori (with Cartesian certainty) by that person at that time.[125]) Secondly, Kripke seems to be supposing that, whereas 'might turn out' expresses (or can express) epistemic possibility, 'might have turned out' expresses metaphysical possibility. Kripke is not, I think, simply stipulating that he shall use 'might turn out' to express epistemic possibility, and 'might have turned out' to express metaphysical possibility. (Otherwise he would not describe 'Hesperus might not have turned out to be Phosphorus' as 'inaccurate', and 'Gold might have turned out to be a compound' as 'loose and inaccurate'; 'N & N', 333.) Instead, his view seems to be that, while 'might turn out' can properly be used to express epistemic possibility, 'might have turned out' can only loosely and improperly be so used.[126]

I have my doubts about this. Given that 'Hesperus and Phosphorus might turn out to be different planets' can express the epistemic possibility, relative to some person or persons, and the present time, of Hesperus and Phosphorus being different planets, it is very natural to suppose that 'Hesperus and Phosphorus might have turned out to be different planets' can express the epistemic possibility, relative to some person or persons, and some past time, of Hesperus and Phosphorus being different planets. 'Might' is most naturally understood as expressing epistemic rather than metaphysical possibility (especially if epistemic possibility is construed as compatibility with what is known); and 'might have' is naturally understood as expressing some non-epistemic variety of possibility.[127] Still, I think that 'might have' can express

---

[125] At 'N & N', 307, Kripke appears to construe epistemic possibility in what I have called the natural way: 'The four-colour theorem might turn out to be true and might turn out to be false.... Obviously, the "might" here is purely "epistemic"—it merely expresses our present state of ignorance or uncertainty.'

[126] Also see 'I & N', 157 n. 15, where Kripke says that if someone insists that this lectern could have turned out to be made of ice, 'what he really means' is that *a lectern* could have looked just like this one, and have been placed in the same position as this one, and yet have been made of ice.

[127] Compare (a) 'I might not exist now' with (b) 'I might not exist tomorrow' and (c) 'I might never have existed'. Though it is easy to think of circumstances in which (b) or (c) might be asserted with propriety, the same cannot be said for (a). I take it this is because (a) and (b) are most naturally understood as meaning (respectively) something like 'For all I know, I do not exist now', and 'For all I know, I will not exist tomorrow', while (c) is naturally understood as not making any claims about my knowledge.

either present or past epistemic possibility, as well as metaphysical possibility. Compare (i) 'I wonder where my keys are; I might have left them in the office' with (ii) 'It was irresponsible of you not to examine that painting more closely; it might have been a Canaletto.' In asserting (i), I would (normally) be asserting that the past-tense proposition *I left my keys in the office* is (now) epistemically possible for me. In asserting (ii), I would (normally) be asserting that the proposition *That painting is a Canaletto* was epistemically possible for you, at the time when you failed to examine the painting closely. I could (truthfully) assert (ii) even if you and I both know (now) that that painting (the one you didn't examine) wasn't a Canaletto, and even if, as a matter of metaphysical necessity, any painting that isn't a Canaletto couldn't have been a Canaletto—as long as, at the time you failed to examine the painting closely, it was, for all you knew, then a Canaletto.

At any rate, Kripke's solution to the puzzle about the necessity or otherwise of identity statements involving proper names does not depend on which expressions express metaphysical possibility and which expressions express epistemic possibility. The crucial point is that metaphysical impossibility is one thing and epistemic impossibility is something else: the rigidity of both terms in a (true) identity statement guarantees that its falsity is (at least weakly) metaphysically impossible, but not that its falsity is epistemically impossible. As Kripke sees it, the failure of Russell, or Marcus, or Quine to solve the puzzle under discussion is due to their failure to distinguish metaphysical impossibility and epistemic impossibility (a priori excludability). Once this distinction is drawn, we can agree with Russell and Marcus that the formal argument set out at the beginning of this section is sound, and agree with Quine that 'Hesperus' and 'Phosphorus' are names and it is an empirical question whether or not they are names of the same thing ('I & N', note 13).

As Kripke notes, many philosophers have wanted to link necessity with apriority. There are various ways in which one might do this. Most strongly, one might maintain that

(1) Every necessary truth is known a priori.

This thesis is far from obvious, inasmuch as there may be necessary truths— say, truths of mathematics—that are not known by anyone and a fortiori not known a priori by anyone. One might more cautiously maintain that

(2) Every known necessary truth is known a priori.

This again is unobvious. Perhaps there is some necessary truth of the form *n is a prime number* that someone knows a posteriori but no one knows a priori. No one has calculated or proven that the number *n* is prime; but a reliable computer has produced the statement *n is prime*. If someone has knowledge of the computer's output, and justified confidence in the computer's reliability, then she can know a posteriori that *n* is prime even if nobody knows a priori that *n* is prime (see 'N & N', 261).

Now, whether or not the necessary truth *n is prime* is in fact known a priori by anyone, it presumably could be known a priori by someone. A proponent of the link between necessity and apriority might accordingly say: every necessary truth is know*able* a priori. But, Kripke points out, this is hardly obvious. There may be mathematical truths that, in spite of being necessarily true, could not be known a priori (or known at all): Goldbach's conjecture might be one such ('N & N', 261–3). Kripke mentions that someone might object here that, if Goldbach's conjecture is true, it is knowable a priori, inasmuch as an infinite mind could prove it by checking each even number and ascertaining that that number is indeed the sum of two primes. But, Kripke replies, it is not clear that it is genuinely possible for a person (with an infinite mind) to acquire knowledge of Goldbach's conjecture by ascertaining, for each even number, that that number is the sum of two primes ('I & N', 151).

In light of this, the proponent of the link between necessity and apriority might say that

(3) Every knowable necessary truth is knowable a priori

or more cautiously say that

(3′) Every known necessary truth is knowable a priori.

Kripke finds the notion of a priori knowability obscure ('N & N', 260, 766). But, he holds, inasmuch as we restrict a priori knowledge to the 'standard, human sort', it seems not to be true that every known necessary truth is knowable a priori ('N & N', 766). And, he thinks, true identity statements such as *Hesperus = Phosphorus* are examples of known necessary truths not accessible to a priori knowledge (at least, of any standard, human sort): 'we do not know *a priori* that Hesperus is Phosphorus, and are in no position to find out ... except empirically' ('N & N', 308).[128]

So, Kripke thinks, if (as is more or less standard) we understand 'a priori truth' as 'a truth that it is possible to know a priori' (where 'possible' means something like 'humanly possible'), it does not seem that all necessary truths are a priori truths. To put it another way, there seem to be necessary a posteriori truths: these include not just *Hesperus = Phosphorus*, but also such theoretical identifications as *Gold = the element with atomic number 79*, *Heat = molecular motion*, statements subsuming one kind under another, such as *Gold is an element*, and compositional statements, such as *Water is (partly) composed of hydrogen* (cf. 'N & N', 263, 331–3). I have not found in Kripke an explicit definition of 'a posteriori truth', but presumably, since Kripke thinks of an a priori truth as one that can be known independently of experience, he thinks of an a posteriori truth as one that cannot be known independently of experience. (Obviously, an a posteriori truth cannot be

---

[128] See also 'I & N', 141, where Kripke says that no amount of a priori ratiocination on the part of astronomers could have enabled them to figure out that Hesperus = Phosphorus.

defined as a truth that can be known through experience, lest a priori truths turn out to be a posteriori truths.)

Because all this is very persuasive, Kripke succeeded in making the necessary a posteriori almost uncontroversial. Thus it is (nowadays) often assumed without argument, in discussions of the mind–body problem, that statements identifying mental and physical properties, or mental and physical events, could still be necessarily true, even if they could not be known a priori. In light of this consensus, it is worth saying that a case can be made (indeed, has been made) for explaining away Kripke's (putative) cases of the necessary a posteriori.

In a number of works Robert Stalnaker has expounded and defended a possible-worlds account of propositions and of belief.[129] Here a proposition should be understood as the kind of thing that is or could be the object of belief, disbelief, doubt, entertainment, and so on; as the kind of thing that is true or false (and could be necessarily true or necessarily false, contingently true or contingently false); and as the kind of thing that is 'expressed' or 'denoted' by sentential complement of the form 'that such-and-such'. (For example, in 'Amanda believes that pigs are clever', the sentential complement, 'that pigs are clever,' expresses or denotes the proposition *that pigs are clever* that Amanda is said to stand in the believing relation to.) For Stalnaker, a proposition may be thought of as a function from possible worlds into truth-values.[130] For instance, the proposition *that the gravitational constant is* 6.67×10^{-11}, may be thought of as a function which assigns truth to possible worlds in which the gravitational constant is 6.67×10^{-11} and falsity to possible worlds in which it is not.

According to Stalnaker, the belief state of a person may be represented by a set of possible worlds (intuitively, the set of possible worlds in which everything the person believes is true). What it is for a person to believe that P is for P to hold at each of the possible worlds belonging to the set that represents the person's belief state (equivalently, for P to hold at all of a person's 'doxastic alternatives').[131]

On this picture, the acquisition of information is the elimination of possibilities. Before I have a view on whether or not it is raining, my belief state comprises possible worlds in which it is raining and possible worlds in which it is not. After I've gone to the window (and, perhaps, leaned out of it), it comprises possible worlds of the first sort, or possible worlds of the second sort, but not both. The more I learn, the more possibilities I rule out: if I could somehow attain omniscience, I would have only one doxastic alternative left in my belief state (the actual world **G**).

[129] See R. Stalnaker, *Inquiry* (Cambridge, Mass.: MIT Press, 1987), and 'Semantics for Belief', *Philosophical Topics*, 15 (1987), 177–90.

[130] That propositions might be thought of as functions from possible worlds into truth-values was, Stalnaker thinks, originally suggested by Kripke in the early 1960s.

[131] See Stalnaker, *Inquiry*, 69. As Stalnaker notes, an early version of this sort of account of belief states is found in J. Hintikka, *Knowledge and Belief* (Ithaca: Cornell University Press, 1962).

Suppose we try to apply this account to what goes on when a person (call her S) learns that Hesperus is Phosphorus. Given that 'Hesperus' and 'Phosphorus' are rigid, it will be necessary that Hesperus is Phosphorus.[132] In that case there won't be any possible worlds in which Hesperus is not Phosphorus, and a fortiori there won't be any such worlds among S's doxastic alternatives before, but not after, she discovers that Hesperus is Phosphorus.

One might conclude that the possible-worlds account of belief (and of information) is unworkable, because it cannot accommodate cases in which the information acquired is necessarily true (or, for that matter, cases in which the misinformation acquired is necessarily false). Alternatively, one might argue that the possible-worlds account can, after all, give an account of such cases—albeit a not entirely straightforward one. Stalnaker takes the latter tack.

Suppose S says to herself (either out loud, or *in foro interno*), 'Hesperus is Phosphorus'. Her utterance expresses a necessarily true proposition P (as Stalnaker thinks of it, a constant function assigning truth to each possible world). But it might have expressed a (different) necessarily false proposition P'. Suppose S had inhabited a world in which there were two different planets, one of which appeared in the evening here (where Hesperus appears in the actual world) and was called 'Hesperus', and the other of which appeared in the morning there (where Phosphorus appears in the actual world) and was called 'Phosphorus'. Then S's utterance 'Hesperus is Phosphorus' would have expressed a false—indeed a necessarily false—proposition.

Now consider the proposition that is true at a possible world just in case the utterance 'Hesperus is Phosphorus' expresses a true proposition at that world. Stalnaker calls this proposition the *diagonal proposition* of the propositional concept for 'Hesperus is Phosphorus'.[133] The diagonal proposition for 'Hesperus is Phosphorus' is, like P, true, but, unlike P, only contingently true. (P is true in every world in which Hesperus is Phosphorus—that is, in every possible world; the diagonal proposition (call it D) is true in all the worlds in

---

[132] I neglect complications about whether it is strictly or only weakly necessary that Hesperus is Phosphorus.

[133] As Stalnaker thinks of it, a propositional concept is a function from possible worlds into propositions—that is, a function from possible worlds into functions from possible worlds into truth-values. Suppose that there are only two possible worlds, G and H. Suppose also that the utterance 'Hesperus is Phosphorus' expresses a (necessarily) true proposition at G and a (necessarily) false proposition at H. Then we can represent the propositional concept associated with 'Hesperus is Phosphorus' as follows:

	G	H
G	T	T
H	F	F

where T and F stand for truth and falsity, and each horizontal line represents the value of the propositional concept for the argument written to the left of the line. Each horizontal line represents the horizontal proposition expressed by 'Hesperus is Phosphorus' at a world; the diagonal from top left to bottom right represents the diagonal proposition true at a world just in the case horizontal proposition the utterance expresses at that world is true at that world. (For more details, see R. Stalnaker, 'Assertion', *Syntax and Semantics*, 9 (1978), 315–32.)

which the *utterance* 'Hesperus is Phosphorus' expresses a (necessary) truth, and false in all the worlds in which 'Hesperus is Phosphorus' expresses a (necessary) falsehood.) When S has no view about whether Hesperus is Phosphorus, there are possible worlds in her total belief state in which **D** is true, and possible worlds in her total belief state in which **D** is false. After she has learned that Hesperus is Phosphorus, the only possible worlds left in her belief state are ones in which **D** is true. So, Stalnaker concludes, we can hold onto the possible-worlds account of belief and information, as long as we suppose that what S learns, when she learns that Hesperus is Phosphorus, is not the necessary proposition **P**, but the contingent proposition **D**.

If the proposition that someone comes to know when she learns that Hesperus is Phosphorus is **D** rather than **P**, then there is no such thing as the necessary a posteriori proposition *that Hesperus is Phosphorus.* The sentential complement 'that Hesperus is Phosphorus', as it occurs in the sentence 'It is necessary that Hesperus is Phosphorus', expresses or denotes a necessarily true proposition (**P**). That same sentential complement, as it occurs in 'S knows a posteriori that Hesperus is Phosphorus', expresses or denotes a proposition that is knowable only a posteriori (**D**). But (for Stalnaker) the idea that there is a single proposition that is both necessary and a posteriori rests on a conflation of **P** with **D**.[134] Moreover (Stalnaker thinks), the strategy can be generalized to explain away Kripke's other examples of the necessary a posteriori.

A defender of the necessary a posteriori might object that we can explain away the necessary a posteriori only if we make the unnatural and unmotivated assumption that a sentential complement such as 'that Hesperus is Phosphorus' denotes one proposition in a context like *It is necessary that ——* and another proposition in a context like *S has learned that ——.* But is the assumption unmotivated? Kripke himself notes that 'There is a very strong feeling that leads one to think that, if you can't know something by a priori ratiocination, then it's got to be contingent' ('N & N', 306).

I am not sure this is quite the right way of putting it: do we really have a strong feeling that all necessary truths are knowable (never mind whether a priori or a posteriori)? But there is, I think, a strong feeling that if we can ascertain the truth-value of a proposition through empirical investigation, but not without empirical investigation, then that proposition is contingent. And this feeling would seem to constitute a motivation for supposing that in contexts such as *S has learned empirically that ——,* and *It is knowable only a posteriori that ——,* 'that Hesperus is Phosphorus' denotes a contingent proposition.

Also, Stalnaker would say, it is intuitively plausible that to learn that Hesperus is Phosphorus is to rule out certain possibilities. A defender of the necessary a posteriori can accept this claim, and insist that the possibilities ruled out are epistemic rather than metaphysical. But the characterization of

---

[134] For another elaboration and defence of this claim, see P. Tichy, 'Kripke on Necessity A Posteriori', *Philosophical Studies*, 43 (1983), 225–41.

belief and its objects (propositions) in terms of metaphysical possibilities fits into the larger project that motivates Stalnaker's account—the attempt to give a naturalistic account of intentionality.[135] If we characterize belief and its objects in terms of epistemic possibilities rather than metaphysical ones, we presuppose intentional notions in our account of belief. Hence, Stalnaker argues, the possible-worlds account of belief and propositions, unlike an account of belief and propositions in terms of epistemic possibilities, offers the prospect of an explanation of intentionality.[136]

The possible-worlds account of belief seems to me attractive but problematic. To use one of Stalnaker's examples, suppose that O'Leary believes that Hesperus is Mars. It seems that I could truly say:

(*a*) That Hesperus is Mars is impossible (true in no possible world), but O'Leary believes it just the same.

On Stalnaker's account, it is hard to see how (*a*) could be literally and straight-forwardly true. For, on that account, the proposition that is impossible (the proposition 'horizontally' expressed by some utterance of 'Hesperus is Mars') is different from the proposition that O'Leary believes (the proposition 'diagonally expressed' by that same utterance). But the occurrence of 'it' in (*a*) seems to imply that one and the same proposition is both true at no possible world and believed by O'Leary. Perhaps not; perhaps the 'it' works more like a so-called 'pronoun of laziness' (however exactly that is). But suppose that O'Leary studies astronomy and modal logic, and comes to accept that *Hesperus isn't Mars* and *Necessarily Hesperus isn't Mars* are both true. It seems as though he could then truthfully say:

(*b*) What I see now is impossible (true at no possible world) I used to believe was true.

(*b*) certainly seems to imply that one and the same proposition is necessarily false and was the object of O'Leary's belief. Perhaps Stalnaker would say that the truth that O'Leary is trying to state is something along the lines of:

(*b'*) The utterance which I now see expresses a necessarily false proposition I used to think expressed a true proposition.

So Stalnaker can provide a truth in the neighbourhood of (*b*). But (I want to say), it looks as though (*b*) itself is (strictly and literally) true, and inasmuch as the possible-worlds account of belief is incompatible with (*b*)'s (strict and literal) truth, that tells against the account.

---

[135] See the first chapter of *Inquiry* for a statement of Stalnaker's naturalism about intentionality and an explanation of how that naturalism motivates a possible-worlds account of belief and information.

[136] 'If belief and desire are to be explained in terms of naturalistic relations such as indication and tendency-to-bring-about, then the possibilities used to individuate propositions must be the ones relevant to these relations, and these clearly must be genuine, rather than epistemic, possibilities' (Stalnaker *Inquiry*, 25).

My purpose here is not to offer (much less argue for) a definitive verdict on the possible-worlds account of belief. I want only to suggest that a certain tendency should be resisted. It is natural to suppose that the rigidity of proper names, the necessary truth or falsity of identity statements flanked by proper names, and the necessary a posteriori are all part of the same package. There is indeed a very plausible line of reasoning from the rigidity of proper names to its being necessary or impossible that, but not determinable a priori whether, Hesperus is Phosphorus. But however plausible the line of reasoning is, it is not irresistible: as Stalnaker has argued, the rigidity of proper names does not by itself settle the question of whether some necessary truths are knowable only a posteriori. Kripke does not show sympathy for, or even discuss, Stalnaker's way of accepting the rigidity of proper names, and rejecting the necessary a posteriori. As we shall see, though, he agrees that the rigidity of proper names does not by itself definitively settle the question of whether there are necessary a posteriori truths.

As Kripke notes, it is natural to suppose that, if something isn't true in all possible worlds, you cannot ascertain its truth by a priori ratiocination: you have to investigate empirically whether you are in a world in which it is true, or a world in which it is false ('N & N', 263). Nevertheless, Kripke argues, considerations involving rigidity seem to show that we can have a priori knowledge of contingent truths. The line of thought that originally suggested this conclusion to Kripke is the following: suppose someone introduces a rigid designator a into the language via the following ceremony: 'Let a designate the only object that in fact has property F with respect to any actual or counterfactual situation.' It seemed clear to Kripke that

if a speaker did introduce a designator into a language that way, then in virtue of his very linguistic act, he would be in a position to say, 'I know that Fa', but nevertheless 'Fa' would express a contingent truth (provided that F is not an essential property of the unique object that possesses it). (N & N, 14)

For example, suppose a is '1 metre', F is *being a length of stick S has at time t_0*, and the reference formula is: let '1 metre' (rigidly) designate the length that in fact is the length S has at time t_0 (more idiomatically, the length that S in fact has at time t_0). Then, Kripke asks:

What is the *epistemological* status of the statement 'Stick S is one meter long at t_0', for someone who has fixed the metric system by reference to stick S? It would seem that he knows it *a priori*. For if he uses stick S to fix the reference of the term 'one meter', then as a result of this kind of 'definition' (which is not an abbreviative or synonymous definition), he knows automatically, without further investigation, that S is one meter long. On the other hand, even if S is used as the standard of a meter, the *metaphysical* status of 'S is one meter long' will be that of a contingent statement, provided that 'one meter' is regarded as a rigid designator: under appropriate stresses and strains, heatings or coolings, S would have had a length other than one meter even at t_0... So in this sense, there are contingent *a priori* truths. ('N & N', 275)

But—it is natural to object—the person who fixes the reference of a via the description 'the F' cannot know that a is F unless he knows that F is uniquely exemplified: if F isn't uniquely exemplified, the attempt to fix the reference of a will misfire, and Fa will be either false or undefined. Moreover—the objection goes—if 'Fa' expresses a contingent truth about the external world, the speaker cannot know a priori that F is uniquely exemplified.

Applying this to Kripke's example, the reference-fixer cannot know that 1 metre is the length of stick S at time t_0 unless he knows that *being a length of stick S at time t_0* is uniquely exemplified. And he cannot know a priori that *being a length of stick S at time t_0* is uniquely exemplified (otherwise, he could know a priori that stick S exists at t_0).

Because Kripke does not discuss this objection explicitly, I am not sure how he would respond to it. For our purposes, what matters is that the objection does not (clearly) apply to all of the cases of (allegedly) a priori knowledge of contingent truths about the external world discussed by Kripke. Kripke suggests that when a reference-fixer introduces a name a to rigidly designate the unique individual that is (actually) F, he can know a priori that the statements 'a exists' and 'Exactly one thing is F' are materially equivalent, even if the biconditional 'a exists if and only exactly one thing is F' is only contingently true ('N & N', n. 33). I am not sure about this. To use another example discussed by Kripke, suppose that Leverrier introduces the name 'Neptune' to rigidly designate the planet (actually) causing these perturbations in the orbits of those planets. For all Leverrier knows a priori, there is no unique planet causing the relevant perturbations, and his attempt to fix the reference of 'Neptune' misfires. Now, we might say that, if his attempt misfires, then 'Neptune exists' is false, or we might say that, if his attempt misfires, then 'Neptune exists' is undefined (neither true nor false). If we say the latter, then we shall say that, for all Leverrier knows a priori, 'Neptune exists' and 'Exactly one planet causes the perturbations in the orbits of these planets' are not materially equivalent.

Still, Kripke could say, just in virtue of his reference-fixing stipulation, Leverrier can know (a priori) that if some (one) planet is the cause of the relevant perturbations, Neptune exists. If it is claimed that the introducer of the expression '1 metre' could know a priori that 1 metre is the length of stick S at time t_0, it may be objected that he could not know that a priori, because he could not know a priori that his attempt to fix the reference of '1 metre' by reference to the length of stick S at time t_0 had succeeded. It is not clear that the same objection can be made to the claim that Leverrier can know a priori that if some (one) planet is the cause of the relevant perturbations, Neptune exists. For there is no obvious reason to suppose that a priori knowledge of this conditional would entail a priori knowledge that the attempt to fix the reference of 'Neptune' had succeeded.[137] Even so,

---

[137] We can imagine someone who disagrees with Leverrier: she doesn't believe in the (alleged) perturbations or the (alleged) planet causing them. She might say to Leverrier: 'Of course, *if* there is some (one) planet that is the cause of the relevant perturbations, then Neptune exists.'

as Kripke emphasizes, it is only contingently true that if some (one) planet is the cause of the relevant perturbations, Neptune exists; there are Neptune-less possible worlds in which some other planet (Saturn, say) is the cause of the relevant perturbations. Similarly, suppose it is suggested that simply in virtue of having made the right reference-fixing stipulation, Leverrier knew a priori the contingent truth that Neptune is the planet causing such-and-such perturbations. Then it may be objected that Leverrier could not know that contingent truth a priori, because he couldn't know a priori that his attempt to fix the reference of 'Neptune' succeeded, because he couldn't know a priori that the relevant perturbations existed, or were caused by just one planet.[138] It is not clear that the same objection can be made, if it is suggested that, simply in virtue of having made the right reference-fixing stipulation, Leverrier knew a priori the contingent truth that, if some planet is the cause of the relevant perturbations, Neptune is; or knew a priori the contingent truth that, if Neptune exists, it is the cause of the relevant perturbations.

Applying this to the case of the metre stick, let us suppose that the reference-fixer for '1 metre' is not in a position to know a priori that 1 metre is the length of stick S at time t_0 (equivalently, that stick S is 1 metre long at time t_0), because he is not in a position to know a priori that his attempt to fix the reference of '1 metre' succeeded, because he is not in a position to know a priori that S exists at t_0. It still could be true that the reference-fixer is in a position to know a priori that

(a) if there is such a thing as the length of stick S at time t_0, then there is such a thing as (the length) 1 metre;

(b) if there is such a thing as the length of stick S at time t_0, then 1 metre is the length of S at t_0;

and

(c) if there is such a thing as (the length) 1 metre, 1 metre is the length of S at t_0.

Assuming that our reference-fixer can know a priori both that S exists at t_0 if and only if there is such a thing as the length of S at t_0, and that 1 metre is the length of S at t_0 if and only if S is one meter long at t_0, it follows that, if she can know a priori that (b) and (c), she can also know a priori that

(d) if S exists at t_0, then S is 1 metre long at t_0;

and

(e) if there is such a thing as the length 1 metre, then S is 1 metre long at t_0.

---

[138] Leverrier could no more know a priori that Neptune is the cause of the perturbations, it might be said, than an astronomer of the 19th century could know a priori that Vulcan is a cause of such-and-such features of the orbit of Mercury.

Although Kripke introduces a number of cases that seem to him to be cases of the contingent a priori, he is actually more tentative about the existence of contingent a priori truths than he is about the existence of necessary a posteriori truths. Thus early on in *Naming and Necessity* (263) Kripke says that he will argue that there *are* necessary a posteriori truths and that there are *probably* contingent a priori truths. And at note 26 of the same work Kripke writes:

If someone fixes a meter as 'the length of stick S at t_0', then in some sense he knows a priori that the length of stick S at t_0 is one meter, even though he uses this statement to express a contingent truth. But, merely by fixing a system of measurement, has he thereby *learned* some (contingent) *information* about the world, some new *fact* that he did not know before? It seems plausible that in some sense he did not, even though it is undeniably a contingent fact that S is one meter long. So there may be a case for reformulating the thesis that everything a priori is necessary so as to save it from this type of counterexample.

How might we reformulate the thesis that everything a priori is necessary? Taking a cue from Kripke's talk of '(contingent) information about the world', we might try:

> No a priori knowledge gives its possessor any information about how the world is contingently,

where by 'the world' we could understand 'the external world', so as to accommodate those philosophers who think that contingent truths such as *I exist* are known a priori.

As Kripke concedes, there is a feeling that the item of knowledge Kripke takes to be contingent a priori—the thing that the reference-fixer comes to know simply by virtue of fixing the reference of, say, '1 metre'—does not actually give the reference-fixer any (contingent) information about the world. This may not be immediately clear: under normal circumstances, someone who came to know that, if this stick exists at t_0, it is 1 metre long at t_0, or that, if there is such a thing as the length 1 metre, then this stick is 1 metre long at t_0, would thereby acquire some (contingent) information about the world. Imagine, though, that you have been told that there is just one stick behind the house, but have no information on how long the one stick behind the house is, if indeed there is (just) one. Imagine, further, that you accept the testimony of the person who told you that there is just one stick behind the house, and say to yourself: let 1 *schmetre* (rigidly) designate the only length that is in fact a length of the only object that is in fact a stick behind the house. (In other words, let 1 *schmetre* designate, with respect to any possible world, the length which is actually the length of the thing which is actually the only stick behind the house.) Suppose that, as Kripke suggests, the reference-fixer for '1 metre' is in a position to know a priori that, if there is such a thing as the length of S at t_0, then the length of S at t_0 is 1 metre. Then it is also true that the reference-fixer for '1 schmetre' is in a position to know a priori that, if there is just one stick behind the house, the length of the stick behind the house is

1 schmetre.[139] Suppose we say, in Kripkean fashion, that 'in virtue of your very linguistic act', you are in a position to know 'automatically and without further investigation' that, if there is just one stick behind the house, the stick behind your house is 1 schmetre long. Be that as it may, on the face of it, it remains true that you have no information about how long the stick behind the house is, or about how long the stick behind the house is, if indeed there is just one stick behind the house. Suppose that, after you have been told that there was just one stick behind the house, but before you introduce the name '1 schmetre', someone asks you how long the stick behind the house is, if there is just one stick behind the house. You could (cautiously, given your belief that there is just one stick behind the house, but truthfully) reply: 'I have no information about how long the stick behind the house is, if indeed there is just one such stick.' And you surely do not acquire information about how long the stick behind the house is, if indeed there is just one such stick, simply by introducing the name '1 schmetre' in the way described above.[140]

Absent empirical investigation, the person who attempts to fix (and, let us suppose, succeeds in fixing) the reference of the name '1 schmetre' still has no information about the length of the stick behind the house, or about the length of the stick behind the house, if there is just one stick behind the house. The act of reference-fixing doesn't put her in a position to have that sort of information about the (external) world. But if it doesn't put her in a position to have that sort of information about the (external) world, it is difficult to see what other sort of information about the (external) world it might put her in a position to have. Even if the act of reference-fixing puts the reference-fixer in a position to know a priori a contingent truth (namely, that if there is just one stick behind the house, it is a schmetre long), knowledge of that truth does not seem to give its possessor any (contingent) information about the external world.

The point seems to generalize to other Kripkean examples of the contingent a priori. Suppose that the police have found N's (dead) body, and believe that N has been murdered, and has exactly one murderer. Suppose they have no information about who murdered N, if indeed someone did. They introduce the name 'Jack' via the following ceremony: 'Jack' shall rigidly designate the

---

[139] The reasoning would go: you know a priori that if there is just one stick behind the house, then there is such a thing as the length of the stick behind the house. And you know a priori that if there is such a thing as the length of the stick behind the house, the length of the stick behind the house is 1 schmetre. So you know a priori that if there is just one stick behind the house, then the length of the stick behind the house is 1 schmetre. So you know a priori that if there is just one stick behind the house, the stick behind the house is 1 schmetre long.

[140] Contrast this with a case in which you are told that there is just one stick lying on the ground in your back garden—a stick that does not extend into any of the adjoining gardens. If you check the surveyor's report for your house to determine the area of your back garden, and on that basis calculate that the longest stick that could fit in your garden (if laid on the ground) is n metres long, you thereby acquire (contingent) information about how long the stick lying on the ground in your back garden is, if indeed there is just one stick on the ground in your back garden (it is no more than n metres long).

(unique) murderer of N. Suppose we say that, in virtue of the way they introduced the name 'Jack', the police are in a position to know a priori that, if anyone is the murderer of N, Jack is. The fact remains that the introduction of the name 'Jack' does not give the police any information about who murdered N, if anyone did. (It would be a cruel joke if, after introducing the name 'Jack', the police called N's spouse to tell her they had new information about who murdered N, if anyone did.) And if the introduction of the name 'Jack' does not put its introducers in a position to have any new information about who (if anyone) murdered N, it is hard to see what other information about the (external) world it might put them in a position to have. So, on the supposition that the introduction of the name 'Jack' puts the police in a position to know a priori the contingent truth that, if N was murdered, Jack murdered N, knowledge of that truth does not seem to give its possessor any (contingent) information about the external world.

Now there is an obvious tension in accepting all of the following claims:

(1) Simply in virtue of an act of reference-fixing, the reference-fixer is in a position to come to know—and come to know a priori—things like *if S exists at t_0, then S is 1 metre long at t_0.*

(2) Truths like *if S exists at t_0, then S is 1 metre long at t_0* are contingent.

(3) A priori knowledge never gives its possessor information about how the (external) world is contingently.

After all, unlike, say, the preposition *I exist*, the preposition *if S exists at t_0, then S is 1 metre long at t_0* seems not only to be contingent, but also to imply something about how the external world is. It accordingly seems to provide information about how the external world is contingently to anyone who knows it. So it looks as though the defender of (3)—our attempt to reformulate the thesis that everything a priori is necessary—is going to have to challenge (1), or (2), or both.

One strategy for holding onto (2) and (3) and rejecting (1) has been employed by Alvin Plantinga, Keith Donnellan, and Keith Hossack (*inter alios*).[141] The basic idea is that in the sorts of cases discussed by Kripke one does not, simply as a result of reference-fixing, come to know a contingent truth a priori; instead, one comes to know that a certain sentence expresses a contingent truth, without knowing a priori the contingent truth it expresses. For example, in virtue of fixing the reference of the name 'Neptune', Leverrier comes to know that the sentence 'If exactly one planet is the cause of such-and-such perturbations, then Neptune exists' expresses a contingent truth,

---

[141] See A. Plantinga, *The Nature of Necessity* (Oxford: Clarendon Press, 1974), 8–9 and n. 1; K. Donnellan, 'The Contingent A Priori and Rigid Designators', in P. French, T. Uehling, and H. Wettstein (eds.), *Contemporary Perspectives in the Philosophy of Language* (Minneapolis: University of Minnesota Press, 1979); and K. Hossack, 'Implicit Definition, Descriptive Names, and the Contingent A Priori', MS, 2000. Many other philosophers have either advanced a similar line, or expressed sympathy for, say, Donnellan's advancement of it; see e.g. Tichy, 'Kripke on Necessity A Posteriori', 226.

without knowing the contingent truth that sentence expresses (without coming to know that, if exactly one planet is the cause of such-and-such perturbations, then Neptune exists). Similarly, simply in virtue of fixing the reference of '1 metre', the reference-fixer comes to know that 'If S exists at t_0, then S is a metre long at t_0' expresses a truth, without knowing the contingent truth that sentence expresses (without knowing that, if S exists at t_0, then S is 1 metre long at t_0). (1) is accordingly false, and we can hold on to both (2) and (3).

The obvious consideration in favour of the Plantinga–Donnellan-*et-alii* view is that, although it seems very odd that reference-fixing could afford the reference-fixer knowledge of how the (external) world is contingently, it seems less odd that reference-fixing could afford him knowledge about which sentences in his language express contingent truths about how the external world is contingently.[142] The drawback of the view is that it doesn't obviously mesh with our (or at least my) intuitive judgements about what we may properly say people know or believe in actual or hypothetical cases.

Suppose that Leverrier introduces the name 'Neptune' to rigidly designate the cause of such-and-such perturbations, saying, '*Neptune* is the cause of such-and-such perturbations.' The sentence he utters expresses a truth. And perhaps Leverrier knows that the sentence he utters expresses a truth. But, for Donnellan, Leverrier doesn't believe, any more than he knows, the truth that the sentence expresses. As Donnellan sees it, the situation is (in some ways) analogous to what happens if, in complete ignorance of what 'oblateness' means (and of what oblateness is), I happen to see in *Scientific American* the sentence 'The oblateness of Mars is .003.' When I am subsequently asked (say, on a quiz show) what the oblateness of Mars is, I may answer, 'The oblateness of Mars is .003.' But (Donnellan thinks) I do not thereby express my belief that the oblateness of Mars is .003.

The difficulty is that, although it seems plausible that (in Donnellan's story) I don't believe that the oblateness of Mars is .003, it seems much more doubtful that (after fixing the reference of the name 'Neptune') Leverrier doesn't believe that Neptune is the cause of such-and-such perturbations. Suppose that shortly after fixing the reference of 'Neptune' Leverrier gives a talk on his latest research to a group of astronomers. Suppose that M attended the talk, and M's colleague N did not. If N asks M what the talk was about, M might reply:

> It was very interesting. Leverrier thinks there is an eighth planet—he calls it 'Neptune'. He thinks that Neptune is not only responsible for such-and-such perturbations, but also . . .

For Donnellan, this perfectly natural description of Leverrier's views is a misdescription. Leverrier uttered the words 'Neptune is the cause of such-and-such perturbations,' and he believed that they expressed a truth, but he didn't believe the truth they expressed. When Leverrier said, 'Neptune is the

---

[142] Cf. Donnellan, 'The Contingent A Priori and Rigid Designators', 52–3.

cause of such-and-such perturbations,' he no more believed what he was saying than I do, when on the quiz show I say, 'The oblateness of Mars is .003.' This seems implausible to me—or at least, unobvious. After all, when I say (on the quiz show), 'The oblateness of Mars is .003,' I don't know what I am saying. When in the course of his talk Leverrier says, 'Neptune is the cause of such-and-such perturbations,' it is not at all clear that he doesn't know what he is saying.

To take another example, suppose that one of the detectives in Scotland Yard has introduced the name 'Jack the Ripper' to rigidly designate whichever person (uniquely) killed X, Y, and Z; and suppose that the other detectives at the Yard have acquired the name from its introducer. In that case, Donnellan would say, none of the detectives at Scotland Yard believe that Jack the Ripper killed X, even if they all believe that 'Jack the Ripper killed X' expresses a truth. In the circumstances, though, it would be very odd to say that none of the detectives at Scotland Yard believes that Jack the Ripper killed X, Y, or Z. By contrast, it might be perfectly natural to say something Donnellan would classify as false: 'All of the detectives believe (agree) that Jack the Ripper killed X, though only some of them think he killed A.'

What seems plausible in the Donnellan–Plantinga view is the idea that, when he fixes the reference of 'Neptune', Leverrier in some sense does not thereby know anything over and above the fact that statements such as, say, 'If there is just one planet causing such-and-such perturbations, the cause of those perturbations is Neptune' express truths. What seems doubtful is the claim that we cannot (properly) say of Leverrier, *qua* reference-fixer, that he knows—or even believes—that, if there is just one planet causing such-and-such perturbations, then the cause of those perturbations is Neptune. So a natural move here is to appeal to Stalnaker's idea that, in belief and knowledge statements, sentential complements may express the diagonal proposition of the propositional concept associated with the statement rather than the horizontal proposition of the propositional concept associated with that statement. Then we can say both that Leverrier knows a priori that, if there is just one planet causing such-and-such perturbations, then the cause of those perturbations is Neptune, and that what Leverrier knows a priori is nothing over and above the fact that 'If there is just one planet causing such-and-such perturbations, then the cause of those perturbations is Neptune' (horizontally) expresses some (contingently) true proposition. (After all, the diagonal proposition associated with a statement is true (at a world) just in case the statement horizontally expresses some proposition or other that is true (at that world).)

If we go this (Stalnaker) route, the contingent a priori proposition *If there is just one stick behind the house, it is 1 schmetre long* vanishes. The proposition H 'horizontally expressed' by 'If there is just one stick behind the house, it is a schmetre long' is contingently true, and the diagonal proposition D 'diagonally' expressed by that sentence is known a priori. But H isn't known a priori, and D isn't contingent. In just the same way, the Kripkean contingent a priori proposition *If S exists at t_0, S is a metre long at t_0* vanishes. There is, accordingly, no way to get from (1) and (2) above, to the denial of (3). If (1) is true, it is a truth about the epistemological status of certain diagonal propositions; if (2) is true, it is a truth about the modal status of a certain horizontal proposition; so, on any construal of (1) and (2) on which they both come out true, they don't rule out (3).

To forestall a possible misunderstanding: the horizontal proposition (horizontally) expressed by 'If S exists at t_0, then S is a metre long at t_0' is, of course, a possible object of belief, and of knowledge. In fact, we can attribute knowledge of that horizontal proposition to someone by saying that she knows that, if S exists at t_0, then S is a metre long at t_0.

For example, suppose that, long after t_0, stick S is acquired by someone other than the reference-fixer. S's new owner does not know that S is the standard metre (or, perhaps, the 'retired' standard metre). She does, however, know that S is a metre long (having measured it in the usual way). Suppose S's new owner subsequently learns that S has always had, and will always have, the length it has now, and concludes that, if S exists at t_0, then S is a metre long at t_0.[143] The conclusion she draws is the proposition horizontally expressed by 'If S exists at t_0, then S is a metre long at t_0', and we can attribute knowledge of that proposition to her by saying that she has learned that, if S exists at t_0, then S is a metre long at t_0.

So the idea is not that a statement such as 'So-and-so knows that, if S exists at t_0, then S is a metre long at t_0' always attributes knowledge of the proposition diagonally expressed by the sentential complement in that statement. It is instead that 'So-and-so knows a priori that, if S exists at t_0, then S is a metre long at t_0' is true only if the proposition (a priori) knowledge of which it attributes to so-and-so is the diagonal proposition diagonally expressed by the sentential complement in that statement rather than the horizontal proposition horizontally expressed by the sentential complement in that statement. We might say truly of one person (the reference-fixer for the term 'metre') that he knows a priori that, if S exists at t_0, then S is one metre long at t_0, and say of someone else (S's new owner) that she knows a posteriori that, if S exists at t_0, then S is one metre long at t_0. But, on the view being considered, we don't have a case of one person knowing a priori what another person knows a posteriori: the objects of knowledge, as well as the knowers, and the mode of knowing, differ.

Someone may ask: given that in the two knowledge attributions we have the same sentential complement, what motivates the assumption that we have different objects of knowledge? Why haven't we simply got a case like the one in which a mathematician knows a priori that the number n is prime (having proven that it is), and I know a posteriori that n is prime (having heard from the mathematician that it is)?

---

[143] I ignore complications about tense, which I take it do not affect the point at issue.

Well, as Kripke would concede, and as I have tried to bring out, there is a strong feeling that what the reference-fixer for '1 schmetre' knows, independently of experience and simply by virtue of linguistic fiat, provides him with no information about the length of the stick behind the house, if indeed there is just one stick behind the house. Thus, assuming (*pace* Donnellan) that we can truly say that the reference-fixer knows a priori that, if there is just one stick behind the house, then the stick behind the house is a schmetre long, that knowledge attribution is not a specification of the reference-fixer's information about the length of the stick behind the house, if indeed there is just one stick behind the house. In the same way, what the reference-fixer for 'metre' knows, independently of experience and simply by virtue of linguistic fiat, provides him with no information about how long S is at t_0, if indeed S exists at t_0. (This would be more obvious if the reference-fixer for 'metre' fixed the reference of 'metre' with respect to an unseen stick S.) So, assuming (*pace* Donnellan) that we can truly say that the reference-fixer knows a priori that, if S exists at t_0, then S is a meter long at t_0, that knowledge attribution is not a specification of the reference-fixer's information about the length of S at t_0, if indeed S exists at t_0. On the other hand, when we attribute to S's new owner the knowledge that, if S exists at t_0, then S is a metre long at t_0, that *is* a specification of S's new owner's information about the length of S at t_0, if indeed S exists at t_0. This is a motivation for saying that the sentential complement that appears in the two knowledge attributions 'The reference-fixer knows a priori that, if S exists at t_0, S is a metre long at t_0' and 'S's new owner knows a posteriori that, if S exists at t_0, then S is a metre long at t_0' does not express the same proposition in the first knowledge attribution as it does in the second. The case is unlike the one mentioned above, in which there is no reason to suppose that the sentential complement that appears in the two knowledge attributions 'The mathematician knows a priori that *n* is prime' and 'I know that *n* is prime' expresses one proposition in the first knowledge attribution, and a different one in the second.

To recap: there is a motivation for rejecting the contingent a priori, if the rejection of the contingent a priori is the thesis that nothing we know a priori provides us with information about how the (external) world is contingently. We can uphold this thesis, in the face of the sort of cases discussed by Kripke, if, with Donnellan, we refuse to allow that in those cases the reference-fixer knows or believes the truths expressed by the statements he knows a priori are true. If, on the other hand, we find Donnellan's way of rejecting the contingent a priori difficult to reconcile with what we are wont to say about what people believe or know in actual or hypothetical cases, we can still reject the contingent a priori, if we say that in the sorts of cases discussed by Kripke, what the reference-fixer knows a priori is not a contingent (horizontal) proposition, but a necessary (diagonal) one.

We have already seen that diagonalization can be used to get rid of necessary a posteriori propositions, as well as contingent a priori propositions. In both cases we distinguish a horizontal proposition **H** and a diagonal proposition **D**, and insist that **H** has the modal property (being necessarily true, or being contingently true), while **D** has the epistemological property (being knowable only a posteriori, being knowable a priori). In both cases we have to make the (initially unobvious, at least) supposition that sentential complements can express different propositions, depending on whether they are embedded in an alethic context (such as 'It is necessary (contingent) that ——') or an epistemic context such as 'It could be known independently of experience (could only be known through experience) that ——'). In both cases our making that initially unobvious supposition enables us to hang onto (what Kripke grants is) an initially attractive belief (that empirical investigation is necessary for ascertaining the truth-value of a proposition (whose truth-value is ascertainable) only if that proposition is contingent; that reference-fixing is not a way of finding out how the (external) world is contingently.

As I noted earlier, in *Naming and Necessity* Kripke is less confident that there are contingent a priori truths than he is that there are necessary a posteriori truths. It is my impression that far more philosophers have accepted Kripkean arguments in favour of the necessary a posteriori than have accepted Kripkean arguments in favour of the contingent a priori.[144] Now Kripke's arguments for the necessary a posteriori and the contingent a priori are symmetrical (they both turn crucially on the rigidity of proper names). And, by appealing to the idea that sentential complements (typically) express a horizontal proposition in alethic contexts, and sometimes express a diagonal proposition in doxastic and epistemic contexts, we can accept the rigidity of names, and avoid both the necessary a posteriori and the contingent a priori. So why have Kripkean arguments on behalf of the necessary a posteriori found more favour than Kripkean arguments on behalf of the contingent a priori?

I'm not sure. Perhaps part of the answer is this: if we accept Kripke's arguments for the necessary a posteriori, we must embrace the idea that there is necessary information to which our only possible access is empirical. If we accept Kripke's arguments for the contingent a priori, we must embrace the idea that linguistic stipulation can by itself afford us access to information about how the world is contingently. Both of these claims are at least initially surprising, but the former claim is perhaps easier to learn to live with than the latter.

We have already seen that in *Naming and Necessity* Kripke stops short of committing himself to the contingent a priori, although he at least comes very close to committing himself to the necessary a posteriori. Caution about the contingent a priori reappears in 'A Puzzle About Belief', and is extended to the necessary a posteriori. Early on in that piece Kripke writes:

According to Mill, a proper name is, so to speak, *simply* a name... If a strict Millian view is correct, and the linguistic function of a proper name is completely exhausted by

---

[144] The impression is shared by Salmon; see *Reference and Essence*, 78.

the fact that it names its bearer, it would appear that proper names of the same thing are everywhere interchangeable not only *salva veritate* but also *salva significatione*: the proposition expressed by a sentence should remain the same no matter what name of the object it uses... If Mill is completely right, not only should 'Cicero was lazy' have the same *truth value* as 'Tully was lazy', but the two sentences should express the same *proposition*, have the same content... If such a consequence of Mill's view is accepted, it would seem to have further consequences regarding 'intensional' contexts. Whether a sentence expresses a necessary truth or a contingent one depends only on the proposition expressed and not on the words used to express it. So any simple sentence should retain its 'modal value' (necessary, impossible, contingently true, or contingently false) when 'Cicero' is replaced by 'Tully' in one or more places, since such a replacement leaves the content of the sentence unaltered... The situation would seem to be similar with respect to contexts involving knowledge, belief, and epistemic modalities. Whether a given subject believes something is presumably true or false of such a subject no matter how that belief is expressed; so if proper name substitution does not change the content of a sentence expressing a belief, coreferential proper names should be interchangeable *salva veritate* in belief contexts. Similar reasoning would hold for epistemic contexts ('Jones knows that... ') and contexts of epistemic necessity ('Jones knows *a priori* that... ') and the like. ('APAB', 104–5)

As we have seen, Kripke neither accepts nor rejects the soundness of the argument for the interchangeability of co-referential proper names in belief contexts sketched in this passage. As he sees it, it is an open and vexed question whether belief contexts are 'Shakespearean' (that is, allow substitution of co-referential proper names *salva veritate*): 'Philosophers have often, basing themselves on Jones' and similar cases, supposed that it goes virtually without saying that belief contexts are not 'Shakespearean'. I think that, at present, such a definite conclusion is unwarranted' ('APAB', 136).

As Kripke notes in the longer of the two passages just cited, the same sort of argument that can be used to defend the interchangeability *salva veritate* of co-referential proper names in belief contexts can be used to defend the interchangeability *salva veritate* of co-referential proper names in what Kripke calls 'contexts of epistemic necessity'. If it would be premature to dismiss the argument for the interchangeability *salva veritate* of co-referential proper names in belief contexts, it should also presumably be premature to dismiss the argument for the interchangeability *salva veritate* of co-referential proper names in contexts of epistemic necessity. If, however, co-referential proper names are interchangeable *salva veritate* in contexts of epistemic necessity, then S knows a posteriori that, if Hesperus exists, then Hesperus = Phosphorus just in case S knows a posteriori that, if Hesperus exists, then Hesperus = Hesperus. Moreover, it is knowable only a posteriori that, if Hesperus exists, then Hesperus = Phosphorus just in case it is knowable only a posteriori that, if Hesperus exists, then Hesperus = Hesperus. This puts a question mark over the claim that *If Hesperus exists, then Hesperus = Phosphorus* is knowable only a posteriori, since there is nothing obvious about the claim that *If Hesperus exists, then Hesperus = Hesperus* is knowable only a posteriori.

Also, Kripke's own case for the contingent a priori depends on the premiss that statements like *If Jack the Ripper exists, then Jack the Ripper is a murderer* are—*pace* Donnellan *et alii*—knowable a priori (by the introducer of the name 'Jack the Ripper'). If, however, the introducer of the name 'Jack the Ripper' is in a position to know a priori that, if Jack the Ripper exists, then Jack the Ripper is a murderer, then surely the introducer of the name 'Hesperus' is (*inter alios*) in a position to know a priori that, if Hesperus exists, then Hesperus = Hesperus. So, if we accept the interchangeability of co-referential proper names in contexts of epistemic modality, we should conclude that, if Kripke's case for the contingent a priori succeeds, then his case for the necessary a posteriori fails. At the risk of belabouring the point: if Kripke's case for the contingent a priori succeeds, then statements like *If Jack the Ripper exists, then Jack the Ripper is a murderer* are knowable a priori. If such statements are knowable a priori, then *If Hesperus exists, then Hesperus = Hesperus* is knowable a priori. And, assuming that co-referential proper names are interchangeable in contexts of epistemic modality, *if Hesperus exists, then Hesperus = Hesperus* is knowable a priori only if it is not after all true that *If Hesperus exists, then Hesperus = Phosphorus* is knowable only a posteriori.

We can run the same argument in the opposite direction, to show that, assuming the interchangeability of co-referential proper names in contexts of epistemic modality, if Kripke's case for the necessary a posteriori succeeds, then his case for the contingent a priori fails. From the premiss that it is knowable only a posteriori that, if Hesperus exists, then Hesperus = Phosphorus, and interchangeability, we may conclude that it is knowable only a posteriori that, if Hesperus exists, then Hesperus = Hesperus. Surely it is no more knowable a priori that, if Jack the Ripper exists, Jack the Ripper is a murderer, than it is knowable a priori that, if Hesperus exists, then Hesperus = Hesperus. So it is not after all knowable a priori that, if Jack the Ripper exists, then Jack the Ripper is a murderer.

The moral would seem to be that, until puzzles about the interchangeability or non-interchangeability *salva veritate* of co-referential proper names in contexts of epistemic modality are sorted out, the question of the coincidence or otherwise of the necessary–contingent distinction, and the a priori–a posteriori distinction, remains wide open. Strict Millianism looks initially as though it should be congenial to the necessary a posteriori and the contingent a priori, since, as we have seen, Kripkean arguments for the necessary a posteriori and the contingent a priori turn on rigidity, and names are rigid if (though not necessarily only if) they are 'Millian' (simply referential). But, if names are simply referential, that suggests that names should be interchangeable in contexts of epistemic modality, which in turn raises difficulties for Kripke's case for the necessary a posteriori and the contingent a priori.

To forestall a possible misunderstanding: even pending a resolution to Kripke's puzzle about belief, we can say that (i) 'Hesperus = Phosphorus' expresses a necessary truth, but empirical inquiry was needed to ascertain that it does. And we can say that (ii) 'If Jack the Ripper exists, he was a murderer' expresses a contingent truth, but no empirical inquiry was needed (on the part of the introducer of the name 'Jack the Ripper') to ascertain that it does. But it is one thing to accept (i) and (ii), and another to embrace the necessary a posteriori and the contingent a priori (as these terms are usually understood). As we have already seen, Donnellan accepts (ii), but is no friend of the contingent a priori. *Pari ratione*, one could accept (i) without being a friend of the necessary a posteriori. Someone who embraces the necessary a posteriori (on the usual understanding of the term) holds that there are truths that are necessary but can only be the objects of a posteriori knowledge. She will hold, say, that S knows that Hesperus = Phosphorus only if S knows a posteriori that Hesperus = Phosphorus, even though it is necessary that Hesperus = Phosphorus. Likewise, someone who embraces the contingent a priori holds that there are truths that are contingent but may still be the objects of a priori knowledge. She will hold, say, that the introducer of the name 'Jack the Ripper' knows a priori that, if Jack the Ripper exists, he is a murderer, even though it is contingent that, if Jack the Ripper exists, he is a murderer. Such claims go beyond (i) and (ii), and do so precisely in a way that is problematic, if in fact contexts of epistemic modality are Shakespearean.

It seems, then, that Kripke's views on the necessary a posteriori and contingent a priori have been subject to a certain degree of misunderstanding. Kripke is often represented as a firm believer in both the necessary a posteriori and the contingent a priori. He certainly appears to champion the necessary a posteriori, and express (strong) sympathy towards the contingent a priori in *Naming and Necessity*. But in 'A Puzzle about Belief' Kripke says that he realized, even at the time of *Naming and Necessity*, that there are very delicate issues concerning the necessary a posteriori and the contingent a priori—issues that call into question, for example, the claim that astronomers knew a priori that Hesperus = Hesperus even before they discovered a posteriori that Hesperus = Phosphorus (see 'APAB', nn. 10 and 44). Certainly, his views as expressed in 'A Puzzle About Belief' do not commit him to the claim that there are necessarily true propositions (objects of knowledge, contents of belief) that can only be known a posteriori, or the claim that there are contingently true propositions (objects of knowledge, contents of belief) that can be known a priori. Instead of seeing Kripke as a champion of the necessary a posteriori and the contingent a priori, it would be better, I think, to see him as challenging the (once orthodox) view that all and only necessary truths are knowable a priori.[145]

---

[145] This is not the only area in which philosophers have failed to appreciate fully how aporetic Kripke's (considered) views are. In the context of the mind–body problem Kripke is often thought of as an out-and-out property dualist, even though (in *Naming and Necessity*) his last words on the subject are 'I regard the mind–body problem as wide open and extremely confusing.' That Kripke is not infrequently interpreted as holding somewhat more definite views than he actually holds is unsurprising, in light of the fact that *Naming and Necessity* is a (very lightly edited) transcript of a series of talks given without either a written text or notes, and it is often very difficult in a talk to put in all the nuances of one's view without making the audience lose sight of the view itself.

3, THE ESSENTIALITY OF ORIGIN AND CONSTITUTION

Suppose that we understand *essentialism* as the view that individuals have some of their properties essentially.[146] According to one sort of essentialism, which we may call *hypoessentialism*, all the essential properties of an individual are trivially essential to it. The properties trivially essential to an individual will include properties that nothing could fail to have (if it existed), such as *being self-identical, being round or not round*, and *being massive if heavy*. They will also include some properties that the individual could not fail to have if it existed, and nothing else could have (*being this very individual*), together with properties such as *being this very individual or round* and *not being identical to that individual*. If necessity is understood in terms of possible worlds, they will also include such properties as *being round in* **G** or *being visible in* **H**. But the properties trivially essential to an individual will not include qualitative properties like *being round*, or their complements. For the hypoessentialist, although a particular individual couldn't have been any other individual, it could have been *just like* any other individual. (And, although an individual could not have been 'self-distinct', it could have completely unlike the way it actually is.) It follows that, for the hypoessentialist, an individual does not essentially belong to any kind, unless belonging to that kind sets no limits on how that individual is qualitatively. An individual can essentially belong to the kind (or pseudo-kind) *thing*, because belonging to the thing-kind does not require or preclude being any particular way qualitatively. But an individual will not essentially belong to the kind *lion*, or *proton*, at least as long as we (plausibly) suppose that being a lion, or being a proton, requires or precludes having certain qualitative properties. Hypoessentialism has been considered sympathetically by Terence Parsons and Robert Stalnaker, and defended by Pavel Tichy.[147]

At the other end of the spectrum of essentialist views from hypoessentialism is what we may call *hyperessentialism*: the view that all of an individual's properties are essential to it. Although I do not know of any contemporary defenders of hyperessentialism,[148] it appears to have been Leibniz's view.

---

[146] The view just characterized is actually essentialism *about individuals*, and not essentialism as such. As we shall see, Kripke accepts (moderate) essentialism about individuals. He also accepts essentialism about properties, since he does not hold, for example, that heat might have been light, or even that heat might have been just like light.

[147] See T. Parsons, 'Essentialism and Quantified Modal Logic', in L. Linsky (ed.), *Reference and Modality* (Oxford: Oxford University Press, 1971); Stalnaker, 'Anti-Essentialism'; and Tichy, 'Kripke on Necessity A Posteriori', esp. 241. What I call 'hypoessentialism', and count as a (weak) form of essentialism, Stalnaker calls 'anti-essentialism'. The issue is (at least partly) terminological: whether or not that view counts as a form of essentialism depends on whether 'property' is construed broadly enough to cover what I call 'trivially essential properties'.

[148] Some champions of trans-world identity have argued that counterpart theorists are in fact committed to hyperessentialism, but counterpart theorists reject this claim. (See Ch. 3, sect. 2.)

At least as long as we construe 'property' broadly enough, hyperessentialism entails that no individual exists in more than one world. (Suppose that some individual i exists at both the actual world **G** and some alternative world **H**. Let P_G be a proposition true at **G** and at no other world. Then i has the (extrinsic) property of *being such that P_G* at **G**, but not at **H**; so i (actually) has at least one property accidentally.) But one could formulate a more nuanced version of hyperessentialism according to which all of an individual's (qualitative or non-qualitative) intrinsic properties are essential to it. This sort of hyperessentialism would allow for trans-world individuals—though only ones whose intrinsic character does not vary from world to world.

As we have seen, Quine holds that quantified modal logic indefensibly presupposes 'an invidious distinction' between the properties an individual has essentially and the properties it has accidentally. Philosophers who share Quine's view that this distinction is problematic—but do not want to abjure *de re* modality altogether—have a motivation for embracing either hypoessentialism or hyperessentialism.[149] After all, if hyperessentialism (in its less nuanced form) is presupposed, then there is no problem about how to sort an individual's properties into essential and accidental ones: they are all essential. If hypoessentialism is presupposed, there is some sorting to be done; but it is at least arguable that there won't be any insuperable difficulties about doing the sorting. Certainly the kind of question Quine found bewildering (should we say that this mathematical cyclist is essentially rational and accidentally bipedal, or vice versa?) will have a straightforward answer: our mathematical cyclist will be accidentally rational, and accidentally bipedal.

Hypoessentialism and hyperessentialism contrast with *moderate essentialism*, according to which individuals have accidental as well as essential properties, and non-trivial as well as trivial essential properties. Untutored intuition clearly recommends moderate essentialism over hyperessentialism: we ordinarily suppose that there are (qualitative) properties that an individual acquires at some point in its existence, but might never have acquired. Untutored intuition also appears to favour moderate essentialism over hypoessentialism. We ordinarily suppose that there are some (qualitative) properties an individual could not lose without going out of existence. The book you are now reading could lose its current shape and acquire a somewhat different one (it could, for example, become warped). But, it seems, it could not lose all its distinctively bookish qualitative properties and acquire all the distinctive qualitative properties of a heap of ashes. Of course, if you commit it to the flames, you could turn the book into a heap of ashes. But, on the face of it, turning into a heap of ashes (or something just like a heap of ashes) is not something a book could do and survive, in the way that turning 20 is something a teenager could do and survive. Similarly, a gold ring could turn into a gold sphere (if it were melted down, and the molten

---

[149] Cf. Stalnaker, 'Anti-Essentialism', in P. French, T. Uehling, and H. Wettstein (eds.), *Midwest Studies in Philosophy* (Minneapolis: University of Minnesota Press, 1979), 343–4.

gold put in the right sort of mould), but turning into a gold sphere is not something a gold ring could do and survive. (It is something that a ring-shaped bit of gold could do and survive.)

Given his views on the necessity of identity, we know that Kripke thinks some form of essentialism is true.[150] Given the evidential weight Kripke puts on intuitions, and his unwillingness to grant Quine that the distinction between accidental and essential properties is problematic, we would expect him to take a dim view of both hypoessentialism and hyperessentialism. In fact, both Kripke's modal arguments against descriptivist theories of names, and his arguments in favour of the contingent a priori, depend on the falsity of hyper-essentialism. (Those arguments work only if, say, Moses might have spent all his days in Egypt, Aristotle might never have gone in for philosophy, stick S might have had some length other than 1 metre at t_o, and so on.) Moreover, Kripke is committed to the view that individuals have non-trivial essential properties. So Kripke is a moderate essentialist; but it is worth underscoring just how cautious a moderate essentialist he is.

To start with, in *Naming and Necessity* and 'Identity and Necessity', Kripke does not formulate—much less endorse—any fully general principles about the non-trivial essential properties of any individuals whatever. He does for-mulate some general principles concerning the non-trivial essential properties of any material objects, to the effect that material objects have their origin and original constitution essentially ('N & N', n. 57). But he introduces those prin-ciples by saying that they are *suggested* by reflection on particular examples, and he stops short of unequivocally endorsing them. Moreover, essentialists have traditionally supposed that, for each individual, there is at least one (nat-ural or artificial) kind to which that individual essentially belongs. Kripke for-mulates no general principles about the essentiality of kind membership, although he commits himself to certain claims about the essentiality of mem-bership in certain particular kinds.[151] Again, some essentialists (such as Duns Scotus) would argue that individuals—or at least material objects—do not have purely qualitative essences; other essentialists (such as Leibniz) would argue that they do. Kripke formulates no general principles addressing this question, although he does briefly discuss a particular instance of it.

Where particular cases are concerned, Kripke is often non-committal about whether an individual has a property essentially. For instance, he writes that, if Nixon is human, then 'we might not imagine' that he is only accidentally animate, and 'perhaps' he is not only essentially animate, but essentially human ('N & N', 268). In the same passage he avers that there might be an argument to show that Nixon has a purely qualitative essence, and leaves the

---

[150] Also see 'N & N', n. 12, for an explicit endorsement of essentialism.

[151] Even here Kripke sometimes expresses himself guardedly, for example, saying that '(roughly) *being a table* seems to be an essential property of a table' ('N & N', n. 57). On the other hand, he says without hedging that the bits of gold in the sample used to fix the reference of the term 'gold' are essen-tially gold ('N & N', 328).

question open. Even in cases where Kripke thinks that an individual has a certain property essentially, he often expresses his views rather tentatively. For example, at 'N & N', 269, he asks whether a table that is actually (originally) composed of molecules might not (ever) have been composed of molecules. He answers, 'certainly there is some feeling that the answer must be "no" '. And at 'N & N', 321, Kripke writes:

The molecular theory has discovered, let's say, that this object here is composed of molecules. . . . Now imagine an object occupying this very position in the room which was an ethereal entelechy. Would it be this very object here? It might have all the appearance of this object, but it seems to me that it could not ever be *this thing*. We can imagine having discovered that it wasn't composed of molecules. But once we know that this is a thing composed of molecules—that this is the very nature of the substance of which it is made—we can't then, at least if the way I see it is correct, imagine that this thing might have failed to have been composed of molecules.

The occurrence of the qualifying phrases 'it seems to me', and 'at least if the way I see it is correct' suggest that Kripke is less confident about (this instance of) the essentiality of constitution than he is about, say, the necessity of identity or the rigidity of proper names. Similarly, although Kripke thinks that a table with a certain origin could not have had a different origin, he appears to acknowledge that there is room for doubt about this ('N & N', n. 57).

The general principle about origin that Kripke says is suggested by reflection on particular cases is the following: 'If a material object has its origin from a certain hunk of matter, it could not have had its origin in any other matter' (ibid.).

On the face of it, this principle falls short of saying that material objects essentially originate from the matter they actually originate from: it does not rule out that a material object having its origin in this matter might have originated from an ethereal entelechy, or might have always been there, and thus might not have had any origin at all. Kripke has, however, often been interpreted as holding, not just that a material object originating from a certain hunk of matter could not have originated from different matter, but also that a material object originating from a certain hunk of matter could not but have originated from that hunk of matter. This interpretation is not unnatural, given that Kripke says that examples suggest that an object has its origin (and not just its non-origin) essentially.[152]

---

[152] Suppose that in world H God makes table T out of nothing at all. In world H', God creates the very same chunk of wooden matter that constitutes T at H (same wood, same molecules, same atoms...). But in H' God simultaneously creates some other wooden matter surrounding the chunk of wooden matter that in H constitutes T. He then wills away the surrounding matter, bringing into existence a table T'. If (like me) you think that T and T' are the same table, then you think that God could create the same table *ex nihilo* (from no matter at all) or *ex ligno* (from this chunk of (originally surrounded) wooden matter). In that case you will reject (the necessitation of) the principle of origin for material objects, since you will think that a (hypothetical) table made *ex ligno* might instead have been made *ex nihilo*. But you may still accept (the necessitation of) the principle of non-origin, since—for all

Is either the weaker (essentiality of non-origin) principle or the stronger (essentiality of origin) principle true? A first thought, suggested by some examples of Strawson's, is that neither principle is true, because the same building might have been made from different sets of bricks, and the same ship might have been made from different batches of steel.[153]

A second thought is that there is something suspect about the supposition that, say, the *QE II* could have been made from a completely different batch of steel. Suppose the *QE II* was actually made from a batch of steel B. And suppose that at some other possible world, a ship was made from a completely different batch of steel B′. Why should we think that the ship in the other world—the one B′ comes to constitute—is the *QE II*? Well, it might be said, suppose the other-worldly ship is made by the shipbuilders who built the *QE II*, to the exact same specifications as those of the *QE II*. Wouldn't that be enough for the other-worldly ship to be the *QE II*? It seems not. For the shipbuilders might make two ships to the same specifications as the *QE II*—one from B (the batch of steel the *QE II* was actually made from), and one from the completely different batch of steel B′. Since the two other-worldly ships are different from each other, they cannot both be the *QE II*. Moreover, if either of those ships is the *QE II*, it is the ship that is not only made to the *QE II*'s actual specifications, but also made from what the *QE II* is actually made from. So *being made by the same builders to the same specifications* cannot after all be sufficient for being the *QE II*.

Still, the inessentialist about origin might say, none of this shows that the *QE II* couldn't have been made from a completely different batch of steel. If the builders had made a ship from B′ to the specifications of the *QE II*, and no other ship to those specifications, then B′ would have constituted the *QE II*. If instead the builders had made two ships to the specifications of the *QE II*, one from B and one from B′, then B would have constituted the *QE II* (just as it actually did), and B′ would have constituted a numerically different ship just like the *QE II*.

Now, this entails that whether or not a batch of steel B originally constitutes a particular ship could depend on what happens to a completely different batch of steel; and the essentialist about origin may find this idea unmotivated and objectionable. In fact, though, there appear to be independent reasons for supposing that whether a bit of matter constitutes a particular ship could depend on what happens to a completely different batch of matter. It seems at least initially plausible that some reassembled planks would constitute the same ship those planks originally constituted if but only if there weren't a different set of planks, which planks had gradually replaced the planks that

---

the example shows—any table made from a particular bit of matter could not have been made from any other particular bit of matter.

[153] P. F. Strawson, 'Maybes and Might Have Beens', in A. Margalit (ed.), *Meaning and Use* (Dordrecht: D. Reidel, 1979).

originally constituted the ship and were subsequently reassembled.[154] The reassembled planks have, as it were, some claim to constitute the ship they originally constituted. In the absence of any other planks with a stronger claim, they do constitute that ship. But, if there are replacing planks, then the replacing planks have a stronger claim to constitute that ship than the reassembled planks do.

Someone may have different views about the relative strengths of the claims of the replacing and reassembled planks to constitute the ship the reassembled planks originally constituted. If, however, she thinks that the reassembled planks have what Salmon calls a 'dominant' claim to constitute that ship, and the replacing planks have a 'recessive' claim, she should still conclude that which ship some planks constitute can depend on what happens to a completely different set of planks. For she should think that whether or not some replacing planks constitute a certain ship could depend on whether or not the replaced planks have been reassembled (in the right way). Similarly, someone who thinks the replacing and reassembled planks have equally strong claims to constitute the original ship should conclude that which ship some planks constitute may depend on extrinsic factors. The only obvious way to avoid this conclusion is to suppose (implausibly, to my mind) that either the replacing planks or the reassembled planks have no claim at all—even in the absence of the other—to constitute the ship we started with.

The essentialist about origin might say that, even if there are reasons to think that whether or not some matter constitutes a ship could sometimes depend on what happens to some completely different matter, there is no reason to think that whether or not B′ constitutes the *QE II* could depend on whether or not B has been made into a ship with the same specifications as the *QE II*. But, the inessentialist about origin could reply, we can imagine the head of the company that built the *QE II* (truthfully) saying: 'We built the *QE II* from German steel, but if we'd known how much cheaper Italian steel was, we would have built her from steel made in Italy.' This certainly suggests that a different batch of steel could have constituted the *QE II*. But it also seems plausible that if at another world B and B′ both constitute ships built to the exact same specifications as the *QE II*, then B (the batch of steel the *QE II* was actually made from) and not B′ constitutes the *QE II* at that world. So, the inessentialist could say, in the *QE II* case, just as in the ship of Theseus cases, we have dominant and recessive constitution claims, and dependence of constitution on extrinsic factors.[155]

---

[154] I assume that the plank-replacement leaves the structure of the ship just as it was before the planks were replaced, and that the reassembled planks are reassembled in just the same way that they were originally assembled. For more on the idea that which ship a bit of matter constitutes may depend on extrinsic factors, see Salmon, *Reference and Essence*, 224–7, and C. Hughes, 'Same-Kind Coincidence and the Ship of Theseus', *Mind*, 106 (1997), 53–69.

[155] If this is right, then one and the same bit of matter could, in Nathan Salmon's phrase, have more than one potential table 'in' it (and, more surprisingly, more than one potential table of this size, shape,

Here is another line of thought suggesting that an artefact could have been made of completely different matter from the matter it was actually made from. Suppose someone carves a wooden duck paperweight out of a block of wood. Suppose further that the paperweight subsequently falls into a stream and undergoes petrification, so that the wood constituting it is gradually replaced by calcite. Could the wooden paperweight (as opposed to the wood constituting the paperweight) survive its petrification? I don't see why not. The paperweight would no longer have its original matter, of course. But lots of things—living beings, inanimate objects such as rivers, and artefacts—can survive the loss of their original matter, and I don't see why there would be a special problem about supposing that a petrified paperweight could survive the loss of its original wooden matter (as long as it retained its original size and shape, and the ability to hold down papers).[156]

Suppose that some day there is a craze for objects made of petrified wood. Since petrified wood is scarce, artefacts made of it are very expensive. One day an enterprising artisan discovers a way of artificially (and expeditiously) petrifying wood. Using this technique, she gets rich making (and selling) petrified wood tables. First she makes a bunch of table parts from wood (say, four legs and a top). Then she puts the parts together to make a table. Finally, she immerses the table in a vat, and artificially petrifies it. Suppose the artisan has made some wooden table parts, and put them together to get a table T, which she subsequently petrifies. She might instead have made the same wooden table parts, petrified them, and then put them together (in the same way) to get a table T'. Must T be a different table from T'? It is not at all obvious to me that it must. In one case, the artisan makes some table parts, puts them together to make a table, and petrifies the table. In the other, the artisan makes the exact same table parts, petrifies them, and then puts them together (in the same way) to get a table. Why aren't these just two different ways of making the same table from the same set of table parts?[157]

---

composition, and so on). At 'N & N', n. 56, Kripke sets out an argument purporting to show that the essentiality of (non-)origin holds in a certain class of cases. If Salmon is right, that argument presupposes a principle in the neighbourhood of the claim that there is just one potential table in a given bit of matter (*Reference and Essence*, 208). I shall not discuss Kripke's argument here, both because it presents formidable exegetical difficulties, and because Kripke writes in the preface of the revised edition of *Naming and Necessity* that the footnote now seems to him to have problems requiring further discussion.

[156] Venetian buildings rest on wooden poles driven through the mud to the less marshy soil below (there are at least 100,000 such poles under Santa Maria della Salute). The reason that so many (very old) Venetian buildings are still standing is that, rather than rotting away, the poles on which the buildings rest have undergone a process of mineralization. Even though the poles driven into the soil hundreds of years ago are no longer wooden (or at any rate no longer completely wooden), they are still there, doing the job they were meant to do.

[157] Someone might say, 'Because T is, at its first moment of existence, made of wood, while T' is, at its first moment of existence, made of calcite.' True enough: but if this sort of difference in constitution is compatible with numerical identity across time, why isn't it compatible with numerical identity across worlds?

If, however, T and T' are the same table, then our (hypothetical) table is made from wood in one possible world, and made from calcite in another. This is compatible with the essentiality of non-origins principle, since the table is only hypothetical; but it is inconsistent with the necessitation of that principle. Suppose, though, that we start with an actually existing wooden table T, made from these wooden table parts. Suppose also that in an alternative world H, we petrify those parts, and assemble them in just the same way that we actually assembled T's parts, bringing into existence a (calcite) table T'. If T is T', then the essentiality of non-origin, and not just its necessitation, is false. Again, it seems plausible that a table existing at a later time could be a table existing at an earlier time, in spite of the fact that the table existing at the earlier time is (at the earlier time) made of wood, and the table existing at the later time is (at the later time) made of calcite. And it seems plausible that some to-be-assembled table parts existing at an earlier time could be some to-be-assembled table parts existing at a later time, in spite of the fact that the parts existing at the earlier time are (then) made of wood, and the parts existing at the later time are (then) made of calcite. So why couldn't a table existing in an alternative possible world be a table existing in this world, in spite of the fact that the table existing in the alternative world is (at that world) made from these parts (put together in this way) at a time when the parts are made of calcite, and the table existing in the actual world is (in the actual world) made from the same parts (put together in the same way) at a time when they are made of wood?

Just as (some) artefacts can survive the loss of their original matter, so can (at least some) living beings.[158] As far as I know, there is no reason to think that gametes couldn't survive the (natural or artificial) loss of their original matter, as long as the replacement of original matter by new matter occurred gradually, and as long as the structure of the gametes was unaffected by the replacement. Suppose that I actually come from gametes G_1 and G_2, and that G_1 and G_2 are actually (respectively) constituted by portions of matter M_1 and M_2. Suppose also that at an alternative possible world H, G_1 and G_2 gradually lose all their original matter, and come to be constituted of portions of matter M_1' and M_2'. Finally, suppose that in H, after G_1 and G_2 have acquired their new matter, G_1 fertilizes G_2 (just as it did in the actual world). The result of the fertilization in H is a human being who came from the same gametes I actually came from, and is (let us suppose)—at least at his first moment of existence—exactly like me. That human being, I want to say, is not just *like* me; he *is* me. If this is true, then I might have come from matter different from the matter I actually came from. So if humans (or non-human animals) are (animate) material objects, they appear to constitute a counter-example to the principle of the essentiality of non-origins formulated by Kripke.

---

[158] In the next few paragraphs I draw on some examples from my 'The Essentiality of Origin and the Individuation of Events', *Philosophical Quarterly*, 44 (1994), 26–45.

So do things like waterfalls or lakes, unless we refuse to count them as material objects. We could make a waterfall by modifying the terrain around a river (say, by using explosives to create a vertical drop where there hadn't been one). Depending on when exactly we modified the terrain, the waterfall would be 'made from'—and initially made of—different portions of water. But there seems no reason to suppose that, if the terrain had been modified a bit later rather than a bit earlier, the result would have a been different waterfall. Which bit of water a waterfall is made from, and made of at its first moment of existence, seems just as inessential to it as what water is made from at subsequent moments of its existence. *Mutatis mutandis*, the same goes for lakes. In Ham there is an artificial lake called 'Ham Lake'. I do not know how Ham Lake was made, but perhaps it was like this: people dug a large hole. Then they diverted water from the (almost) adjoining Thames to fill the hole, thereby bringing the lake into existence. Suppose they filled the hole with a portion of Thames water on Tuesday morning. Would they have brought a different lake into existence if they had filled it with a different portion of Thames water on Tuesday afternoon? It seems not. On the face of it, Ham Lake could have been made from, and initially made of, any old portion of water, just as one and the same portion of water could have been made into any old lake.

The moral would seem to be this: in the case of some kinds of material objects, it is difficult to imagine objects of that kind coming from, or being made from, matter different from the matter they actually came from or were made from. For instance, it is difficult (at least for me) to imagine the circumstances under which this very ice-cube, or this very icicle, could have been made from different (aqueous) matter. An ice-cube made from different water, it seems, would *eo ipso* be a different ice-cube. In the case of other kinds of material objects (tables, organisms) it is harder, but not obviously impossible, to imagine circumstances under which objects of that kind come from or are made from different matter: in the case of other kinds of material objects (waterfalls, lakes) it is on the face of it easy to imagine circumstances under which material objects of that kind come from or are made from different matter.

More generally, as Aristotle notes, different kinds of material objects have different kinds of essences. An ice-cube (at least arguably) has its material origin (or at any rate a certain sort of non-origin) essentially. A lake has its material origin only accidentally, although it appears to essentially stand in certain (spatial) relations to its surroundings. (It is hard to imagine Ham Lake having come into existence on Pluto, millions of miles away from the Thames Valley. It is also hard to imagine 'moving' Ham Lake (as opposed to the water constituting it) to Pluto, and leaving all its current surroundings on earth.) A human being has both his material origin and its (original and subsequent) surroundings only accidentally. (The same gametes made respectively of this and that matter might have been made respectively of this other and that other matter; the gametes that got together in these surroundings might have got together in completely different surroundings.) Perhaps there are material objects (mountains?) that have both their material origin and their surroundings essentially. In light of all this variety, it does not look as though we can move from intuitions about the essentiality of origin or non-origin of material objects of a particular kind—say, tables—to a general principle of the essentiality of origin or non-origin meant to apply to any material object whatever, any more than one could move from intuitions about the essentiality of surroundings for material objects of a particular kind (say, lakes) to a general principle of the essentiality of surroundings meant to apply to any material object whatever.

At note 57 of *Naming and Necessity* Kripke writes that 'in addition to the principle that the *origin* of an object is essential to it, another principle suggested is that the *substance* of which it is made is essential'. Although Kripke does not explicitly formulate an essentiality of substance principle, he has in mind a principle according to which a material object must originally (rather than subsequently) be made of the substance it is actually (originally) made of. He also appears to have in mind a principle according to which a material object is essentially originally made of the kind of substance (say wood, or balsa wood) it is actually originally made of, and not a principle according to which a material object is essentially made of the particular bit of a particular kind of substance it is actually made of (ibid. and 'I & N', 152). The essentiality of original 'substantial makeup' (as Kripke calls it) does not in any obvious way entail the essentiality of origin. The former principle concerns particular kinds of substance, whereas the latter concerns particular bits of matter. And it might be true, for example, that a human being actually originally made of the substance(s?) a fertilized egg is made of could not but have been originally made of those substances, even if it is false that a human being actually originally made from a particular bit of matter could not but have been originally made from that particular bit of matter. Conversely, if a particular bit of matter could only accidentally constitute a bit of a substance of a certain kind, perhaps it could happen that a material object was essentially made from the matter it was actually made from, but not essentially originally made of the substance that bit of matter (only accidentally) constituted at the first moment of the existence of the object. If, on the other hand, it is assumed that a bit of matter essentially constitutes the substance(s) it actually constitutes, there is a natural route from the essentiality of origin to the essentiality of original substantial make-up: for the matter a thing is made from is the matter that thing is actually (originally) made of.

Many of the alleged counter-examples to the essentiality of origin discussed so far (e.g. the one involving gametes, and the one involving a table that might have been created either *ex nihilo* or *ex ligno*; see note 152 above) raise no difficulty for the essentiality of original substantial make-up. Indeed, I am inclined to think that it is harder to find (apparent) counter-examples to the essentiality of original substantial make-up than it is to find (apparent) counter-examples to the essentiality of origin (or non-origin).

If, however, the same table that is actually made from wood might have been made from petrified wood, then the same table that is actually initially made of wood might have been initially made of calcite, and that table's original substantial make-up, as well as its origin, is accidental.

Furthermore, according to the *OED*, 'river' can either mean 'a copious stream of water' or 'a copious stream of something'. When a volcano erupts, a stream of molten rock comes up from the inside of a mountain and flows down the mountain's slope until it solidifies. On a planet much hotter than earth the molten rock might not solidify, and copious streams—that is to say, rivers—of molten rock might be enduring features of the landscape. Such rivers would not necessarily always have the same 'substantial make-up': one and the same river might be constituted of different kinds of molten rock at different times. On a suitably hot planet there might be not just rivers but also 'waterfalls' (that is, 'molten-rock-falls') and lakes made of molten rock. If—as I have argued—an ordinary waterfall or lake could have been made from, and initially made of, a different portion of water than the water it was actually made from, and initially made of, then it seems that a molten-rock-fall, or molten-rock-lake, could have been made from, and initially made of, a different kind of substance than it was actually made from and initially made of.

Finally, if, as Strawson suggests, the same ship might have been made from a different batch of steel, it is not clear on what grounds we can rule out the idea that the same ship might have been made from a different batch of a different kind of steel, or a different batch of some other kind of metal. I do not mean to suggest here that there couldn't be such grounds; but I am at a loss to see just what they would be.

None of this suggests that material objects never have their original substantial make-up essentially. At least so long as 'material object' is construed broadly, material objects include bits of gold, and it is at least initially plausible that a bit of gold could not but have been originally (and indeed, subsequently) made of gold. Similarly, it is at least initially plausible that an ice-cube is essentially originally made of ice; that a milk-shake is essentially originally made of milk; that an (ordinary) cloud is essentially originally made of water vapour. And perhaps a human zygote is essentially originally made of the substances it is actually originally made of. Be that as it may, given how varied material objects are, it is doubtful that they all have their original substantial make-up—any more than their material origin—essentially.

3

Identity, Worlds, and Times

At 'Naming and Necessity', 266, Kripke notes that some philosophers have found the notion of identity across worlds problematic, on the grounds that we cannot legitimately make judgements of identity across worlds unless we have a criterion for trans-world identity. If, for example, we judge that someone in an alternative world is the same person as Nixon, then, they have thought, we need a criterion saying, in non-trivial terms, under what conditions a person P in a world **H** is the same person as a person P′ in a (different) world **H′** (in other words, a criterion that provides us with non-trivial necessary and sufficient conditions for the identity of person P in world **H** with person P′ in world **H′**). But—these philosophers have maintained—it is very difficult to provide a criterion of identity across worlds for persons, or material objects, or the like.

Kripke answers that, although it is very difficult to come up with a set of necessary and sufficient conditions for the identity across time of, say, persons, or ships, this is not generally thought to show that the notion of identity across time is illegitimate. In any case, Kripke avers, the worry about trans-world identity just sketched depends on a certain sort of picture of what possible worlds are, and of our access to them:

One thinks, in this picture, of a possible world as if it were like a foreign country. One looks upon it as an observer. Maybe Nixon has moved to the other country, or maybe he hasn't, but one is given only qualities. One can observe all his qualities, but of course, one doesn't observe that someone is Nixon. ('N & N', 266–7)

But, Kripke protests,

A possible world isn't a distant country that we are coming across, or viewing through a telescope...A possible world is *given by the descriptive conditions we associate with it*...'Possible worlds' are *stipulated*, not *discovered* by powerful telescopes. There is no reason why we cannot *stipulate* that, in talking about what happened to Nixon in a certain counterfactual situation, we are talking about what would have happened to *him*. ('N & N', 267)

Don't ask: how can I identify this table in another possible world, except by its properties? I have the table in my hands, I can point to it, and when I ask whether *it* might

have been in another room, I am talking, by definition, about *it*. I don't have to identify
it after seeing it through a telescope. ('N & N', 273)

There seem to be two points here. First, possible worlds are stipulated, not
discovered. Secondly, it is perfectly all right for us to specify possibilities
(counterfactual situations, possible worlds) in terms that make reference to
particular (actually existing) individuals; we needn't think of such a specification as a promissory note to be redeemed by a specification in purely qualitative terms.

There is something initially puzzling about the claim that possible worlds are
stipulated, and not discovered with the aid of telescopes. For one thing, 'stipulates' is ordinarily followed by a sentential complement, rather than a direct
object; one stipulates *that* such-and-such. So when Kripke says that possible
worlds are stipulated, not discovered, what exactly is it that Kripke thinks we
stipulate (that) rather than discover (that)? Not, surely, that there are possible
worlds, or that there are possible worlds in which, say, Nixon lost the 1968 election. Kripke holds that facts about what is possible (or impossible) are not of the
right sort to be stipulated: one could not stipulate that, say, gold might not have
had atomic number 79, or that Elizabeth II might have been the (biological)
daughter of Harry Truman, any more than one could stipulate that Nixon could
not have lost the 1968 election. And one could not stipulate that Nixon might
have lost the 1968 election any more than one could stipulate that he could not
have: that Nixon could have lost the 1968 election is something whose truth is
independent of our stipulations, just as that Nixon could not have lost the 1968
election is something whose falsity is independent of our stipulations. What we
can stipulate, for Kripke, is that we are considering, or entertaining, or talking
about, a possible situation in which this particular individual (say, Nixon) has a
certain property. At least, we can do that if there are possible worlds in which
Nixon has that property: if the property in question is one that Nixon could not
have had, the attempted stipulation misfires.

Presumably, though, those philosophers who insist that there is a problem
about trans-world identity would not deny that, *if* there are possible worlds in
which Nixon is F, *then* we can stipulate that we are entertaining, or discussing,
possible worlds in which *Nixon*, and not somebody else, is F. As we shall see,
some of those philosophers have disagreed with Kripke about whether, in a
large range of cases, we can be confident that there are possible worlds in
which Nixon has a certain property, so that we can stipulate that we are considering or discussing a possible world where that man has that property,
without fear that the stipulation will misfire. If this is right, then the fundamental issue between Kripke and those who think trans-world identity is
problematic does not (ultimately) concern the legitimacy of certain stipulations. All the parties to the dispute agree that we can unproblematically stipulate that, say, we are considering a possible world in which *Nixon* lost the
1968 election, if but only if it is unproblematic that there are possible worlds

in which Nixon lost that election. The question is whether the existential claim—and others like it—are unproblematic.

Kripke's second point is that we should not assume that possible worlds are initially 'given' in a purely qualitative form, and must be specifiable in purely qualitative terms. On the view Kripke is opposing, it is not supposed to be problematic that there are possible worlds in which someone with a five-o'clock shadow, an oddly shaped nose, and so on, loses the 1968 election. But it is supposed to be problematic that in one of those worlds the man with the five-o'clock shadow, oddly shaped nose, etc. losing the 1968 election is Nixon. Why the asymmetry? Most of us 'intuit that' there are possible worlds in which someone with such-and-such qualitative properties loses the 1968 election. But then, most of us intuit that there are possible worlds in which Nixon loses the 1968 election. There is no obvious reason to think that our intuitions about whether such-and-such qualitative properties are co-exemplifiable are somehow more trustworthy than our intuitions about whether this particular individual could have had (or lacked) this particular property.

As I have already noted in Chapter 1, Kripke does not argue in *Naming and Necessity* that possible worlds cannot be (completely) specified in purely qualitative terms, or argue that individuals do not have purely qualitative essences. For all Kripke says there, a (complete) purely qualitative description of a possible world might settle whether or not that world contains Nixon—if there are purely qualitative necessary and sufficient conditions for being Nixon. For all Kripke says, a (complete) purely qualitative description of a possible world might settle every question about whether this or that actually existing individual exists at that world (if every actually existing individual has a purely qualitative essence). Kripke's point is simply that one's confidence that, say, there is a possible world in which Nixon lost the 1968 election is perfectly legitimate, even if one does not have a view about what purely qualitative conditions (if any) are necessary and sufficient for being Nixon.

In fact, depending on how exactly we understand the notion of a qualitative property, Kripke may hold views that are incompatible with the idea that (all) material objects have purely qualitative essences. In his unpublished lectures 'Time and Identity' Kripke appealed to a (subsequently) much-discussed example to argue that two things can be in all the same places at all the same times, and be made of all the same matter at all the same times. Suppose, for example, that a plant dies before it grows past its stem. Then the plant and the stem will have occupied all the same places at all the same times, and have been made of all the same places at all the same times (forget about the roots). Nevertheless, Kripke maintains, the plant and the stem will be distinct: for it will be true of the plant, but not the stem, that, had things been different, it would have grown past its stem. More generally, Kripke holds, two things can be discernible with respect to modal and dispositional properties, and hence distinct, even though they are, throughout their existence, in exactly the same places, and made of exactly the same stuff. Although Kripke does not explicitly define the notion of a (purely) qualitative property, it seems plausible that he has in mind properties such as *being round*, and not modal or dispositional properties (or sortal properties such as *being a plant* and *being a stem*) of the sort with respect to which (Kripke holds) the plant and the stem are discernible. If the latter sort of properties are not qualitative, and if Kripke is right about the plant and the stem being distinct, then it cannot be true that the plant and the stem have purely qualitative essences. For the plant and the stem (being distinct individuals) will have distinct essences, in spite of having all the same qualitative properties.

Let us grant that there is no reason to suppose that worlds are initially 'given' purely qualitatively, or must initially be specified or described in purely qualitative terms. There remains a question about whether the picture Kripke opposes is, as he maintains, behind the worries about trans-world identity prevalent at the time Kripke wrote *Naming and Necessity*.

At least some proponents of quantified modal logic who have thought there was a problem about trans-world identity have done so because they accepted at least a good deal of the picture of possible worlds and our access thereto that Kripke argues against. Thus, in setting out what he calls 'the problem of cross-identification', Hintikka writes:

Each possible world contains a number of individuals . . . with certain properties and with certain relations to each other. We have to use these properties and relations to decide which member (if any) of a given possible world is identical with a given member of another possible world. Individuals do not carry their names in their foreheads; they do not identify themselves.[159]

It is not so clear that the opponents of quantified modal logic who thought there was a problem about trans-world identity have done so because of a commitment to the picture of possible worlds at issue. In 'Propositional Objects' Quine raises the following worries about trans-world identity:

How is Catiline to be identified in . . . various possible worlds? Must he have been named 'Catiline' in each, in order to qualify? How much can his life differ from the real life of Catiline without his ceasing to be our Catiline and having to be seen as another man of that name? Or again, how much can the [great] pyramid differ from the real one? . . . Is it sufficient, for its identification in other worlds, that it have been built by Cheops? How much then can his life differ from the real life of Cheops without his ceasing to be our Cheops?[160]

Given the way this passage starts, one might initially think that Quine is presupposing just the picture that Kripke challenges: we start out with not-yet-identified individuals in other worlds, having such-and-such properties,

---

[159] J. Hintikka, 'The Semantics of Modal Notions', in D. Davidson and G. Harman, (eds.), *Semantics of Natural Language* (Dordrecht: D. Reidet, 1972), 400. See also D. Kaplan, 'Trans-world Heir Lines', in M. Loux (ed.), *The Possible and the Actual* (Ithaca: Cornell University Press, 1979).

[160] W. V. O. Quine, 'Propositional Objects', in Quine, *Ontological Relativity and Other Essays* (New York: Columbia University Press, 1969), 153.

and then ask whether or not one of those individuals is Catiline, or the Great Pyramid.

In fact, though, Quine's fundamental worry about trans-world identity is expressed in the second sentence of the passage: how different from how he actually is might Catiline have been? Quine thinks there is no good (non-arbitrary) way of answering this question, and thus no way of saying, of some other-worldly individual, whether he is the same as or different from Catiline. Bracketing the question of whether this worry is a genuine one, it does not, in any way I can see, presuppose the picture of possible worlds and our access thereto that Kripke argues against. On the face of it, one could perfectly well hold that there is no good answer to the question 'How different from how he actually is might Catiline have been?' without presupposing that possible worlds are like foreign countries or far-away planets whose only 'observable' features are qualitative.

In 'Propositional Objects' Quine suggests, rather than argues, that there is a problem about trans-world identity. In an article that appeared two years earlier[161] Chisholm offers an argument, which begins as follows:

Suppose that Adam and Noah are trans-world individuals. Then, it would seem, there is a possible world containing Adam and Noah in which Adam's life-span is, say, a year closer to Noah's actual life-span, and Noah's life-span is a year closer to Adam's actual life-span. And there is another possible world containing Adam and Noah in which Noah's life-span is two years closer to Adam's actual life-span, while Adam's life-span is two years closer to Noah's actual life-span. Continuing in this way, we get to a world in which Adam has Noah's actual life-span, and Noah has Adam's actual life-span. If there is a possible world in which Adam and Noah have each other's actual life-spans, then, it seems, there are possible worlds in which they also have each other's names, birth-dates, physical characteristics, psychological characteristics, and so on. But now it begins to look as though there is a possible world in which Adam and Noah, as it were, occupy each other's (actual) roles. Adam might have been qualitatively just the way Noah actually is, and also 'fitted into the world' in just the way Noah actually does. And Noah might have been qualitatively just the way Adam actually is, and also 'fitted into the world' in just the way Adam actually does. Moreover, it seems that there could be two different possible worlds that were discernible only inasmuch as in one world, Adam fills the Adam-role, and Noah fills the Noah-role, whereas in the other Adam fills the Noah-role, and Noah the Adam-role.

Some essentialists (for example, hypoessentialists such as Pavel Tichy) would find nothing troubling here. But Chisholm thinks there is a problem about supposing that there is a pair of worlds that differ only with respect to which member of {Adam, Noah} fills the Adam-role and which member fills the Noah-role. Where W^1 is the actual world (where Adam fills the Adam-role and Noah fills the Noah-role), and W^n is the (qualitatively indiscernible) world where Adam and Noah have switched roles, Chisholm writes:

Now we must ask ourselves: how is one to tell the difference between the two worlds W^1 and W^n? ... If W^1 and W^n are two different possible worlds, then of course there are

[161] R. Chisholm, 'Identity Through Possible Worlds: Some Questions', *Nous*, 1 (1967), 1–8.

infinitely many others, equally difficult to distinguish from each other and from W^1 and W^n. For what we have done to Adam and Noah, we can do to any other pair of entities.[162]

In this passage we do seem to have something like the 'foreign country' picture of possible worlds and our access to them that Kripke rejects. It is as though we are looking at W^1 and W^n from far away and attempting to find their distinguishing features. Kripke would of course say that—supposing for the sake of argument that there are the possible worlds W^1 and W^n—we know perfectly well, without telescopes, what distinguishes W^1 and W^n: in W^1 Adam is the person who lives for 930 years, and so on, and Noah is the person who lives for 950 years and so on; in W^n it's the other way round.

Kripke would in fact deny that there is any such possible world as W^n. If there were, then there would be a possible world in which Noah—who actually came from these particular gametes—came from no gametes, and Adam—who actually came from no gametes—came from Noah's actual gametes. As Kripke sees it, there are some pairs of properties of Adam and Noah that they could not exchange, because there are some properties that Adam couldn't lack and Noah couldn't have, and some properties that Noah couldn't lack and Adam couldn't have.

Chisholm considers this sort of move:

If we accept this doctrine of essential properties, we may say, perhaps, that the property of living for just 930 years is not essential to Adam and therefore that he may inhabit other possible worlds without living for just 930 years in each of them. And so, too, perhaps, for having a name which, in English, ends with the letter 'm'. But, we may then go on to say, somewhere in the journey from W^1 to W^n, we left the essential properties of Adam (and therefore Adam himself) behind. But where? What *are* the properties that are essential to Adam? Being the first man? Having a name which, in English, begins with the first letter of the alphabet? But why *these* properties? . . . It seems to me that even if Adam does have such essential properties, there is no procedure at all for finding out what they are. And it also seems to me that there is no way of finding out whether he *does* have any essential properties.[163]

This passage is reminiscent of the passage from Quine cited a bit earlier.

The idea seems to be that there is no way of knowing which properties of an individual, if any, are essential to it. If that is so, there are going to be a great many unresolved questions about trans-world identity and, in that sense, a problem about trans-world identity. Again, this problem appears independent of the 'foreign country' picture of possible worlds.

---

[162] R. Chisholm, 'Identity Through Possible Worlds: Some Questions', *Nous*, 1 (1967), 1–84.

[163] Ibid. 6. In this passage Chisholm uses the term 'essentially' in a non-standard way. As Chisholm understands the term, an individual has a property essentially just in case necessarily something has that property if and only if it is that individual.

Chisholm and Quine think there is a problem about trans-world identity; Kripke does not.[164] I am inclined to think that, at the most fundamental level, this difference is not attributable to the fact that, for Kripke, we begin with actual identified objects and ask what might have been true of those objects; whereas, for Chisholm and Quine, we start with real, purely qualitatively characterized worlds, and then try to find a criterion to settle questions of trans-world identification (cf. 'N & N', 273). At the most fundamental level, I think, the difference is attributable to the fact that Kripke has robust intuitions—in a large range of cases, at least—about how different from how it actually is an individual might have been, whereas Chisholm and Quine have fewer and weaker intuitions. Thus Quine attempts to support the claim that there are puzzles about how different Catiline might have been by asking whether he could have lacked the name 'Catiline'. He would not ask that question if he thought that *being named 'Catiline'* was obviously accidental to Catiline. But, Kripke would say, it is obviously accidental to him. Again, Quine would not ask whether all there is to the essence of the Great Pyramid is being a pyramid made by Cheops if he thought the answer was obviously no. And Kripke would say the answer is obviously no: otherwise there would only be one possible pyramid that Cheops could have built—and there are worlds in which Cheops builds two or more. Similarly, Chisholm considers the possibility that *being the first man* is an essential property of Adam; he presumably would not do so if he thought that property was obviously accidental to Adam. But, Kripke would say, it is obviously accidental, since God might have brought Adam into the world along with (at the same time as) a different man, Adam$_2$.

To recap: as Kripke says, 'The question of essential properties . . . is supposed to be equivalent (and it is equivalent) to the question of identity across possible worlds' ('N & N', 266). So, if there is a problem about essence and accident, then there is a problem about trans-world identity. And at least some of the most influential philosophers who have thought that there was a problem about essence and accident have not thought so (simply, or even primarily) because of their antecedent commitment to the 'foreign country' model of possible worlds.

Instead, they have thought that (primarily) because they have not taken themselves (or the rest of us) to have strong or stable intuitions about essences, and have not seen how beliefs about essences could be defended if not by appeal to intuition.[165]

---

[164] The claim about Kripke will be qualified in the next section. It might be better to say that Kripke thinks there isn't the problem about trans-world identity that Quine, or Chisholm, or Hintikka have supposed there is.

[165] Kripke would agree that, in defending the claim that an individual has a certain property essentially, we may ultimately have nothing more to appeal to than that, upon reflection, that is the way things seem to us ('N & N', 313). So Kripke would agree with Chisholm that, in one sense, there is no 'procedure' for finding out which of an individual's properties are accidental and which are essential.

2, COUNTERPARTS

I have never crossed the Himalayas, though I might have done. So there is a non-actual (or, if you prefer, a non-actualized) possible world (or possible state of the world) whose existence is a sufficient condition for the truth of 'I might have crossed the Himalayas'. At that possible world someone crosses some mountains. Is that person me, and are those mountains the Himalayas? Or are they (non-actual) individuals different from me and from the Himalayas?

According to David Lewis, the latter suggestion is the better one:

Where some would say that you are in several worlds, in which you have somewhat different properties and somewhat different things happen to you, I prefer to say that you are in the actual world and no other, but you have counterparts in several other worlds. Your counterparts resemble you closely in content and context in important respects. They resemble you more closely than do the other things in their worlds. But they are not really you. Indeed we might say, speaking casually, that your counterparts are you in other worlds, that they and you are the same; but this sameness is no more a literal identity than the sameness between you today and you tomorrow. It would be better to say that your counterparts are *the men you would have been*, had the world been otherwise.[166]

For Lewis, what an individual might have done (or might have been) is what its counterparts do (or are). (When the possibility is unrealized, the counterparts will all be in other worlds.[167])

Kripke emphatically rejects counterpart theory:

[For Lewis] if we say, 'Humphrey might have won the election' (if only he had done such-and-such), we are not talking about something that might have happened to *Humphrey* but to someone else, a 'counterpart'. Probably, however, Humphrey could not care less whether someone *else*, no matter how much resembling him, would have been victorious in another world. Thus Lewis' view seems to me even more bizarre than the usual notions of transworld identification that it replaces. ('N & N', n. 13)

Even more objectionable is the view of David Lewis. According to Lewis, when we say 'Under certain circumstances Nixon would have gotten Carswell through', we really mean 'Some man, other than Nixon but closely resembling him, would have gotten some judge, other than Carswell but closely resembling him, through.' Maybe that is so, that some man closely resembling Nixon could have gotten some man closely resembling Carswell through. But *that* would not comfort either Nixon or Carswell, nor would it make Nixon kick himself and say, '*I* should have done such and such to get Carswell through.' The question is whether under certain circumstances Nixon *himself* could have gotten *Carswell* through. ('I & N', 148)

---

[166] D. Lewis, 'Counterpart Theory and Quantified Modal Logic', *Journal of Philosophy*, 65 (1968), 113–26; repr. in Lewis, *Philosophical Papers*, 2 vols. (Oxford: Oxford University Press, 1983–6), vol. i; the citation is from *Philosophical Papers*, 27.

[167] Cf. D. Lewis, *Counterfactuals* (Cambridge, Mass.: Harvard University Press, 1973), 40.

Kripke appears to be saying here that, on Lewis's view, if we say 'Humphrey might have won the election', what we really mean is: 'Someone different from Humphrey, but resembling Humphrey in certain ways, might have won the election': winning the election isn't (on Lewis's view) something that might have happened to Humphrey; it is something that might have happened to one of his counterparts. Similarly, if we say 'Under certain circumstances, Nixon would have got Carswell through', what we really mean is 'Under certain circumstances, someone different from Nixon, but resembling him in certain ways, would have got someone different from Carswell, but resembling him in certain ways, through.'[168] But, Kripke thinks, it is obvious that, if we say 'Humphrey might have won the election' or 'Nixon would have got Carswell through', we mean what we say: we mean that *Humphrey* might have won the election, or that *Nixon* would have got Carswell through.

It is obvious; but why exactly is this a problem for Lewis? Lewis holds that it is Humphrey's counterpart, rather than Humphrey himself, who wins the election (in another world); and it is Nixon's counterpart, rather than Nixon himself, who gets (a counterpart of) Carswell through (in another world). But he does not hold that it is Humphrey's counterpart, and not Humphrey himself, who *might have* won the election; and he does not hold that it is a counterpart of Nixon, and not Nixon, who *might have* got Carswell through. For Lewis, it is true that Humphrey himself would have won under different circumstances, precisely inasmuch as it is true that someone different from Humphrey (his counterpart) wins in an alternative possible world.[169]

If Kripke's objection to Lewis is that, on Lewis's account, 'Humphrey might have won the election' is not genuinely about Humphrey, then the objection misses the target. But it is perhaps uncharitable to suppose that that is Kripke's principal objection in the passages just cited. On a natural way of reading them, Kripke's main point is that Lewis's counterpart-theoretic truth-conditions for *de re* modal predications are wrong-headed, with regard to what they do and do not make reference to. Suppose that (in alternative possible circumstances) some man who is not Nixon, but resembles Nixon in certain ways, gets through some man who is not Carswell, but resembles Carswell in certain ways. Be that as it may, the question remains (as Kripke puts it, 'the question is'): could Nixon *himself* have got Carswell *himself* through? As Kripke sees it, the question of whether, in different circumstances, Nixon might have got Carswell through is just the question of whether there are alternative possible circumstances in which Nixon gets Carswell through; it is not the question of whether there are alternative possible circumstances in

---

[168] Compare: 'When we say that my great-aunt died last Thursday, and we buried her on Saturday, what we really mean is that she died last Thursday, and we buried *her remains* (or *her body*) on Saturday.'

[169] See Lewis, postscript B to 'Counterpart Theory and Quantified Modal Logic', in *Philosophical Papers*, i. 42; and his *Counterfactuals*, 40: 'Ripov's honest counterparts make him someone who might have been honest.'

which someone who is not Nixon, but resembles him in certain ways, gets through someone who is not Carswell, but resembles him in certain ways.

Like a great many other philosophers, I find this last claim very plausible.[170] But, a champion of counterpart theory might understandably protest, it is not so much an argument against counterpart theory as a denial thereof.[171] And, she might continue,

> Two objections to counterpart theory might be extracted from the passages of Kripke cited above, namely:
>
> (A) The question of whether Nixon might have got Carswell through is the question of whether Nixon *himself* might have got Carswell *himself* through—and not the question of whether someone who is not Nixon (however much he might resemble Nixon) might have got through someone who is not Carswell (however much he might resemble Carswell)
>
> and
>
> (B) The question of whether Nixon might have got Carswell through is the question of whether there is a possible world in which Nixon *himself* gets Carswell *himself* through—and not the question of whether there is a possible world in which someone like Nixon in certain ways gets someone like Carswell in certain ways through.
>
> (A) is not a genuine objection to counterpart theory at all, since it is perfectly consistent with it.[172] (B) is not an argument against counterpart theory, but the conclusion of such an argument. Hence Kripke hasn't provided any argument against counterpart theory.

I agree with the hypothetical defender of counterpart theory to this extent: the passages of Kripke cited above, taken on their own, do not obviously constitute a decisive objection to counterpart theory. On the other hand, we shouldn't neglect the context of those passages. In both 'Identity and Necessity' and *Naming and Necessity*, by the time Kripke criticizes counterpart theory, he has already argued that we are perfectly within our rights to stipulate that we

---

[170] See e.g. R. Adams, 'Theories of Actuality', *Nous*, 8 (1974), 211–31; Plantinga, *The Nature of Necessity*, ch. 6, esp. p. 116; N. Salmon, *Reference and Essence*, app. I, esp. n. 16.

[171] It would be different, the defender of counterpart theory might say, if the opponent of counterpart theory could describe a possible situation in which a *de re* modal predication of the form *x might have been F* was true, but its counterpart-theoretic translation was false (or vice versa). That would be a (non-question-begging) argument against counterpart theory. But in neither *Naming and Necessity* nor 'Identity and Necessity' does Kripke attempt to describe possible situations in which a *de re* modal predication and its translation into counterpart theory have different truth-values. He does not, for example, say: if things had gone this way, then it would be true that some counterpart of Nixon might have got some counterpart of Carswell through, but false that Nixon might have got Carswell through.

[172] Because counterpart theory does not require that the counterpart relation be transitive, a thing can have a counterpart that has an F-ish counterpart without itself having an F-ish counterpart. So, the counterpart theorist would say, it could be true that some (other-worldly) thing resembling *x* in the right sort of ways to be its counterpart might be F, and false that *x* itself might be F.

are talking about or have in mind a possible circumstance or situation in which *Humphrey* (and not just someone like Humphrey in certain ways) wins the election, or a possible situation in which *Nixon* (and not just someone like Nixon in certain ways) gets *Carswell* (and not just someone like Carswell) through. Given that Nixon was once a child, we are perfectly entitled to stipulate that we are talking about or have in mind a past situation in which Nixon himself is a child. In exactly the same way, Kripke thinks, given that Nixon might have been the president of Students for a Democratic Society (the SDS), we are entitled to stipulate that we are talking about or considering a possible situation in which Nixon himself is the president of the SDS (equivalently: a possible situation in which the president of the SDS is Nixon).[173]

Moreover, Kripke holds, a possible world is not 'a distant country that we are coming across, or viewing through a telescope' ('N & N', 267); it is just a (complete) possible situation. So if we are entitled to stipulate that we are talking about a possible situation in which Humphrey (himself) wins the election, or Nixon (himself) loses it, we are likewise entitled to stipulate that we are talking about a possible world in which Nixon (himself) loses the election:

> Why can't it be part of the *description* of a possible world that it contains *Nixon* and that in that world *Nixon* doesn't win the election? It might be a question, of course, whether such a world *is* possible. (Here it would seem, *prima facie*, to be clearly possible.) But, once we see that such a situation is possible, then we are given that the man who might have lost the election or did lose the election in this possible world is Nixon, because that's part of the description of the world . . . There is no reason why we cannot *stipulate* that, in talking about what would have happened to Nixon in a certain counterfactual situation, we are talking about what would have happened to *him*. ('N & N', 267)

If, however, we may stipulate that we are talking about a possible world in which Humphrey himself wins, or Nixon himself loses the election (equivalently: that we are talking about a possible world in which the man who wins the election is Humphrey, or the man who loses the election is Nixon), then there are such worlds. And if there are possible worlds in which, say, Nixon himself loses the election, then the existence of a world in which Nixon himself loses the election is surely a necessary and sufficient condition for the truth of *Nixon might have lost the election*. So an account of the truth-conditions of *de re* modal predications in terms of trans-world identity is correct. Moreover, a counterpart-theoretic one—at least, any counterpart-theoretic one that includes Lewis's second postulate, according to which nothing exists in more than one possible world—is wrong. For on any such theory *In the actual world Nixon exists* and *In some possible world Nixon himself loses the election* jointly

---

[173] In the temporal case it is a matter of indifference whether we speak of a past situation in which Nixon was a child, or a past situation in which Nixon is a child. In exactly the same way, Kripke would say, it is a matter of indifference whether we speak of a possible situation in which Nixon would be president of Student for a Democratic Society (the SDS), or a possible situation in which Nixon is president of the SDS.

entail *Nixon loses the election*. But if a trans-world identity account of the truth of *de re* modal predications is right, then *In the actual world Nixon exists* is equivalent to *Nixon actually exists*, and *In some possible world Nixon himself loses the election* is equivalent to *Nixon might have lost the election*; and *Nixon actually exists* and *Nixon might have lost the election* do not jointly entail *Nixon loses the election*.

Here is a slightly different way to bring out the grounds of Kripke's anti-pathy to counterpart theory: given that Humphrey is the man who lost the (1968 American presidential) election, the counterpart theorist will want to say that there is no possible world in which *that man* wins the election (although there are possible worlds in which some man resembling him in such-and-such ways wins the election). Now, a possible world is just a 'complete' or com-pletely determinate possible circumstance. And if there are no completely determinate possible circumstances in which that man (Humphrey) wins the election, then there are no possible circumstances in which that man wins the election. If, however, there are no possible circumstances in which that man wins the election, how can that man's winning the election be a possibility? How can it be true that that man might have won the election? The counter-part theorist answers: it is true that that man might have won the election, inasmuch as there are possible circumstances in which some counterpart of that man wins the election. But, Kripke objects, on the understanding of counterparthood advanced by the counterpart theorist, there being possible circumstances in which some counterpart of that man wins the election amounts to—is nothing over and above—there being possible circumstances in which someone who is not that man, but resembles him in such-and-such ways, wins the election. And how could the existence of possible circum-stances in which someone as like as you please to that man wins the election, *in the absence of possible circumstances in which that man himself wins the elec-tion*, be sufficient for the truth of *that man might have won the election*? Supposing that it could is like supposing that the existence of future circum-stances in which someone as like as you please to this man wins the election, *in the absence of future circumstances in which that man himself wins the elec-tion*, is sufficient for the truth of *this man will win the election*.[174]

I think Kripke succeeds in showing that there is something deeply counter-intuitive about counterpart-theoretic truth-conditions for *de re* modal

---

[174] Note that this argument is again directed at a version of counterpart theory which denies that things exist in more than one world: it does not refute a version of counterpart theory that takes no stand on the existence of trans-world individuals. As we have seen, Kripke does not want to rule out explicitly the possibility that individuals have purely qualitative essences. So, I presume, he would not want to rule out that some qualitatively defined counterpart relation is necessary and sufficient for identity. Thus he would not, I take it, want to say that a version of counterpart theory that does not include Lewis's second postulate necessarily gives the wrong truth-conditions for *de re* modal predica-tions. He would nevertheless think it wrong-headed to bring a counterpart relation into the truth-conditions for *de re* modal predications, since he thinks that the recourse to a counterpart relation, rather than identity, is motivated by the 'faraway planet' picture of possible worlds (cf. 'N & N', n. 13).

predications. But defenders of counterpart theory have advanced various arguments in its favour. If (at least some of) those arguments were cogent, a counterpart-theoretic account of modality would be defensible, in spite of its (initial) counter-intuitiveness. Since—as we shall see—Kripke concedes that at least some of the considerations adduced in favour of counterpart theory are not easily dismissed, it is worth having a look at the kind of arguments Lewis and others have advanced on behalf of counterpart theory.

Suppose that, like David Lewis, you accepted something like the 'foreign country' or 'faraway planet' picture of possible worlds that Kripke rejects. That is, suppose you thought that the actual world was the universe we inhabit, and other possible worlds were other universes spatio-temporally unconnected to our own. Then, it might seem, you would embrace counterparts, and abjure trans-world individuals, on the grounds that an individual can inhabit at most one universe. Kripke himself says that counterpart theory is the most reasonable view to hold for one who takes the foreign-country view of worlds: 'No one far away on another planet can be strictly identical with someone here' ('N & N', 147). And various other philosophers have concurred with Kripke.[175]

But why exactly couldn't one thing be in each of two (spatio-temporally unconnected) universes? After all, if (as Lewis supposes) the two universes don't share any times or places, this wouldn't be a case of one and the same thing being in two different places at the same time; it would be a case of one and the same thing being in two different places at two (temporally unrelated, and a fortiori) different times. Even if no one who is *now* far away on another planet can be strictly identical with someone here *now*, nothing prevents someone who is far away on another planet at another time from being strictly identical to someone here now.[176]

---

[175] See e.g. Michael Loux's introduction to Loux (ed.), *The Possible and the Actual* (Ithaca: Cornell University Press, 1979), 63–4, and W. Lycan, 'Two—No Three—Concepts of Possible Worlds', *Proceedings of the Aristotelian Society*, 94 (1990), 221. Interestingly, although Lewis offers a number of considerations in favour of counterpart theory in *Counterfactuals* (39–41), none involve his conception of worlds as universes. In 'Individuation by Stipulation and Acquaintance', *Philosophical Review*, 92 (1983), 20–4, and in *On the Plurality of Worlds* (Oxford: Basil Blackwell, 1986), he does suggest that someone who thinks of possible worlds as universes has reason to believe in counterparts, lest properties that are not intuitively relations to worlds (or world–time pairs) turn out to be such relations. As Bigelow and Pargetter argue (in *Science and Necessity* (Cambridge: Cambridge University Press, 1990), 168–9), this line of reasoning seems less than conclusive: someone who thinks possible worlds are universes might hold that it is at least as counter-intuitive to deny that one and the same thing can exemplify roundness at different times and in different worlds as it is to suppose that *being round* is a relation between individuals and world–time pairs. Thus Bigelow and Pargetter consider it an open question whether someone who thinks of possible worlds as universes should believe in counterparts or trans-world individuals.

[176] Loux writes: 'It is easy to see why Lewis is so anxious to deny that a single individual can exist in more than one world ... If all possible worlds are equally real ... then you and I can exist in but a single world; we cannot, so to speak, inhabit two worlds at once' (*The Possible and the Actual*, 47). But why does it follow from the fact that we cannot inhabit two worlds *at once* that we cannot inhabit two worlds?

Sydney Shoemaker holds that an individual persists through time only if there are causal connections between the earlier stages of that individual's life (or, in the case of an inanimate object, existence) and the later stages; qualitative similarity plus spatio-temporal continuity, in the absence of causal connections, is insufficient for identity through time.[177] In support of this contention, he describes a case in which an absent-minded deity first decrees that a stone tablet will disappear into thin air at just before noon, and then—having forgotten all about his disappearance decree—also decrees that a (qualitatively indiscernible) stone tablet will appear at exactly noon. In this case, Shoemaker says, there are two different stone tablets (the a.m. tablet and the p.m. tablet), even though, to anyone not apprised of the deity's activity, it would look as though there were just one. The reason the a.m. tablet and the p.m. tablet cannot be identified is the absence of any (and, a fortiori, of the right sort) of causal relations between the a.m. stages of the existence of the tablet *t* there before noon, and the p.m. stages of the existence of the table *t* there from noon on.

If the identity of this individual at this time with that individual at that (distinct) time requires that there be causal connections between a this-timely stage of this individual's existence and a that-timely stage of that individual's existence, then individuals existing at different times in (spatio-temporally) unconnected universes are identical only if there are causal connections between events in (spatio-temporally) unconnected universes.

Perhaps the idea of causal relations obtaining between events occurring in (spatio-temporally) unconnected universes is incoherent; perhaps not. But it is very doubtful that someone who thinks that alternative possible worlds are universes spatio-temporally unconnected to the one we inhabit can countenance the idea that there are causal connections between events in such universes. For if there are causal links between, say, what is happening now in this universe and what is happening in some disjoint and unconnected universe, then surely the other universe is not an alternative possible world (a *merely* possible world); it is a part of the actual world (spatio-temporally unconnected to, though causally connected to, the part we inhabit).[178]

So there is a line of thought suggesting that, if the actual world is our universe, and alternative possible worlds are universes not (spatio-temporally) connected to our own, then the individuals existing in those alternative worlds are distinct from the individuals existing in the actual world. But I am uncertain how cogent it is. Shoemaker's thought-experiment seems to show that even if

---

[177] S. Shoemaker, 'Identity, Properties, and Causality', in P. French, T. Uehling, and H. Wettstein (eds.), *Midwest Studies in Philosophy* (Minneapolis: University of Minnesota Press, 1979). Different arguments for the same view are offered by Michael Slote in 'Causality and the Concept of a Thing', ibid.

[178] Most medieval philosophers held that God was the aspatial and atemporal first cause of the universe. Surely no one could (coherently) maintain both that the existence of the universe is the effect of a divine choice, and that God and the relevant choice-event were merely possible individuals and events found only in alternative possible worlds.

tablet T is spatio-temporally continuous and contiguous with tablet T′, it does not follow that T = T′. But one can accept this claim without conceding that the cross-time identity of x at t with y at $t′$ requires causal relations between the t-stage of x (or, for the three-dimensionalist, the t-stage of x's career) and the $t′$-stage of y (or the $t′$-stage of y's career). After all, some philosophers would maintain that it is logically possible for a thing and all its parts to go out of existence at an earlier time, and then, after an interval of non-existence, come back into existence, in spite of the absence of any causal relations between the phase of the thing's career just before its existence and the phase of the thing's career after it came back into existence.[179] Their intuition is that, just as a thing could (uncausedly) pop into existence or (uncausedly) pop out of existence, a thing that had (uncausedly, say) ceased to exist could (again, uncausedly) pop back into existence after an interval of non-existence. Why not? Perhaps the worry is that nothing could ground the fact that we have the same entity before and after the interval of non-existence. But why couldn't the identity of the pre-demise and post-demise thing be a primitive fact not grounded in anything else?

In sum: it is not clear that the identity of x existing at t with y existing at $t′$ presupposes the existence of causal relations between the t-stage of x's existence and the $t′$-stage of y's existence. And, if we can have non-causally mediated cross-time identity between two things in the same universe, it is unclear that we couldn't have it between two things in different universes.

That said, whether or not one and the same thing can exist in many disjoint universes, difficulties arise if we suppose both that alternative possible worlds are disjoint universes and that we exist in more than one possible world.

When we speak of someone's welfare, we might have in mind either their current welfare or what we might call their welfare *sub specie aeternitatis*. Current welfare is, roughly, a matter of how good your life is now (how good the current part of your life is), while welfare *sub specie aeternitatis* is a matter of how good your life as a whole is. (If you agree with Solon that we cannot tell whether or not people are happy until they have died, you are probably thinking of happiness as something like welfare *sub specie aeternitatis*.) However things might be with your current welfare, your welfare *sub specie aeternitatis* depends on what happens to you not just now, but at other times: the good or bad things that happen to you at past or future times, as well as the good or bad things that happen to you now, make a difference to your welfare *sub specie aeternitatis*. (From now on, for the sake of brevity, I will mean by 'welfare' 'welfare *sub specie aeternitatis*': when I have in mind current welfare I will always say 'current welfare'.)

Your welfare (broadly construed) is a matter of how good your life, con-sidered as a whole, is. This will be true if, but only if, you only have one life. If you have had many different lives in the past (say, punctuated by periods of

---

[179] For more on the possibility of 're-existence', see my 'Starting Over', in A. Bottani and P. Giaretta (eds.), *Individuals, Essence, and Identity* (Boston: Kluwer, 2002).

temporary non-existence), or will have many different lives in the future, any-thing good or bad that happens in any of your lives is relevant to your welfare.

A being that persisted through a series of incarnations would have many dif-ferent (and disjoint[180]), but spatio-temporally connected lives. A being that existed in a plurality of (non-spatio-temporally connected) universes would have many non-spatio-temporally connected, as well as different and disjoint, lives. And the good or bad things that happened to that being in any of those lives would be relevant to that being's welfare: anything good or bad that happens to someone—anywhere and 'anywhen'—is relevant to that someone's welfare.

That something good (or bad) happens to me at some past or future time is in and of itself relevant to my welfare. In just the same way, presuming that I exist in various 'alternative' (unconnected) universes, that something good (or bad) happens to me at some alternative universe is in and of itself relevant to my welfare. But the fact that something good (or bad) might in different circumstances have happened to me does not seem in the same way in and of itself relevant to my welfare. My welfare depends on what will happen to me, or did happen to me, even if it is not happening to me now; but my welfare does not depend on what might happen to me, unless it actually happens to me (now or 'elsewhen'). And this spells trouble for the view that I am a trans-world individual and alternative possible worlds are alternative universes. After all, on that view, the fact that such-and-such might in different circum-stances have happened to me just is the fact that in an alternative universe such-and-such does happen to me. The proponent of that view accordingly identifies a fact that is intuitively in and of itself relevant to our welfare with a fact that is intuitively not in and of itself relevant to our welfare.

Here a defender of the view under attack might object that, just as we have distinguished between current welfare and welfare from a temporally more inclusive point of view, we must distinguish between actual welfare and wel-fare from a modally more inclusive point of view. The idea would be that, just as my current welfare depends only on how good the current part of my life is but my welfare *sub specie aeternitatis* depends on how good my whole life is, my actual welfare depends only on how good my actual life is (or actual lives are), while my welfare from a modally non-parochial perspective depends on how good all my (actual or non-actual) lives are.

Well, yes. If alternative possible worlds are alternative (unconnected) uni-verses, and if I exist in more than one possible world, then, just as what did or will happen to me but isn't happening to me now is in and of itself relevant to my welfare most broadly construed, so too what might have happened to me but never actually did and never actually will happen to me is in and of itself relevant to my welfare most broadly construed. If you accept the antecedent of this last conditional, you should accept its consequent; but that doesn't make the consequent any less difficult to believe.

---

[180] As I use the term, things are disjoint if and only if they share no parts.

It may be worth underscoring why the difficulty brought to light here need not arise for either someone who takes possible worlds to be abstract entities and accepts trans-world identity, or someone who takes possible worlds to be universes but insists that no individual exists in more than one possible world. Someone who thinks of possible worlds as abstract entities need not (and presumably should not) think that *such-and-such happens to me in possible world W* implies *such and such happens to me* (full stop). She can say that, just as something that happens in a story, or in a dream, doesn't happen (full stop) unless it also actually happens, something that happens in an alternative possible world **W** doesn't happen (full stop) unless it also actually happens. Thus the good or bad things that happen to me in alternative possible worlds don't happen to me unless they actually happen to me; so the good or bad things that happen to me in alternative possible worlds, but not in the actual one, do not (partially) determine my welfare. Someone who, in Lewisian fashion, takes alternative possible worlds to be something like 'other dimensions of a more inclusive universe' ('N & N', note 13) is by contrast committed to saying that things that happen in alternative possible worlds happen (full stop), even if they don't actually happen. (The things that happen in alternative possible worlds are, as it were, things that happen *very* far away.) But someone who agrees with Lewis that individuals in different worlds are at most counterparts (and never identical) can say that the things that, so to speak, happen to me in alternative possible circumstances in fact happen to somebody else (to the person-I-would-have-been, and not to me). So, he can say, the good or bad things that (loosely speaking) happen to me in alternative possible worlds do not (partially) determine my welfare.

In sum, there is a line of thought that takes us from the premiss that alternative possible worlds are universes spatio-temporally unconnected to our own to the conclusion that you and I are not trans-world individuals, so that anything existing at an alternative possible world is at best a counterpart of you or me. But are alternative possible worlds universes unconnected to our own?

It seems very natural to suppose that they are not. After all, (even David Lewis admits that) a possible world is a 'complete' or 'saturated' possibility—a (complete) way for the world to be, a (complete) state for the world to be in. As Stalnaker has argued, it does not look as if any way the world might have been—including the way the world actually is—can be identified with the world that might have been or is that way: 'The way the world is is the world' sounds like what in the old days would have been called a category mistake.[181] (So does 'The state the world is in is the world'.) And if the way the world actually is is not the world, neither is it the universe (the largest scattered object surrounding us), since the universe is the world, or perhaps a proper part of the world.

---

[181] See R. Stalnaker, 'Possible Worlds', in Loux (ed.), *The Possible and the Actual*, 228.

In a similar vein, Kripke writes:

The 'actual world'—better, the actual state, or history of the world—should not be confused with the enormous scattered object that surrounds us. The latter might also have been called 'the actual world', but it is not the relevant object here. Thus the possible but not actual worlds are not phantom duplicates of the 'world' in this other sense. (*N & N*, 19–20)

To put the Kripke–Stalnaker point slightly differently, inasmuch as a possible world is a state for the world to be in, or a way for the world to be, a possible world is not a world, any more than a fake diamond is a diamond, or petrified wood is wood.[182] And, as long as we are supposing that the actual world is a possible world, the actual world is not a world either. For Lewis, philosophers such as Kripke and Stalnaker, who take possible worlds to be abstract entities, don't really believe in possible worlds: they believe in surrogates for possible worlds, or (as he calls them) 'ersatz possible worlds'. But, if possible worlds aren't worlds, then it is one thing to say that abstract entities are ersatz *worlds*, and quite another to say that they are ersatz *possible worlds*. Features that would preclude a thing's being anything more than an ersatz world might be perfectly compatible with (or even required by) being a genuine possible world.

To be sure, it is natural to construe 'the actual world' as (rigidly) designating the world, in much the way that 'the actual capital of Liguria' (rigidly) designates the capital of Liguria, and natural to suppose that alternative possible worlds are the same sort of thing as the actual world. If we do both of these things at once, we may conclude that possible worlds are worlds. If, however, Kripke and Stalnaker are right, we can only get that conclusion by equivocating on 'the actual world'. To guard against this sort of equivocation, Kripke suggests that it might be better to speak of 'possible states of the world' or 'counterfactual situations' rather than 'possible worlds' ('N & N', n. 15).

There is a complication here. We might naturally say that, corresponding to a set containing two (human) sperm cells and two (human) egg cells, there are four possible humans (at most two of whom will ever actually exist). In this context, a possible human is clearly not a different kind of thing from a (genuine) human. In much the same way, if a theologian says that there are many possible worlds that God could have created *ex nihilo*, in that context a possible world is not a different kind of thing from a (genuine) world. It is not an abstract entity—a (maximal) way things might have been; it is a universe, or something very like a universe. But possible worlds in what we might call the theologian's sense (maximal *creabilia*) are not possible worlds in the modal

---

[182] At least, a possible world is not a world in any of the more or less ordinary senses of 'world'. Given how many philosophers and logicians use 'world' to cover possible and, in some cases, impossible worlds, there is probably now a philosopher's sense of 'world' which means something like '(complete) way for things to be'.

logician's sense (maximal *actuabilia*).[183] After all, for each of the many possible worlds that God might have created, there are many (maximal) ways that world could have been. That's why knowing which possible world, in the theologian's sense, is actual falls well short of knowing which possible world, in the modal logician's sense, is actual.

The thesis that possible worlds are universes gives rise to another difficulty. The proponent of a (standard) possible-worlds account of modality holds that there is a possible world in which there is a golden mountain if and only if there might have been a golden mountain. So someone who offers a (standard) account of modality in terms of possible worlds, and identifies possible worlds with universes, will hold that there is a universe containing a golden mountain if and only if there might have been a golden mountain. If, in addition, she thinks that there isn't actually a golden mountain but there might have been, she will have to say that there is a possible world in which there is a golden mountain but there isn't actually a possible world in which there is a golden mountain. After all, by her lights, if there actually were a possible world in which there is a golden mountain, there would actually be a universe containing a golden mountain. And if there actually were a universe containing a golden mountain, there would actually be a golden mountain: no universe can be 'more actual' than any of its contents.

More generally, whenever she thinks that there aren't actually Ks (or F-ish Ks), but there might have been, she will have to say that there is a possible world in which there are Ks (or F-ish Ks), but there isn't actually a possible world in which there are Ks. (Think of it this way: inasmuch as she doesn't believe there actually are Ks, she doesn't think there actually is a universe or possible world in which there are Ks; inasmuch as she believes there might have been Ks, she believes there is a universe or possible world in which there are Ks.)

The difficulty, as Stalnaker and others have noted,[184] is that it is difficult to believe that there are more things than there actually are. (Compare: Can we believe that more things are true than are actually true?)

These considerations are not guaranteed to move a 'universalist' about possible worlds; they do not move David Lewis. As he sees it, we get the best systematic account of modality by making some unobvious, not initially attractive suppositions: that (complete) possibilities are universes, and that *actualia* are, as it were, only the tip of the iceberg of being. It is beyond the scope of this work to consider Lewis's detailed, subtle, and imaginative arguments in defence of this claim;[185] my aim has only been to argue that universalism is—initially at least—a very difficult view to believe. It is accordingly

---

[183] Cf. *Summa Theologiae* Ia. 45. 4, *responsio*, where Aquinas says that it is things, rather than states of things, that God creates.

[184] Stalnaker, *Inquiry*, 48–9, and Plantinga, *The Nature of Necessity*, 132 *et passim*.

[185] See Lewis, *On the Plurality of Worlds*.

an initially unpromising source of support for a counterpart-theoretic account of modal predication.

Considerations of a very different sort—involving identity and extrinsicality—might be thought to support counterpart theory. Suppose that in the actual world **G** a certain ship S is built from a set of planks, sinks on its maiden voyage, and gradually falls to bits on the sea floor. Suppose that in an alternative possible world **H** the same set of planks is put together in the same way at the same time and place by the same shipwright. The resulting ship does not sink on its maiden voyage, but all its original planks come to be dispersed. They are subsequently reassembled by the same shipwright who originally assembled them, in just the way they were originally assembled. Is the ship resulting from the reassembly S?

Well, it depends. Suppose that in **H** the dispersal of the original planks went hand in hand with the (structure-preserving) replacement of original planks by new planks, in the course of ordinary maintenance and repairs. Then, it appears, the ship whose planks were gradually replaced is ship S, and the ship resulting from the reassembly of the original planks is a different ship (a new ship—a duplicate of the original ship—made from original ship's original planks). Suppose, on the other hand, that in **H** the plank-dispersal was not accompanied by plank-replacement; the sailors took the ship to bits in order to transport it overland more easily, intending all the while that the shipwright would put the ship back together when they got to the next sea. Then, it seems, the ship resulting from the reassembly of the original planks is the ship S.[186]

A counterpart theorist might say: whether or not the ship in **H** resulting from the reassembly of the old planks is the ship S would have been, had **H** been actual, depends on whether there is another ship in **H** that has a stronger claim to be the ship S would have been under those circumstances. This is perfectly compatible with counterpart theory, according to which whether an individual i in an alternative world is a counterpart of an actually existing individual i' depends, not just on the intrinsic properties of i and i', but also on the relational or extrinsic properties of i—in Lewis's words, on i's context as well as its content. Suppose, on the other hand, that individuals are transworld, so that the ship S would have been in such-and-such circumstances is literally identical to S. Then it cannot be that whether a certain ship in another world is the ship S would have been depends on whether in that world there is something else that has a stronger claim to be the ship S would have been. So—the counterpart theorist might conclude—inasmuch as the would-have-been relation has extrinsic determinants, it is not identity: the ship S would have been, had **H** been actual, is not S, but a counterpart of it.

---

[186] This assumes that the replacing planks have a dominant claim to constitute the original ship, and the reassembled planks a recessive one (see Ch. 2, sect. 3). The reader who thinks that the reassembled planks have a dominant claim may modify the story accordingly.

The trouble with this argument—as Salmon has pointed out[187]—is that the champion of trans-world identity (henceforth, 'the (trans-world) identitarian') will deny that there is a particular ship in **H** which is the ship S would have been if but only if nothing else in **H** has a better claim to be the ship S would have been. Instead, the identitarian will say, there is a set of planks in **H** which constitutes the ship S would have been (that is, constitutes S) at a given time if but only if no other set of planks in **H** has a better claim to constitute S at that time. This has the (at least initially surprising) result that which ship some planks constitute has extrinsic determinants, but, as we have already seen in Chapter 2, section 3, the identitarian may find that acceptable (and independently plausible).

A different line of reasoning purporting to favour counterparts over trans-world individuals takes as its starting point the idea that two different things cannot have all the same parts at all the same times. Why should this principle favour counterparts over trans-world individuals? Suppose that individuals are trans-world, so that an individual might have been F if and only if that individual is F in some world. Then i and i' must be distinct as long as there is some property F such that i, but not i', might have been F. (If there is a world **W** and a property F such that i is F in **W**, but i' is not F in **W**, then i and i' are discernible, and thus distinct.) We have already seen, though, that there seem to be cases in which it is true both that this thing and that thing have all the same parts at all the same times, and that this thing but not that thing might have been F. Kripke's example involves a plant that never 'grows past' its stem: the plant and the stem have all the same parts at all the same times, but it is true of the plant, and not of the stem, that it might have grown past its stem. Various other examples have been discussed in the literature.[188]

Actually, one could believe in trans-world individuals, and also think that different things cannot have all the same parts at all the same times, as long as one believes in few enough kinds of individuals. Peter van Inwagen holds just this pair of views, because he doesn't think there are such things as stems, or bodies, or lumps of clay, or statues... only simples and living things.[189] But those of us who are ontologically more generous must choose between the view that things with all the same parts at all the same times are identical, and the view that individuals exist in more than one possible world.

---

[187] *Reference and Essence*, app. 1.

[188] See A. Gibbard, 'Contingent Identity', *Journal of Philosophical Logic*, 4 (1975), 187–221, and D. Lewis, 'Counterparts of Persons and Their Bodies', *Journal of Philosophy*, 68 (1971), 203–11; repr. in Lewis, *Philosophical Papers*, vol. i. In Gibbard's example a (human-shaped) statue and a lump of clay come into and go out of existence together, and thus have all the same parts at all the same times; the lump, but not the statue, might have survived being squeezed into a ball while the clay was still soft. In Lewis's example a person and his body come into and go out of existence together, and thus have all the same parts at all the same times; the person, but not the body, might have switched bodies with someone else. (A reader who doubts the possibility of body-switching may substitute: the body, but not the person, might have outlasted the person.)

[189] See P. van Inwagen, *Material Beings* (Ithaca: Cornell University Press, 1989).

Still, given a modicum of ontological generosity, won't we have to deny that (permanent) co-constitution is identity, whether we accept an identitarian or a counterpart-theoretic account of modal predication? If i, but not i', might have been F, then (for the counterpartist) i, but not i', has an F-ish counterpart, so i and i' are distinct.

Since Lewis wants to say that (permanent) co-constitution is identity, he suggests a modification of his original counterpart theory which allows him to say that. It involves multiple and intersecting counterpart relations, and works like this: one and the same (world-bound) individual can have different sets of counterparts, under different counterpart relations, corresponding to different sortals. For example, it can happen that one and the same world-bound individual is both a person and a body, and has one set of person-counterparts and a different (though intersecting) set of body-counterparts; or that one and the same world-bound individual is both a statue and a lump of clay, and has one set of lump-counterparts and a different (though intersecting) set of statue-counterparts. In a sentence such as 'This statue might have been F' the subject term denotes a world-bound individual and selects a counterpart-relation—in this case, the statue-counterpart relation. 'This statue might have been F' will accordingly be true just in case the individual denoted by 'this statue' has a statue-counterpart that is F. Similarly, 'This lump might have been F' will be true just in case the individual denoted by 'this lump' has a lump-counterpart that is F. If this statue and this lump of clay are (permanently) constituted by all the same parts, then it will be true that this statue = this lump, even though it is true that this lump might have been spherical (because the statue-cum-lump has a lump-counterpart that is spherical), and false that this statue might have been spherical (because the statue-cum-lump has no statue-counterpart that is spherical).[190]

Again, though, these considerations favour counterpart theory only if (permanent) co-constitution is identity. Why suppose that it is? Harold Noonan has argued that, if we don't, then we shall have to say that purely material things having the same material constitution at all times are nevertheless distinct simply in virtue of having different modal, or dispositional, or counterfactual, properties. This last view (Noonan holds) 'is...surely...astonishing':[191] it implies that material objects can be *actually* distinct simply in virtue of the fact that they have different properties in *counterfactual* circumstances. But (Noonan thinks) differences in modal or dispositional properties have to be grounded in differences in 'categorical' (non-modal) properties.

Here someone might object that, if, say, a statue and a lump of clay are discernible with respect to modal or counterfactual properties, they will also be discernible with respect to some categorical properties, such as *being a statue*

---

[190] Cf. Lewis, 'Counterparts of Persons and Their Bodies'.

[191] See H. Noonan, 'The Closest Continuer Theory of Identity', *Inquiry*, 28 (1985), 203, and 'Constitution Is Identity', *Mind*, 102 (1993), 133–45.

and *being a lump*.[192] Indeed, it might be maintained, the discernibility of the statue and the lump with respect to modal or counterfactual properties is grounded in their discernibility with respect to *being a statue* and *being a lump*. (It is because the lump is a *lump*, rather than a statue, that it could survive being squeezed into a ball.)

But is it clear that permanently co-composite objects could belong to different kinds? Following Alan Gibbard, let us suppose that our permanently co-composite statue and lump are (respectively) called 'Goliath' and 'Lumpl'. Goliath has everything it takes to be a statue—the right sort of size, shape, origin, and history. But Lumpl has the same size, shape, origin, and history. So why isn't Lumpl a statue too? The identitarian, like the counterpartist, could accept that Lumpl has everything it takes to be a statue—but only at the cost of embracing the (unattractive) view that the sculptor who made Goliath made *two different statues*. The alternative is the (not obviously attractive) view that something with the same composition, size, shape, origin, and history as a statue may yet not be a statue.

Bracketing the question of whether in the Goliath–Lumpl case we have two statues, do we really have two things or two (material) objects? It is not obvious that we do. After all, suppose that Goliath is on the table and weighs 1 pound. Suppose also (to put it neutrally) that anything on the table that weighs 1 pound permanently has the same constitution as Goliath. How many 1-pound things are there on the table? For the (ontologically generous) identitarian, it would seem, there are (at least) two different 1-pound things (two different material objects) on the table: Goliath and Lumpl. But are there? Suppose we take a non-philosopher, show her the table, equip her with a scale, and ask her how many 1-pound things there are on the table. It is difficult to imagine her coming up with any answer other than 'One'. This suggests that the identitarian sees plurality (and difference) where we are naturally disposed to see unity (and sameness).

The identitarian can grant that, when we count how many 1-pound things there are on a table, we (ordinarily) count as though permanently co-composite things were identical. Still, he can say, we should not conclude that permanently co-composite things are (in every case) identical. He might say that we are (ordinarily) prone to under-count, although this tendency can be corrected by the right sort of metaphysical education. (In the case at hand one comes to see that, since Goliath and Lumpl have different persistence conditions, they are two different 1-pound things on the table, so that there are in fact (at least) two 1-pound things on the table.) Alternatively, he might

---

[192] See T. Burge, 'A Theory of Aggregates', *Nous*, 11 (1977), 114. Although Burge does not discuss statues and lumps, or persons and bodies, he considers the case of a positronium atom and a positron–electron pair that have all the same parts at all the same times. After noting that the pair, but not the atom, could have been scattered, he says it would be a mistake to think that the only differences between the positronium atom and the positron–electron pair are modal; the entities in question also differ in kind.

borrow an idea championed by Peter Geach, Harold Noonan, and David Lewis, and say that counting by identity is just one way of counting.[193] While one may count in such a way that things count as one if and only if they are identical, one may also count in such a way that things count as one if and only if they satisfy some weaker condition—e.g. are permanently co-composite. On this view, someone who says 'There is just one 1-pound object on the table' is not in error, and what she says is perfectly compatible with Goliath and Lumpl's being distinct. Of course, if Goliath and Lumpl are distinct, and each is a 1-pound thing on the table, then there are two 1-pound things on the table. That is, there are two 1-pound things on the table (counting by identity). Nevertheless, there is just one 1-pound thing on the table (counting by some relation weaker than identity, as we are wont to do in non-philosophical contexts). Finally, the identitarian might say that, when we count 1-pound things on the table, in that context, 'thing' should be understood in a restricted sense which covers, say, statues, but not bits of clay: given the restriction, there is just one 1-pound thing on the table.

Still, the counterpartist may object, what reason is there to think that, in the case under discussion, someone who says there is just one 1-pound thing on the table is either under-counting, or counting by some relation weaker than identity, or presupposing a restricted sense of 'thing'? Well, the identitarian might reply, suppose we modify the case somewhat, so that on the table there is a 3-pound statue, made of some bronze that existed before the statue did, and everything on the table that weighs 3 pounds currently has the same constitution as the bronze statue. Once again, we take a person innocent of metaphysics, show her the table, equip her with a scale, and ask her how many 3-pound things are on the table. In this case, just as in the original case, she will probably answer 'One'. But, the identitarian will continue, it is clear that the statue is not identical to the bronze it is made of, since the statue and the bronze have different origins and histories. So in the modified case either the person under-counted, or she counted by some relation weaker than identity, or she was using a restricted conception of 'thing'. If that is what is going on in the modified case (involving wholly but only temporarily 'overlapping' objects)—the identitarian may conclude—it is incumbent upon the counterpartist to show that it is not what is going on in the original case (involving wholly and permanently overlapping objects). Failing that, our practices of counting 1-pound objects in a place at a time do not favour the claim that Goliath is Lumpl (or favour counterparts over trans-world individuals).

In response, the counterpartist might deny that the bronze statue and the bronze it is made of are distinct, on the grounds that two (material) things cannot be in the same place at any (much less all) of the same times. In order

---

[193] See P. Geach, *Reference and Generality*, 3rd edn. (Ithaca: Cornell University Press, 1980); Noonan, 'Constitution Is Identity', and D. Lewis, 'Survival and Identity', in Lewis, *Philosophical Papers*, vol. i.

to square this with the fact that the bronze existed before the statue did, the counterpartist would have to countenance a less than straightforward account of tensed predication to go along with her less than straightforward account of modal predication. She might, for example, say that in 'this statue is bronze', 'this statue' denotes a time-bound (as well as world-bound) individual, and selects both a counterpart relation (a way of tracing that (world-bound) individual through logical space) and a 'continuer relation' or 'temporal counterpart relation' (a way of tracing that (time-bound) individual through time). Then, she could say, 'This statue will be (was) F' comes out true just in case the individual denoted by 'this statue' has an F-ish statue-temporal counterpart at a later (earlier) time. And 'This bronze will be (was) F' comes out true just in case the individual denoted by 'this bronze' has an F-ish counterpart at a later (earlier) time. That will allow her to say that the statue and the bronze are identical, even though the bronze existed before the statue did.

It is not surprising, though, that Gibbard, Lewis, and most other philosophers who have maintained that permanent co-constitution is identity have not wanted to say that current co-constitution is identity. For one thing, there is something very odd about the supposition that an individual who has existed for, say, twenty years exists at only one time. For another, on the account of (tensed and modal) predication just sketched, there is no such thing as the history or the origin of an individual, any more than there is such a thing as the essence of an individual: just as Goliath–Lumpl has one essence *qua* statue and another *qua* lump, the bronze statue–bronze will have one origin and history *qua* statue and another origin and history *qua* bronze. And it is very hard to believe that individuals don't have just one origin and just one history.

The moral would seem to be that what might be called 'counting-based' arguments for the identity of permanently co-composite things are inconclusive.

This leaves the rather different argument to the effect that the identitarian must choose between too many statues, and things that have everything it takes to be a statue and yet are not statues. The identitarian might try to meet this argument by supposing that in Goliath–Lumpl-type cases there are two statues counting by identity, but only one counting by the relation we normally count by. But he needn't take this (unappealing) line. The identitarian typically distinguishes the 'is' of identity from the 'is' of constitution, and the 'is' of (what I shall call) co-constitution.[194] In 'Hesperus is Phosphorus' the 'is' expresses the relation of identity. In 'The ring is gold' the 'is' expresses a constitution relation (holding between gold and the ring, and more generally between a (kind of) stuff and a thing). In 'the copse is a bunch of trees' or 'the statue is a piece of bronze' the 'is' expresses a co-constitution relation (say, *being (currently) made of the same matter as*). (Because the 'is' of co-constitution expresses a tensed relation, one and the same copse can be different groups of trees at different times.)

[194] See D. Wiggins, *Sameness and Substance* (Cambridge, Mass.: Harvard University Press, 1980).

Since he recognizes the co-constitutive 'is' as well as the 'is' of identity, the identitarian can say that anything with the same composition, size, shape, origin, and history as a statue is itself a statue—where 'is' expresses co-constitution, not identity.

Someone who thinks permanent co-constitution is identity runs into the following difficulties in arguing for that view. She can start from relatively uncontroversial premisses (say, about how we count material objects, or material objects of a certain kind; or about whether things just like statues are statues). Then—as we have seen—the identitarian can grant the premisses, but insist that the premisses (properly construed) do not entail that permanent co-constitution is identity. Alternatively, she can start from a strong metaphysical premiss, such as the claim that a thing's modal properties supervene on its origin-cum-history-cum-constitution-cum-qualities (so that necessarily any two things that are indiscernible with respect to their origin, history, constitution, and qualities are also indiscernible with respect to their modal properties). Then there won't be any daylight between the premisses and the conclusion; but the identitarian will simply reject the premisses. This is not to say that the modal supervenience principle at issue doesn't have a certain intuitive appeal (I think it does); it is just difficult to know what to say to those who find it intuitively unappealing.

On the other hand, there are principles I find initially plausible that are inconsistent with the modal supervenience principle and the sufficiency of permanent co-constitution for identity. Suppose we call a fact *soft* just in case its being a fact depends (logically) on how the future goes, and call a fact *hard* otherwise. That I will never walk on the moon is a soft fact; that I have lived in Castello is a hard fact. I find it intuitively very plausible that facts about how many material objects there are in a given place at a given time are hard facts: how can what happens in the future make any difference to how many things are here *now*? Suppose, though, that Goliath is Lumpl, and Goliath is on the table right now. That Goliath is the same material object as Lumpl will be a soft fact (since its factuality depends on whether or not some future event puts an end to Goliath without at the same time putting an end to Lumpl). So whether we have n or $n + 1$ material objects on the table now will be a soft fact. This unwelcome consequence is avoided if Goliath and Lumpl are conceived of as different trans-world individuals with difference essences: it will then (I presume) be a hard fact that Goliath and Lumpl have different modal properties, and are two material objects rather than one.

Those who would identify Goliath and Lumpl can respond that there is a fact about how many material objects there are on the table now—on the supposition that we are counting by current total overlap rather than identity.[195]

[195] Notice, though, that if we are counting by current total overlap, then how many material objects there are here and now can no more (logically) depend on the past than it can (logically) depend on the future. But, I have argued elsewhere, how many material objects there are here and now can depend on the past (see my 'Same-Kind Coincidence and the Ship of Theseus').

But isn't there hard fact about how many material objects there are on the table right now, even if we are counting by identity? Aren't Goliath and Lumpl one *and the same*, or not, irrespective of what happens in the future?

One reason the identitarian is likely to think the answer is yes is that he is likely to believe that material objects are 'wholly present', not just at different worlds, but also at different times. Given this (Anselmian)[196] picture, it is hard to see how one could avoid thinking that there are either two (currently over-lapping) material objects on the table or just one, and nothing could change that. By contrast, a counterpartist may well suppose that material objects are present at different times by having different temporal parts that are present at those times, in much the way that events are present at different times by having different temporal parts that are present at different times. Given this picture, it is natural to suppose that there can be soft facts about how many material objects there are here now. Suppose a battle is being fought now. It could be that, depending on how the future goes, that battle will turn out to have been part of one war or two. If so, there won't be a hard fact about how many wars (and how many events) are getting under way now. Similarly, if what is wholly present on the table (or, rather, the current stage of the table) is a temporal part or stage of a (four-dimensional) material object, then there need not be a hard fact about whether that stage is a part of two different four-dimensional material objects (a statue, and a lump), or just one (a statue-cum-lump). So there need not be a hard fact about how many material objects are there on the table right now. This suggests that a serious objection to the thesis that permanent co-constitution is identity can be met if, but only if, material objects are not Anselmian.

Peter van Inwagen has suggested that, in order to defend a four-dimensionalist (or 'temporal parts') account of material objects against a certain objection, the four-dimensionalist must reject an identitarian account of modal predication. I am suggesting that, in order to defend a counterpart-theoretic account of modal predication against a certain objection, the counterpart theorist must reject a three-dimensionalist (or 'temporally part-less') account of material objects. The view that objects are wholly present in different worlds and the view that objects are wholly present at different times are, as it were, made for each other, and someone who rejects one is well advised to reject the other.

We have not yet found any arguments decisively favouring counterparts over trans-world individuals. But we have not yet considered what Lewis (and,

---

[196] The view that material objects are temporal-part-less, and wholly present at all the times they exist, is often associated with Aristotle; and that view certainly does go very naturally with the Aristotelian account of change. As far as I know, though, Anselm was the first philosopher to argue explicitly that material objects and other substances lack temporal parts and are wholly present at every time they exist, unlike processes or events, that are present at different times by having different parts present at those times. See his *Monologion*, 21.

it seems, Kripke) consider the strongest argument for their view, which could be put as follows:

> The identitarian thinks that the person you would have been, had things been different, is just you, and that the would-have-been relation is just identity. (We may speak of the *would-have-been relation* in order not to prejudge whether the relation is identity or some (qualitatively based or non-qualitatively based) counterpart relation.) But the would-have-been relation does not have the right logical properties to be identity.

Why not? Suppose that a bicycle B_0 is made by a bicycle-maker from (small) parts $P_1 \ldots P_n$.[197] It seems that the bicycle-maker could have made that same bicycle from all the same (small) parts except one (as long as she had put the parts together in the same way, in the same place and time, etc.). On the other hand, it seems that no bicycle-maker could have made that very bicycle from a completely different set of (small) parts: a bicycle made from completely different (small) parts would have been a different bicycle.[198] There is a problem, though, about reconciling these two claims. In the actual world **G** our bicycle-maker makes a bicycle B_0 from (small) parts $P_1 \ldots P_n$. Since she might have made B_0 from all the same (small) parts but one, there is a possible world $\mathbf{H_1}$ in which she makes B_0 from $P_1 \ldots P_{n-1}, P'_n$ (where P'_n is distinct from but just like P_n: in what follows, a P' will always be distinct from a P). Now, if she had made B_0 out of $P_1 \ldots P_{n-2}, P_{n-1}, P'_n$ it would then have been true that she could have made B_0 out of $P_1 \ldots P_{n-2}, P'_{n-1}, P'_n$ So, it seems, there is a possible world $\mathbf{H_2}$ in which she makes B_0 from $P_1 \ldots P_{n-2}, P'_{n-1}, P'_n$. Continuing in this way, we reach a possible world $\mathbf{H_n}$ in which our bicycle-maker makes B_0 from a set of parts $P'_1 \ldots P'_n$ completely different from B_0's actual parts $P_1 \ldots P_n$. If there is such a world, then the bicycle-maker could after all have made the bicycle that she actually made from one set of parts from a completely different set of parts. Given that we accept the (intuitively plausible) claim that a bicycle could have been made from a slightly different set of parts, we seem to be stuck with the (intuitively implausible) claim that a bicycle could have been made from a completely different set of parts.[199]

The counterpart theorist can, however, take the better without the bitter. As she sees it, it could have been that B_0 was made from $P_1 \ldots P_{n-1}, P'_n$, since B_0 has a counterpart that was made from $P_1 \ldots P_{n-1}, P'_n$. And it could have been that it could have been that B_0 was made from $P_1 \ldots P_{n-2}, P'_{n-1}, P'_n$, since B_0 has a counterpart that has a counterpart that was made from $P_1 \ldots P_{n-2}, P'_{n-1}, P'_n$. By

---

[197] Assume that all of the parts $P_1 \ldots P_n$ are small, and that the bicycle has no parts disjoint from each of $P_1 \ldots P_n$: even the bicycle's frame can be broken down into many small parts.

[198] Here it is important that each of the bicycle's parts overlap with some of (the small parts) $P_1 \ldots P_n$. If, say, the frame and the handlebars did not overlap with any one of $P_1 \ldots P_n$, it could be argued that the bicycle could have (originally) lacked all of $P_1 \ldots P_n$.

[199] The example is taken from H. Chandler, 'Plantinga and the Contingently Possible', *Analysis*, 37 (1976), 152–9, though, as we shall see, the moral Chandler draws from it is quite different.

the same reasoning, it could have been that it could have been that...it could have been that B_0 was made from $P'_1 \ldots P'_m$, since B_0 has a counterpart that...that has a counterpart that was made from $P'_1 \ldots P'_n$. Still, the counterpartist can say, B_0 could not have been made from $P'_1 \ldots P'_m$, since B_0 has no counterpart made from $P'_1 \ldots P'_n$ (even though it has a counterpart that has a counterpart that...has a counterpart made from $P'_1 \ldots P'_n$). The counterpart relation need not be transitive; in fact, if it is similarity-based, one would expect it to not be (since a lot of little differences can add up to a big difference).

So, the counterpart theorist may say, the bicycle case shows that the would-have-been relation is non-transitive. Corresponding to our sequence of possible worlds $G, H_1 \ldots H_n$, we have a sequence of bicycles $B_0, B_1 \ldots B_n$. Each (non-initial) B_j is the bicycle that B_{j-1} would have been had H_j been actual; but B_n is not the bicycle B_0 would have been had H_n been actual. Since the would-have-been relation is non-transitive, it isn't identity.

All this presupposes that a bicycle could not have been made from a completely different set of parts. That assumption is contestable. But I doubt there is much mileage in trying to block the argument by contesting it. The argument depends on the idea that an individual could have been slightly different along a certain dimension (in this case, the parts it was made from and originally made of), but could not have been too different along that dimension. Even if a bicycle could have been made from a completely different set of parts, there will surely be some individual and some dimension such that that individual could have been slightly different, but not too different, along that dimension. For instance, it seems that a clay statue might have had an (original) shape slightly different from its actual (original) shape. But it could not have had an (original) shape hugely different from its actual (original) shape. A cup made by a potter could have been originally made of slightly more or slightly less matter than the amount of matter it was initially made of. But it could not have been originally made of an amount of matter sixteen times greater, or sixteen times less, than the amount of matter it was actually originally made of. Richmond Park might have a slightly different original location—it might have (originally) extended slightly into the region of space actually occupied by Wimbledon Common—but it could not have originally been entirely within the city limits of Edinburgh. And so on. Denying all these claims would commit us to an appreciable (and unappealing) degree of hypoessentialism.

Similarly, there is not much mileage in blocking the argument by denying the premiss that a bicycle could have been made from all the same parts except one. If we blocked similar arguments in the same way, we would end up with a version of hyperessentialism—at least with respect to the properties an individual has at its first moment of existence.

Of course, all of the kinds of entities appealed to in arguing for the non-transitivity of the would-have-been relation are at least somewhat ontologically controversial: van Inwagen, for one, would deny the existence of every one of them. But anyone who (like van Inwagen) countenances living beings will face problems concerning the transitivity of the would-have-been relation. After all, it seems possible that some day we shall be able to grow cells individually and subsequently 'assemble' them in such a way as to get an organism. Moreover, it seems that the same organism might have been 'assembled' from almost all but not quite all the same cells, but couldn't have been 'assembled' from a completely different batch of cells. That is enough to call into question the transitivity of the would-have-been relation. (The point may be worth emphasizing, inasmuch as van Inwagen once averred that opposition to, or even scepticism about, trans-world identity is simply the result of one or another sort of confusion.) Perhaps we can avoid problems for the transitivity of the would-have-been relation if our ontology includes only simples—say, space-time points *vel similia*. I take it, though, that an ontology this bare is not significantly more attractive than hypo- or hyperessentialism.

In any case, most philosophers who have addressed the problem under consideration have conceded the existence of the entities that give rise to the problem, and rejected both hypoessentialism and hyperessentialism. Indeed Kripke (tentatively) accepts that an artefact could have been made from and originally of slightly different parts or matter, and insists that an artefact could not have been made from or originally of completely different parts or matter. So how would Kripke respond to the argument just sketched? Judging from note 18 of *Naming and Necessity*, surprisingly concessively:

If a chip, or molecule, of a given table had been replaced by another one, we would be content to say that we have the same table. But if too many chips were different, we would seem to have a different one. The same problem can, of course, arise for identity over time.

Where the identity relation is vague, it may seem intransitive; a chain of apparent identities may yield an apparent nonidentity. Some sort of 'counterpart' notion (though not with Lewis' underpinnings of resemblance, foreign country worlds, etc.), may have more utility here. One could say that strict identity applies only to the particulars (the molecules), and the counterpart relation to the particulars 'composed' of them, the tables. The counterpart relation can then be declared to be vague and intransitive. It seems, however, utopian to suppose that we will ever reach a level of ultimate, basic particulars for which identity relations are never vague and the danger of intransitivity is eliminated.

Although Kripke expresses himself rather tentatively ('we would be content to say...we would seem...may seem intransitive...may have more utility'), he seems to concede that there are cases in which what I have been calling the would-have-been relation at least appears to be intransitive, and that a counterpart-theoretic account of modal predication—unlike an identitarian one—can obviously accommodate this apparent fact. If it turned out that an identitarian account of modal predication cannot be squared with the idea that a bicycle could have been made from some different parts but not from completely different parts, or the idea that a table could have been made from an only slightly different portion of wood, but not from a completely different portion of wood, that certainly would strongly favour a counterpart-theoretic account of modal predication over an identitarian one. And inasmuch as the former account clearly and unproblematically accommodates the (apparent) fact at issue and the latter account does not, that favours the former account over the latter one.[200]

Hugh Chandler has suggested, though, that the identitarian can accept that a bicycle could have (originally) lacked some but not all of its original parts (without contesting the transitivity of identity).[201] His idea is that, somewhere in the sequence of worlds that begins with G and ends in H_n, we pass from possible worlds to impossible worlds.[202] H_1 is possible relative to G (that is, everything true at H_1 is possibly true at G). And each (non-initial) H_j is possible relative to H_{j-1}. But H_n is not possible relative to the actual world G (although it is possibly possibly possibly…possible with respect to G). It follows that the relation of relative possibility is not transitive, so that any modal logic at least as strong as S4 (according to which relative possibility is transitive) is too strong (cf. Chapter 2, Section 1). It will not in general be true that $\Box A \supset \Box \Box A$, or that $\Diamond \Diamond A \supset \Diamond A$. Moreover, some possible states of affairs are only contingently possible. Suppose, for the sake of simplicity, that a bicycle with n small parts could have been made from a set of small parts fewer than half of which were different from the small parts that bicycle was actually made from, but could not have been made from a set of small parts half or more of which were different from the small parts the bicycle was actually made from. Given that B_0 was actually made of $P_1 \ldots P_n$, it could have been made from $P'_1 \ldots P'_{(n/2)-1}, P_{(n/2)} \ldots P_n$. Moreover, B_0 could have been made from $P_1 \ldots P_{n-1}, P'_n$. But if it had been, then it wouldn't have been possible for it to have been made from $P'_1 \ldots P'_{(n/2)-1}, P_{n/2} \ldots P_n$ (since that would entail B_0's being made from a set of small parts half of which were different from the small parts it was actually made from).

---

[200] In note 18 Kripke at least suggests that it is the (apparent) vagueness of the would-have-been relation that gives rise to its (apparent) intransitivity. But, as Anil Gupta has noted (see his *The Logic of Common Nouns* (New Haven: Yale University Press, 1980)), even if we supposed there was a perfectly precise answer to the question 'How many of its original parts could this bicycle have lacked and still existed?', the would-have-been relation would still be intransitive—unless the answer was 'none' or 'all'. Although the fact that a bicycle could have (originally) lacked some but not all of its original parts provides at least a prima facie reason to think that the would-have-been relation is intransitive, it is unclear that it provides a (prima facie) reason to think that it is vague (see Salmon, *Reference and Essence*, 240–50).

[201] See Chandler, 'Plantinga and the Contingently Possible'.

[202] Chandler suggests that we do not pass from a (definitely) possible H_j to a (definitely) impossible H_{j+1}; instead we move from possible worlds to impossible worlds 'via a region of indeterminacy' where the worlds are presumably neither definitely possible nor definitely impossible. Thus the relation of relative possibility will have the vagueness, as well as the intransitivity, that some counterpart theorists have wanted to ascribe to the would-have-been relation.

My first reaction to Chandler's proposal was disbelief: it seemed paradoxical that what possibilities there are could depend on which of the possibilities that there are is actualized. Moreover, consider the state of affairs B_0's *being made of* $P'_1 \ldots P'_{n/2}, P_{n/2 + 1} \ldots P_n$. Given our assumptions about how many of its actual parts a bicycle would have to have been made from, this is a (just barely) impossible state of affairs. But it is not, as it were, intrinsically impossible—impossible simply in virtue of being the state of affairs it is. If B_0 had been made from $P_1 \ldots P_{(n/2) - 1}, P'_{n/2}, P_{(n/2) + 1} \ldots P_n$,—as it perfectly well might have been—then B_0's *being made of* $P'_1 \ldots P'_{n/2}, P_{(n/2) + 1} \ldots P_n$ would have been a (just barely) possible state of affairs. So B_0's *being made of* $P'_1 \ldots P'_{n/2}$, $P_{(n/2) + 1} \ldots P_n$ is impossible only because it happens to be just a bit too far from actuality. And how could the (metaphysical) impossibility of a state of affairs be a contingent and extrinsic feature of that state of affairs?

Although I continue to resonate to these intuitions, I am no longer sure how much weight they should be accorded. For the defender of Chandler could say that, however plausible it might be *ex ante* that what is possible does not depend on which possibilities are actualized, or that metaphysical impossibility is an intrinsic feature of a state of affairs, cases involving artefacts show—by appeal to widely shared and robust intuitions—that relative possibility is intransitive, and that both of the (initially plausible) principles just mentioned have counter-examples.

Indeed, two commonly made criticisms of Chandler can be played off against each other. It is often held to be intuitively obvious—or at least very plausible—that (where metaphysical possibility is at issue) whatever is possible is necessarily possible.[203] And it is often said that Chandler's solution to the bicycle problem is ad hoc, inasmuch as Chandler provides no independent reason for thinking that relative possibility is intransitive.[204] Chandler could reply that (i) the bicycle case and its congeners provide (tolerably) clear counter-examples to the transitivity of relative possibility (sufficiently motivating the denial of transitivity), and (ii) counter-examples to the transitivity of relative possibility are relatively hard to come by—which explains the initial appeal of principles such as 'whatever is possible is necessarily possible' (and the related principles discussed in the last paragraph). An analogy: many non-philosophers, and even some philosophers, are initially disposed to regard as intuitively obvious the claim that, if this material object and that material object are in the same place at the same time, then this material object and that material object are identical.[205] But most (contemporary analytic) philosophers deny the claim, and explain its initial appeal in terms

---

[203] See e.g. Plantinga, *The Nature of Necessity*, 51–5.

[204] See Gupta, *The Logic of Common Nouns*, 101.

[205] I put the principle this way, rather than in terms of the impossibility of having two material objects at the same place at the same time, to avoid complications concerning whether we are counting by identity or some other relation.

of our failure to think initially of the kind of cases that are counter-examples
to it. It is open to Chandler to make the same kind of move with respect to
the intuitive appeal of the principles incompatible with his treatment of the
bicycle case.

Here the champion of counterparts might object:

> As Kripke notes, the same sorts of difficulties concerning vagueness and
> intransitivity arise for identity through time that arise for identity across
> worlds: a chain of apparent identities can lead to an apparent
> non-identity. So, for example, a minuscule deformation of the bit of clay
> a statue is made of will result in the (minuscule) deformation of the
> statue, and not in the statue's demise. But a long enough series of minus-
> cule deformations will result in the statue's demise and its replacement by
> something else.[206] How is this possible? We cannot answer this question
> along Chandler's lines (since the temporal analogue of the relative pos-
> sibility relation (the earlier-than-or-simultaneous-with-or-later-than
> relation) is unquestionably transitive. Thus Chandler's approach is un-
> satisfactory, because it does not provide a general solution to a problem
> that comes in both a modal and a temporal version.

It is true that Chandler's approach cannot be brought to bear on the (superficially similar) statue problem. But what would an account providing a
uniform solution to the modal and temporal problems look like? One such
account would hold that identity through time is genuine identity, and avoid
the paradox by maintaining that some of the relevant alleged (cross-time)
identities are false. On this approach, in the statue case described above we
deny that, for all times t and $t + \epsilon$, if the clay statue exists at t, and the clay the
statue is made of still exists at $t + \epsilon$, and is only minimally deformed with
respect to its shape at t, then the statue existing at $t =$ the statue existing at
$t + \epsilon$. At some point, when the statue gets too far away from its original shape,
a slight deformation of the clay it is made of becomes fatal to it. On an alternative account, we would say that an identity statement may be completely
true, or completely false, or something in between. And we would say that
some of the relevant identities are less than completely true: at some point a
slight deformation results in its no longer being completely true that the statue
survives.

If we treat the temporal case along these lines, a parallel treatment of the
modal case would look like this: identity across worlds is genuine identity, and
some of the alleged (cross-world) identities are false. We deny that, for any
worlds $\mathbf{H}$ and $\mathbf{H}'$, if the bicycle existing at $\mathbf{H}$ differs from the bicycle existing
at $\mathbf{H}'$ only with respect to one small (original) part, then the bicycle at $\mathbf{H} =$ the

---

[206] I have switched from bicycles to statues because it is by no means clear that a bicycle could not
survive the total replacement of all its original small parts, even if it could not have been made from a
completely different set of small parts.

bicycle at H'. Alternatively, we say that identities may be completely true, completely false, or something in between, and we deny that for any worlds H and H', if the bicycle existing at H differs from the bicycle existing at H' only with respect to one small part, then it is completely true that the bicycle at H = the bicycle at H'.

I don't find either of these uniform treatments of the modal and temporal problems especially appealing: I am inclined to think the two problems, in spite of their apparent similarity, have different solutions. But, for present purposes, what matters is that the defender of counterpart theory cannot champion either of them, inasmuch as they presuppose that identity across worlds is genuine identity. If the counterpart theorist wants a uniform treatment of the temporal and modal cases, then she is going to need to suppose that there is a cross-time counterpart relation, as well as a cross-world counterpart relation, and that the cross-time counterpart relation, like the cross-world counterpart relation, is non-transitive. On this approach, in the statue case we will have a series of times $t_0 \ldots t_n$, and a series of individuals $i_0 \ldots i_n$, where i_0 is a statue, each i_{j+1} is the individual that i_j will be (when t_{j+1} is present), but i_n is *not* the individual that i_0 will be (when t_n is present). The trouble is that, on this approach, it won't in general be true that, if it will be that it will be that . . . this is F, then it will be that this is F. (For it won't follow from the fact that this thing has a (future) 'temporal counterpart' that has a (future) temporal counterpart . . . that is F, that this thing has a (future) temporal counterpart that is F.) Surely, though, if a statue is going to be going to be spherical (or going to be going to be going to be spherical, or going to be going to be going to be . . . spherical), then it is going to be spherical.

To put essentially the same point in a slightly different way, treating the bicycle case and the statue case in a uniform counterpart-theoretic way is unappealing, because whether or not the would-have-been relation is transitive, the will-be relation certainly is. (It is often objected to (unsophisticated) memory theories of personal identity that, on such theories, it can happen that P_3 at t_3 is the person that P_2 at t_2 will be, and P_2 at t_2 is the person P_1 at t_1 will be, but P_3 at t_3 is not the person that P_1 at t_1 will be. Few philosophers would be tempted to defend the memory theory on the grounds that the will-be relation is intransitive.)

The moral is that the counterpart theorist is in no position to impugn the adequacy of Chandler's solution to the bicycle problem on the grounds that what is wanted is a solution that can also solve the statue problem.

Having considered four lines of argument for counterpart theory, we can take stock. Kripke has argued (to my mind, persuasively) that an identitarian account of modal predication is the 'default' option, in the absence of considerations favouring a counterpart-theoretic one. This rather modest conclusion could be granted by a champion of counterpart theory who thought that there were in fact considerations that overcame the initial presumption in favour of an identitarian account of modal predication. Moreover, although Kripke sometimes speaks as though only confusion would induce one to prefer a counterpart-the-oretic account of modal predication to an identitarian one (see, for example, 'N & N', n. 13), on at least one occasion he recognizes that there might actually be weighty considerations favoring a counterpart-theoretic account of modal predication over an identitarian one (cf. 'N & N', n. 18). It seems, though, that none of the considerations usually adduced in favour of a counterpart-theoretic account of modal predication are likely to persuade, or should persuade, someone initially impressed by the simplicity, straightforwardness, and initial 'intuitivity' of an identitarian account that a counterpart-theoretic account is, all things considered, preferable to it.

3, PERSISTING OBJECTS AND IDENTITY THROUGH TIME

In 1978 Kripke gave a series of lectures on time and identity at Cornell University. (He also presented some of the central themes of these lectures elsewhere, for example in 'Identity through Time', at the 1979 American Philosophical Association conference.) It is regrettable that a series of lectures so packed with important and illuminating ideas has never been published and has exerted influence only by dint of being (much) discussed. In what follows, I shall touch upon two (well-known) themes from those lectures.

Naming and Necessity challenged certain (at the time, widely held) views about trans-world individuals and trans-world identity. Kripke's Cornell lectures similarly challenged certain (at the time, widely held) views about what we might call 'trans-time' individuals and 'trans-time' identity. In particular, Kripke targets two doctrines. The first is:

'Trans-time' (i.e. persisting) objects are non-basic. The more basic things they are 'built up from' are things that exist *at* a time, but not *through* time—things that Quine calls 'momentary objects', and philosophers often call 'temporal stages' or 'time-slices' of objects. Momentary objects do not 'coincide with' (fill exactly the same region of space at the same time as) any object but themselves. A persisting object is a particular kind of construct or 'assemblage' from momentary objects. It is something like a set of momentary objects, or a mereological aggregate of momentary objects, or perhaps a (possibly partial) function from moments of time to momentary objects. At any rate, for each persisting object, there is a corresponding set of momentary objects (the set of 'slices' or 'stages' of that object), and persisting objects that have all the same slices or stages are the same object.

The doctrine just sketched is set out clearly (and considered sympathetically) by Richard Montague:

It is perfectly possible to construct an ontology allowing for physical objects of different sorts, objects that may coincide without being identical. The construction...corresponds, I believe, to Hume's outlook. Let us for present purposes suppose that our *basic objects* have no temporal duration; each of them existing only for a moment; they are accordingly what we might regard as temporal slices of 'ordinary' objects, and might include such physical slices as heaps of molecules at a moment...Then *ordinary objects* or *continuants* (for instance, physical objects, persons) would be constructs obtained by 'stringing together' various basic objects— or more exactly, certain functions from moments of time to basic objects.[207]

David Lewis also endorses something very like the doctrine under discussion, although he disavows any claim that persisting objects are 'constructs' of 'more basic' stages, and doesn't insist on the momentariness of the stages of persisting objects.[208]

In arguing against the doctrine, Kripke asks us to consider a case we have already discussed in previous chapters—the case of a plant that never grows past its stem. The plant and the stem occupy all the same places at all the same times (forgetting, for the sake of the simplicity of the example, the plant's roots). But the plant is not the stem, inasmuch it is true of the plant, but not of the stem, that it might have grown past the stem (might have had parts that extended beyond the stem). The view of persisting objects sketched above cannot accommodate the distinctness of the plant from the stem. After all, if momentary stages can differ only if they do not coincide, and persisting objects can differ only if they don't have all the same momentary stages, it follows that persisting objects can differ only if they do not permanently coincide (even if, as Montague and Lewis suppose, they can differ, in spite of temporarily coinciding).

Given the 'identitarian' account of modal predication it presupposes, Kripke's argument is clearly successful. Incidentally, suppose—as seems to be the case—that events occupying all the same places at all the same times can nevertheless differ with respect to their modal properties.[209] Then an argument analogous to Kripke's can be used to show that, if distinct momentary events do not coincide, then 'time-consuming' (that is, temporally extended) events are not 'assemblages' (aggregates, sets, or what have you) of momentary events.

Defenders of the doctrine Kripke opposes have typically responded to Kripke's sort of argument by rejecting the identitarian account of modal predication it presupposes, and explaining how an alternative account allows one to say both that distinct stages never coincide, and that distinct persisting

---

[207] See Montague, 'The Proper Treatment of Mass Terms in English', 176–7. Montague does not explicitly say in this passage that coincident momentary objects are identical. But in the piece from which the passage is taken his standard example of a momentary object is what he calls a *homaam*— that is, a heap-of-molecules-at-a-moment. And presumably any two homaams that occupy the same place do so at different times.

[208] See Lewis, postscript B to 'Survival and Identity', in *Philosophical Papers*, i. 77. As we shall see, Lewis's version of the doctrine is as vulnerable to Kripke's objection as Montague's, inasmuch as it incorporates the 'same slices, same continuant' assumption. Cf. also Gibbard, 'Contingent Identity'.

[209] See my 'Gli eventi e i loro criteri di identità', in C. Bianchi and A. Bottani, *Significato e Ontologia* (Milan: FrancoAngeli, 2003).

objects never have all the same stages.[210] Since I have already discussed the relative merits of the identitarian account of modal predication and its rivals (in the previous section of this chapter), I shall not say more about that here.

Suppose, though, that someone accepts the identitarian account of modal predication. She might still attempt to hold onto some of the doctrine Kripke attacks, either by giving up the claim that momentary objects coincide with no objects but themselves, or by giving up the claim that persisting objects that differ always differ with respect to their stages.

On the first line, one would distinguish, say, a particular marble-statue-at-a-moment from the bit-of-marble-at-moment that coincides with it, perhaps on the grounds that the statue-at-the-moment, unlike the bit-of-marble-at-the-moment, is essentially an artefact: if the sculptor had not chipped away the (persisting) marble that originally surrounded the (persisting) bit of marble constituting the (persisting) marble statue, the bit-of-marble-at-the-moment would still have existed, but the statue-at-the-moment would not have existed, any more than the persisting statue would have. Having allowed that distinct momentary objects may coincide, one could continue to maintain that persisting objects are 'assemblages' of momentary objects, so that different persisting objects always differ with respect to their momentary stages.

I don't think this will work. For an ordinary persisting object that actually had a certain lifespan might have had a shorter or longer lifespan. So, presuming persisting objects have temporal stages, the same persisting object could have had some but not all of the stages it actually had, or could have had all the stages it actually had, and more; it is not the particular persisting object it is, in virtue of having exactly the stages it actually has. But whether we think of an 'assemblage' of momentary objects as a set of momentary objects, or a mereological aggregate of momentary objects, or a function from times to momentary objects, it seems that an assemblage of momentary objects *is* the particular assemblage it is, in virtue of having precisely the momentary objects it has. So, even if a persisting object has temporal stages, it will be discernible with respect to some modal property from the assemblage of its stages, and will accordingly be distinct from its stages (assuming, as we are, an identitarian account of modal predication).

Someone might attempt to resist this argument by maintaining that persisting objects are assemblages of momentary objects, but those assemblages are not essentially assemblages of exactly the things they are actually assemblages of. I don't think there is much mileage in this. Suppose we say that a persisting object O is a (mereological) aggregate of momentary objects A that might have comprised fewer momentary objects—say, would have comprised no last Thursday momentary objects if it had gone out of existence at the end

---

[210] See (for starters) Lewis's 'Counterparts of Persons and Their Bodies'; Gibbard, 'Contingent Identity'; Gupta, *The Logic of Common Nouns*; and Noonan, 'The Closest Continuer Theory of Identity'.

of last Wednesday, instead of going out of existence at the end of last Thursday, as it actually did. Then (if we are identitarians) we have to say that in an alternative world **H** A exists, and has (at **H**) all of A's actual momentary parts except its last Thursday momentary parts. The difficulty is that in the actual world **G** there is not just A, but an aggregate of momentary objects we may call A minus, which actually has all the parts A actually has, except for A's last Thursday parts. A minus is obviously different from A, since A minus (actually) has fewer parts than A (actually) does. Moreover, A minus exists, not just in **G**, but also in **H**. If, however, both A and A minus exist in **H**, then in **H** we have two different (mereological) aggregates with all the same parts. And surely different aggregates cannot have the same parts in the same possible world, any more than different sets can have the same elements in the same possible world, or different functions can have the same arguments and values in the same possible world. So, we must conclude, even if in **H** the persisting object O has all the same temporal stages it actually has, except for the last Thursday stages it actually has, we cannot say that in **H** the aggregate A has all the same temporal stages it actually has, except for the last Thursday stages it actually has. We cannot say that, because we know that something else existing in **H** (to wit, A minus) has all the temporal stages (in **H**) that A actually has, except for the last Thursday ones, and we know that there is only one aggregate existing in **H** that has (in **H**) all the temporal stages A actually has, except for the last Thursday ones. We can run the same argument whether a persisting object is identified with a mereological aggregate, or a set, or a partial function, or some other kind of assemblage: for assemblages just are the kinds of things that can differ only if they differ with respect to their 'assemblands'.

The moral is that—at least as long as we are identitarians—we cannot identify persisting objects with assemblages of momentary objects, whether or not we assume that distinct momentary objects never coincide. *Mutatis mutandis*, the same goes for time-consuming events. Take Jaegwon Kim's account of events, according to which a (one-place) event consists of an individual's having a property at a time.[211] Such an account of events would clearly allow distinct momentary events (say, events indiscernible with respect to their 'constituent individual' and their 'constituent moment', but discernible with respect to their 'constituent property') to coincide. Someone who had a Kimian conception of momentary events might suppose that the basic events were momentary events, and that time-consuming events were assemblages of momentary events. Her view could not be refuted by appeal to an (event-involving) analogue of the stem–plant case, since that view does not require distinct time-consuming events to differ with respect to where or when they happen. If, however, we make the (intuitively plausible) assumption that a time-consuming event might have gone on for a bit longer, or a bit less long,

---

[211] Cf. J. Kim, 'Events as Property Exemplifications', in M. Brand and D. Walton (eds.), *Action Theory* (Dordrecht: D. Reidel, 1975).   .

than it actually did, then we shall have to deny that time-consuming events are assemblages of momentary events, even if distinct momentary events can coincide.

This leaves the identitarian who wants to save some of the Hume–Montague doctrine with the second of the fallback positions mentioned above: she can give up the claim that different persisting objects always differ with respect to their momentary parts. This would involve saying that persisting objects are wholes whose parts are momentary objects, but not wholes of a sort that are (merely) the sum of their momentary parts. (For, if objects were simply the sums of their momentary parts, there would be no way to get two different objects out of one bunch of momentary parts.)

Fallow deer are members of the set of all the fallow deer there ever were or are or will be. They are also members of the species *Dama dama*. It seems natural to think of both the set and the species as wholes whose parts are deer. It seems, though, that the set cannot be identified with the species.[212] The set is the particular set it is, because it has the members it does; it accordingly could not have had different members than it actually has. The species, on the other hand, is not the particular species that it is because it has the members that it does: *Dama dama* would have existed as long as there had been fallow deer, even if some or many or (perhaps) all of the fallow deer that actually exist (or existed, or will exist) hadn't existed.

It might be that a persisting object was to the corresponding assemblage of its momentary stages as the species of fallow deer is to the set of fallow deer. In each case we might have different wholes of the same parts—one a whole whose existence and identity was at least partly independent of the existence and identity of its parts, and one a whole whose existence and identity was completely dependent upon the existence and identity of its parts.

Suppose someone wanted to defend a chastened version of the Hume–Montague doctrine, according to which persisting objects are wholes, but not mere sums, of momentary objects. She should admit that the analogy between persisting species and objects is in one way imperfect. Different persisting objects (she supposes) can actually have all the same momentary parts, but different species presumably cannot actually have all the same 'animal parts'. Perhaps taxa would provide her with a better analogue than species. After all, it seems that different taxa could have all the same animal parts, if one were subordinate to the other. (Even if the only mammals were guinea pigs, *Cavia cavia* would still be different from *mammal*. For it would be true of *mammal*, but not of *Cavia cavia*, that it could have had as members members of *Mesocricetus auratus*.)

Many philosophers think that persisting objects are mere sums of their temporal parts, and many philosophers deny that persisting objects so much

---

[212] At least, the species and the set cannot be identified by someone who accepts an identitarian account of modal predication.

as have temporal parts.[213] As far as I know, the view that (ordinary) persisting objects have, but are not mere sums of, their temporal parts is not a popular one. But it is not as though one can't imagine how it might be motivated. The analogous view about events—that they have, but are not mere sums of, temporal parts, is attractive, given that events seem to have temporal parts, without having exactly the temporal parts they have essentially.[214] To be sure, it is far less controversial to suppose that, say, there is such a thing as the first half of the great depression than it is to suppose that there is such a thing as the first half of, say, the sun. But various philosophers have argued that, just as we need to suppose that a spatially extended object has spatial parts to explain how there can be 'local variation' in the intrinsic properties of an object (i.e. to explain how one and the same object can be, say, smooth here and bumpy there), we need to suppose that 'temporally extended' objects have temporal parts to explain how there can be 'temporal variation' (change) in the intrinsic properties of an object.[215] If such arguments work, and an identitarian account of modal predication is right, then (ordinary) persisting objects have, but are not the mere sum of, their temporal parts, in just the way that (ordinary) spatially extended objects have, but are not the mere sum of, their spatial parts. And there is no obvious reason to think that the argument that we need to appeal to temporal parts to explain the possibility of intrinsic change works only if some non-identitarian account of modal predication is right.

Suppose that, in the face of the plant–stem argument, a proponent of the Hume–Montague doctrine holds onto the claim that persisting objects are wholes of momentary parts, but gives us the claim that persisting objects are assemblages of their momentary parts. Must she then give up the idea that persisting objects are (in some sense) 'non-basic' and momentary objects are 'basic'? Not obviously. In *Naming and Necessity* Kripke asks:

Does the 'problem' of 'trans-world identification' make any sense? Is it *simply* a pseudo-problem? The following, it seems to me, can be said for it. Although the statement that England fought Germany in 1943 perhaps cannot be *reduced* to any statement about individuals, nevertheless in some sense it is not a fact 'over and above' the collection of all facts about persons, and their behavior over history. The sense in which facts about

---

[213] Proponents of the idea that objects have temporal parts include Armstrong, Lewis, and Heller; opponents include Thompson, van Inwagen, and Merricks.

[214] Admittedly, that events have temporal parts seems less controvertible than that events have momentary (instantaneous) temporal parts: there is a tradition according to which momentary events are not real parts of time-consuming events. Compare: Someone might think that objects have (three-dimensionally) extended spatial parts, but not one- or two-dimensional parts: the Equator, she might say, is not a genuine *part* of the earth, but rather a boundary between genuine parts of the earth. In the same way, momentary events might not be genuine parts of (temporally extended, time-consuming) events; they might instead be boundaries between different genuine parts of a time-consuming event, or between different time-consuming events, or between a time-consuming event and its (temporal) 'surroundings'.

[215] See e.g. David Lewis's discussion of intrinsic change in his *On The Plurality of Worlds*.

nations are not facts 'over and above' those about persons can be expressed in the observation that a description of the world mentioning all facts about persons but omitting those about nations can be a *complete* description, from which the facts about nations follow. Similarly, perhaps facts about material objects are not facts 'over and above' facts about their constituent molecules. We may ask, given a description of a non-actualized possible situation in terms of people, whether England still exists in that situation, or whether a certain nation (described, say, as the one where Jones lives) which would exist in that situation, is England. Similarly, given certain counterfactual vicissitudes in the history of the molecules of a table *T*, one may ask whether *T* would exist, in that situation, or whether a certain bunch of molecules, which in that situation would constitute a table, constitute the very same table *T*. In each case, we ask criteria of identity across possible worlds for certain particulars in terms of those for other, more 'basic' particulars. If statements about nations (or tribes) are not *reducible* to those about other more 'basic' constituents...we can hardly expect to give hard and fast identity criteria; nevertheless, in concrete cases we may be able to answer whether a certain bunch of molecules would still constitute *T*, though in some cases the answer may be indeterminate. I think similar remarks apply to the problem of identity over time; here too we are usually concerned with determinacy, the identity of a 'complex' particular in terms of more 'basic' ones. ('N & N', 271)

Here Kripke countenances a sense of 'basic' in which things of one kind may be more basic than things of another kind, even though things of the other kind are not mere sums or assemblages of things of the one kind. (Tables are obviously not assemblages or mere sums of their constituent molecules.)

In the passage just cited Kripke is open to the possibility that

(A) The (less basic) facts about tables are supervenient on or 'fixed by' the (more basic) facts about molecules: a description of the (physical) world that includes all and only the facts about molecules (and their parts), even though it does not mention facts about tables, is nevertheless a *complete* description, in the sense that it implies all the facts about tables (and all other physical objects).

The analogous claim about persisting objects and momentary objects would be:

(B) The (less basic) facts about persisting objects are supervenient on or 'fixed by' the (more basic) facts about momentary objects: a description of the world that includes all and only the facts about momentary objects, even though it does not mention facts about persisting objects, is nevertheless a *complete* description, in the sense that it implies all the facts about persisting objects.[216]

---

[216] Of course, on one understanding of 'all the facts about molecules', those facts will include the fact that such-and-such molecules constitute this table, and so on; on this understanding it makes no sense to speak of a description of the world that mentions all the facts about molecules but omits the facts about tables. In (A), 'all the facts about molecules' must be understood in a somewhat restrictive way— as including, say, only the facts about what molecules there are, what intrinsic properties they have, where and when they exist, and how they are related to each other—if facts about tables are to follow

In the Cornell lectures Kripke did not challenge (B). Instead he challenged

(C) The (less basic) facts about persisting objects are supervenient upon or 'fixed by' the (more basic) facts about the series of momentary 'holographic' states of the world: a description of the world that includes (just) a complete specification of the series of momentary holographic states, even though it does not mention persisting objects, is a *complete* description, in the sense that it implies all the facts about persisting objects, and, more generally, the whole of history. The history of the world is nothing 'over and above' the series of momentary holographic states of the world.

Here a momentary holographic state of the world is what would be captured by a description that does nothing but specify all the facts about how (purely) qualitative properties[217] are distributed in space at a given moment in history. It will, say, include that there is something small and furry here, something round and shiny there, and so on. But it will not include that there *will be* something small and furry here in two seconds, or that there *was* something round and shiny there two seconds earlier. Nor will it include that there is *this* particular small and furry thing here, or *that* particular round and shiny thing there.[218] Kripke calls states of this kind holographic because they correspond (roughly) to what could be captured by a three-dimensional picture of the universe at instant.

(B) and (C) are quite different ways of filling out the idea that facts about persisting objects are supervenient. (B) says that fixing all the facts about objects of a certain kind (momentary objects) suffices to fix all the facts about objects of a different kind (persisting objects). (C) says that fixing all the facts about the spatio-temporal location of qualitative properties (about which qualitative properties are instantiated where and when) suffices to specify a (complete) history of the world. (B) requires that there be momentary objects—or rather, requires that there be momentary objects if there are

---

from, but not be included in, the set of all the facts about molecules. If we want to leave room for the possibility that the facts about persisting objects will follow from, but be included in, the facts about momentary objects, we must understand 'the facts about momentary objects' in a correspondingly restrictive way.

[217] It might be better to say *local* qualitative properties (cf. Lewis's introduction to his *Philosophical Papers*. I omit this qualification because in the Cornell lectures Kripke did not bring in locality in any explicit way.

[218] In the Cornell lectures Kripke stressed on a number of occasions that a holographic state is 'purely qualitative' and can be captured by a 'qualitative instantaneous description of the world'—one without names or demonstratives for objects. That the description associated with a momentary holographic state will not include singular terms for objects is not always made clear in discussions of Kripke's views on identity through time. For example, in discussing Kripke, Shoemaker says that a momentary holographic state of the world is given by a maximal description of the world at an instant, 'where the description is not such as to imply the existence at any other moment of time of any of the things referred to or quantified over in it' (Shoemaker, 'Identity, Properties, and Causality', 327). This makes it sound as though the description associated with a momentary holographic state may not only quantify over but also refer to 'things' (objects).

persisting ones. (C) requires that there be holographic states if there is history, but it does not require that there be momentary objects if there are persisting ones. A proponent of (C) could perfectly consistently hold that persisting objects persist by being wholly present at more than one time, and there are no momentary objects. So someone could consistently accept (C) and reject (B). It also seems that someone could consistently accept (B) and reject (C), as long as she thought that different possible momentary individuals could have had the same qualities at the same time and place. Suppose you thought that (i) there is a possible world containing only God and the statue-at-a-moment he creates (constituted by this-matter-at-a-moment), and (ii) there is a different possible world containing only God and a different statue-at-a-moment he creates (constituted by a different bit of matter-at-a-moment), where the statue-at-the-moment in the first world and the statue-at-the-moment in the second world, though numerically (and 'materially') different, have exactly the same qualitative properties and spatio-temporal location. Then—even if you accepted (B)—you could (consistently) hold that sometimes possible worlds have different histories involving different momentary individuals, in spite of having the very same distribution of qualities through space and time.

In the Cornell lectures Kripke suggested that (C) at first sight appears to be true: it looks as though, given a complete knowledge of the series of momentary holographic states of the world, we could 'read off' the whole history of the world—and, in particular, all the true identity-across-time statements. Nevertheless, he argued, a certain thought-experiment shows that (C) is false.

It is not clear to me why Kripke thought that (C) initially appears to be true. Consider a possible world—call it H_1—containing nothing but one planet forever circling one star at a constant velocity. The planet, we may suppose, is at place p at time t, back at p in an hour's time, and so on. Isn't there a possible world H_2 in which that same planet forever circles that same sun in that same orbit, at that same velocity, but is exactly a half-orbit away from p at t, and $t + 1$ hour, and $t + 2$ hours, ... (as well as being exactly a half-orbit away from p at $t - 1$ hour, $t - 2$ hours, ...)? I don't see why there shouldn't be such a world. The existence of H_1 and H_2 is compatible with (C), since H_1 and H_2 differ with respect to the distribution of qualities through space and time (the qualities that are instantiated at p at t in H_1 are instantiated at a different place p' (half an orbit away) at t in H_2). But now consider a possible world H_3 containing nothing but two qualitatively indiscernible co-orbital planets forever circling the same star at a constant velocity. Suppose that the two planets are forever π radians (180 degrees) away from each other. If there is a possible world H_2 differing from H_1 in the way described above, then there should also be a possible world H_4 differing from H_3 in the same way—a world where, instead of having *this* planet at place p at ... 9.00, 10.00, 11.00, ..., and *that* planet at place p at ... 9.30, 10.30, 11.30, ... (as happens in H_3), we have *that* planet at place p at ... 9.00, 10.00, 11.00, ..., and *this* planet at place p

at ... 9.30, 10.30, 11.30, ... And if there is such a pair of possible worlds as H_3 and H_4, then there are possible worlds that have different histories, even though they are indiscernible with respect to the distribution of qualitative properties in space and time.

Here is a different sort of case that casts doubt on (C). In the last section we discussed a case of Shoemaker's, in which what looks like persistence turns out not to be. A deity decrees that a tablet will cease to exist at the end of Thursday (its first moment of non-existence will be midnight), and that a new tablet will come into existence, on the very same spot, at precisely midnight. The new tablet is a perfect qualitative duplicate of the old tablet. Thus we have what looks like a case of persistence but is actually a case of replacement. By varying the case, we can raise a different worry about (C).

Suppose there is a possible world—call it H—containing only one persisting object (say, a rock) that pops into existence without a cause, continues to exist for ten minutes, and then pops out of existence, again without a cause. Then, it would seem, there is another possible world H' differing from H only in that, in H', the rock pops out of existence at the end of the fourth minute rather than at the end of the ninth minute. And, it would seem, there is yet another possible world H'' which differs from H only in that the rock that exists in H pops out of existence at the end of the fourth minute, and a new, but qualitatively indiscernible rock pops into existence at the beginning of the fifth minute, and continues to exist for the next five minutes. If there is such a pair of possible worlds as $\{H, H''\}$, then (C) is false, since H and H'' have the same distribution of qualities but different histories.

Setting aside particular (putative) counter-examples to (C), I don't see why one should think that (C) initially appears to be true. To my mind, there is nothing obvious, or even obviously plausible, about the claim that a given actual (or, for that matter, possible) individual with a certain set of qualitative properties, and a certain spatio-temporal location, is the *only possible* individual with those properties and that location. Even if that claim is in fact true, and even if it doesn't initially look false, I don't think it initially looks true. If, however, more than one possible individual could have had the same qualitative properties and spatio-temporal location, there is no obvious reason to think that possible worlds couldn't be indiscernible with respect to the distribution of qualities through space and time but discernible with respect to which individuals exist in them.[219]

---

[219] Unless one accepts a certain sort of reductive theory (e.g. one on which objects just are (partial) functions from moments of times to equivalence classes of compresent tropes), I don't see why one should think that a particular individual is the particular individual it is, merely in virtue of having the spatio-temporal location and qualities it has. See my 'Omniscience, Negative Existentials, and Cosmic Luck'. Kripke rejects the sort of reductive theory at issue: 'I ... deny that a particular is nothing but a "bundle of qualities", whatever that may mean. If a quality is an abstract object, a bundle of qualities is an object of an even higher degree of abstraction, not a particular' ('N & N', 272).

In the preface to the revised edition of *Naming and Necessity* (which appeared after he gave the Cornell lectures) Kripke says this about the existence or otherwise of qualitatively indiscernible but distinct possible worlds:

With respect to possible states of the entire world, I do not mean to assert that there are qualitatively identical but distinct (counterfactual) states. What I do assert is that *if* there is a philosophical argument excluding qualitatively identical but distinct worlds, it cannot be based simply on the supposition that worlds must be stipulated purely qualitatively. What I defend is the *propriety* of giving possible worlds in terms of certain particulars as well as qualitatively, whether or not there are in fact qualitatively identical but distinct worlds. (*N & N*, preface, n. 17)

Kripke seems to suggest here that he does not regard the principle of the identity of qualitatively indiscernible possible worlds as one we should accept in the absence of arguments against it, in virtue of its initial plausibility; in order reasonably to accept it, we would have to be in possession of a good philosophical argument in its favour. And, Kripke seems to be saying, although he knows of no such argument, he would not want to rule out that there is one; he accordingly does not pronounce on whether qualitatively indiscernible possible worlds are ever distinct.

As we have seen, in the Cornell lectures, Kripke suggested that, although (C) looks true (it looks on the face of it as though there is nothing more to history than a series of momentary holographic states), it surprisingly turns out that there are good reasons to think it is false. I am not sure, though, how to fit these remarks together with Kripke's just cited remarks on the principle of the identity of qualitatively indiscernible possible worlds. Although Kripke does not say wherein the qualitative indiscernibility of worlds consists, I take it that he understands qualitative indiscernibility for worlds in such a way that possible worlds are qualitatively indiscernible if and only if they are indiscernible with respect to the distribution of qualities through space and time.[220] If that is so, then the holographic supervenience thesis (C) and the principle of the identity of qualitatively indiscernible worlds seem to come to much the same thing. In that case the principle has as much initial plausibility as the thesis, and an argument against the thesis is an argument against the principle.

Of course, if one construed momentary holographic states in such a way that they included information about which things have which qualitative properties at that moment, then the thesis that there is nothing more to the history of the world than the series of momentary holographic states of the world would be both different from, and initially more plausible than, the

---

[220] If worlds could be qualitatively indiscernible, in spite of the fact that, say, in one world charge was instantiated here and now, and in the other, charge was not instantiated here and now, the principle of the identity of qualitatively indiscernible possible worlds would be an obvious falsehood. And Kripke does not take the principle of the identity of qualitatively indiscernible possible worlds to be an obvious falsehood.

principle of the identity of qualitatively indiscernible possible worlds. But if we think of momentary holographic states as including information about which *persisting* things have which qualities (so that the momentary holographic state for five o'clock on 27 March 1972 will include Nixon's having a five-o'clock shadow), then the facts about cross-time identities *will* supervene on the facts about the momentary holographic states, which is precisely what Kripke wants to argue against. If, in order to avoid this difficulty, we think of momentary holographic states as including information about which *momentary* things have which qualities (so that the momentary holographic state for five o'clock on 27 March 1972 includes this human momentary stage's having a five o'clock shadow), then belief in momentary holographic states commits us to belief in momentary objects. And Kripke's intention, in the Cornell lectures, seemed to be to introduce a notion of momentary holographic state that did not commit one to the existence of momentary objects.

So I am not sure what to think. The holographic supervenience thesis Kripke discussed in the Cornell lectures seems to be very close to the principle of the identity of qualitatively indiscernible worlds.[221] Moreover, as we shall see (in note 222), Kripke's main argument against the holographic supervenience thesis works only if a holographic state can be (completely) specified solely in terms of which qualities are instantiated where at the relevant moment. Thus Kripke's main argument against the holographic supervenience thesis works only if holographic states are understood in such a way as to make the holographic supervenience thesis and the principle of the identity of qualitatively indiscernible worlds substantially the same. Nevertheless, Kripke's take on the thesis (in the Cornell lectures) seemed very different from his take on the principle (in the preface to the revised edition of *Naming and Necessity*). In the Cornell lectures he seems to suppose both that the thesis is initially plausible and that there are cogent arguments against it; in the preface to the revised edition of *Naming and Necessity* he seems to make neither of those assumptions about the principle.

Perhaps the best-known argument from Kripke's Cornell lectures is the argument to the effect that a plurality of possible histories can correspond to the same series of momentary holographic states. Kripke asks us to consider a disk made of uniform and continuous matter. The disk might be rotating, or it might not be rotating. Either way, Kripke maintains, the series of momentary holographic states for the disk could be exactly the same. So different possible histories (in some but not all of which a disk is rotating) can correspond to the same series of holographic states of the world (just imagine that there is

---

[221] If somehow qualities could be distributed not just through time and space but also through some other 'dimension', then perhaps possible worlds could be indiscernible with respect to the distribution of qualities through space and time, without being (qualitatively) indiscernible *sans phrase*.

nothing to the universe except the disk, or that we are considering pairs of possible worlds that differ only with respect to the rotation or otherwise of the disk).[222]

At this point someone might object (indeed, when Kripke gave the lectures at Cornell, some members of the audience did object) that the holographic states of the rotating and non-rotating disks will be different, inasmuch as, at any instant, the disks will have different instantaneous angular velocities. Kripke responded that an instantaneous velocity is not the sort of property whose exemplification will be included in a holographic state. As he put it, nothing about the way the world is at an instant will tell us whether or not a thing is moving then, or what direction it is moving in; the concept of a (linear or angular) instantaneous velocity essentially involves reference to where the thing is at other times during some stretch of time, however small. Since an

---

[222] Notice that this argument is cogent only if momentary holographic states can be specified in purely qualitative (non 'object-involving') terms. If you couldn't completely specify a momentary holographic state without saying that such-and-such qualities were instantiated by *this* (persisting) part of the disk at this time, there would be no reason to think that the series of momentary holographic states for the rotating disk is the same as the series of momentary holographic states for the stationary disk. (Indeed there would be reason to think the opposite.) If you couldn't completely specify a momentary holographic state without saying that such-and-such qualities were instantiated by *this* momentary part of this momentary disk, there would again be no evident reason to think that the series of momentary holographic states for the rotating disk is the same as the series of momentary holographic states for the non-rotating disk. Suppose that, instead of Kripke's disk, we had a disk made partly of gold and partly of silver. And suppose that the momentary part of the momentary disk occupying place p at time t in world H (where the disk revolves) was silver, while the momentary part of the momentary disk occupying place p at time t in world H' (where the disk does not revolve) was gold. Someone might say that the momentary part of the momentary disk that is at p and t in H and the momentary part of the momentary disk that is at p and t in H' are not merely qualitatively discernible (in the sense that the former has different properties in H than the latter has in H') but also numerically distinct. (This would involve supposing that momentary individuals in different possible worlds can have the same spatio-temporal location (in their respective worlds) without being identical.) In the case discussed by Kripke, where the disk is made of uniform, continuous matter, and revolves in H but not in H', someone might likewise say that the momentary part of the momentary disk that is at p at t in H is numerically different from the momentary part of the momentary disk that is at p at t in H', even though the momentary part of the momentary disk that is at p at t in H and the momentary part of the momentary disk that is at p at t in H' are qualitatively indiscernible (in the sense that the former has exactly the same qualitative properties in H that the latter has in H'). (This would involve supposing that momentary individuals in different possible worlds can have the same qualities, as well as the same spatio-temporal location (in their respective worlds) without being identical.) Thus someone could maintain that in Kripke's two possible worlds we have different series of holographic states as well as different histories: the 't-member' of one series (that is, the member of the series of holographic states that is the holographic state of the world at time t) differs from the t-member of the other series with respect to which momentary individual has the qualities instantiated in place p. If it is objected here that momentary individuals in different possible worlds *couldn't* have the same qualities and spatio-temporal location (in their respective worlds) without being identical, that is tantamount to saying that we *can* after all uniquely specify a momentary holographic state without reference to momentary individuals, simply by saying which qualitative properties are instantiated where at that moment. The upshot is that Kripke's disk argument against the holographic supervenience thesis works only if (as Kripke supposes) holographic states are specifiable in non-object-involving terms.

instantaneous velocity is, as it were, a (partly) 'other-timely' property, we can't read off from a holographic state whether or not something has it.

Actually, it seems to me there is room for doubt about whether the instantaneous velocity of an object can be read off from a holographic state. The idea that it obviously cannot might rest on the view that an object's having a given instantaneous velocity at t entails its being elsewhere at $t - \epsilon$ and/or $t + \epsilon$. Bigelow and Pargetter, however, have argued that there is no such entailment.[223] They imagine a situation involving three Newtonian rigid spheres of equal mass, a, b, and c. Sphere a moves with velocity **v** along a line joining the centres of spheres b and c, which are at rest and in contact with each other. Newtonian mechanics tells us that, when a collides with b, the velocity of a will be (instantaneously) handed on to b, and thence to c: so there will be an instant t at which the middle sphere b has velocity **v**, even though it won't be anywhere else at either $t - \epsilon$ or $t + \epsilon$.

Suppose it is granted that, in the three-sphere example, the middle sphere b has a velocity **v** for (just) one instant, even though it isn't anywhere else at $t - \epsilon$ or $t + \epsilon$. Still, it might be said, if the middle sphere b has velocity **v** at that instant, the following counterfactual will be true: were c not there to instantaneously 'absorb' b's velocity, b would be somewhere else at $t + \epsilon$. And, it might be added, there is a logical or conceptual link between b's having the instantaneous velocity it does at the relevant time, and the truth of certain contingent counterfactuals concerning where b would be at $t + \epsilon$. If there is such a link, then the instantaneous velocity of an object is presumably not a property that can be captured by a holographic state.[224]

In arguing that there is no entailment between something's having an instantaneous velocity at a time and its being somewhere else earlier and/or later, Bigelow and Pargetter do not simply rely on hypothetical counterexamples. They also appeal to the idea that a thing's instantaneous velocity at t can explain its being somewhere else at $t + \epsilon$: 'An object will be, say, a little higher a moment from now *because* it is now moving upwards. It will be a little higher because it is *now* moving upwards.... The first-order properties of position are explained by another first-order property of instantaneous velocity'.[225] One can see how this might be so, if instantaneous velocity is what Bigelow and Pargetter call 'a genuine intrinsic property of an object at a time...a further property over and above an object's position, history, and destiny'.[226] But it is unclear how (present) instantaneous velocity can explain (future) change in position, if change in position is, as a matter of conceptual necessity, constitutive of instantaneous velocity.

---

[223] Bigelow and Pargetter, *Science and Necessity*, 67.

[224] Katherine Hawley makes this point in 'Persistence and Non-supervenient Relations', *Mind*, 108 (1999), 58–9.   [225] Bigelow and Pargetter, *Science and Necessity*, 66.

[226] Ibid. 71 and 64.

Now, it is at least arguable that instantaneous velocity can explain counter-factual change in position, just as much as actual change in position. Sphere b would be in a different place at $t + \epsilon$ from the place it is in at t, if not for the velocity-absorbing powers of sphere c. It is at least arguable that this is true because (is explained by the fact that) b has velocity $\mathbf{v}$ at t. (Compare: A certain medicine containing a large amount of arsenic is given to two people, only one of whom has built up a tolerance to arsenic. That the medicine contains a large amount arsenic explains why the one person dies, and also why the other person would have died had she not built up a tolerance to arsenic.)

Suppose, then, that someone thinks the instantaneous velocity of an object cannot be read off from a holographic state, because it essentially involves reference to where that object would under certain circumstances be at other times. She needs either to explain how this is compatible with the idea that instantaneous velocity explains (actual or counterfactual) change in position, or to explain away the appearance that instantaneous velocity explains such changes in position. On the other hand, someone who thinks that instanta-neous velocity can be read off from a holographic state needs to say something about what an object's having an instantaneous velocity could consist in, if it entails nothing contingent about where that object is at other times or would under certain circumstances be at other times.[227]

In fact, Kripke is not unaware of the sort of considerations that might make the Bigelow–Pargetter view of instantaneous motion attractive. At Cornell, in his discussion of Zeno's paradox of the arrow, Kripke noted that we have a deep-set intuition that the past makes a difference to the future only in so far as it makes a difference to the present. As he strikingly put it, if the past were somehow miraculously changed but the present were left (intrinsically) just as is, that change would make no difference to the future. To put the point epistemically, if we wanted to find out how the future will go, and we knew all there was to know about the present, nothing we might subsequently learn about the past would bring to light any new factors influencing how the future will go that had to be factored in before predicting how the future will go, since nothing that happened in the past makes any 'marginal' difference to the future over and above the difference it makes to the present.

Of course, this does not imply that we can always figure out how the future will go on the basis of our knowledge of the present (and our knowledge of the laws). In favourable cases, though, it seems that we can. Suppose two (isolated) spheres are about to collide. If we know that the spheres are isolated, and we know the (current) momenta of the spheres and some mechanics, it seems that we can ascertain what will happen when the spheres collide, without making

---

[227] The quandary about how to think of the relation between facts about the motion of an object and facts about its actual or counterfactual location is of a kind that arises for other fundamental physical properties. On the one hand, there is an inclination to say that there is a conceptual connection between mass and resistance to acceleration; on the other, there is also an inclination to say that more massive objects are harder to accelerate *because* they are more massive.

assumptions about either the past or the future. For example, we needn't make any assumptions about what the momenta of the spheres were five seconds ago, or what the momenta of the spheres will be five seconds hence.

But is this right? As Kripke emphasizes, we cannot tell what will happen when the spheres collide, unless we have their (current) momenta, and hence their (current) instantaneous velocities. If current instantaneous velocity is a (partly) 'other-timely' property, then the fact that the spheres have such-and-such instantaneous velocities is a (partly) 'other-timely' fact. In other words, it is not a 'purely present fact'—not a fact exclusively about how the world is (intrinsically) now. If, however, we can tell what will happen when the spheres collide only if we have their instantaneous velocities, and instantaneous velocity is a partly other-timely property, then it seems we cannot after all ascertain what will happen when the spheres collide, if all we have to go on is information about the present and the laws; we also need to have enough information about the past and/or the future to allow us to ascertain the current instantaneous velocities of the spheres. So, Kripke concluded in the Cornell lectures, the view that instantaneous velocity is a (partly) other-timely property is in tension with the view that, if you want to predict the future and you already have complete information about the 'purely present' factors influencing the future, whatever knowledge about the past or future you might subsequently acquire won't turn up any new factors influencing the future that would have to be factored in in order to ascertain how the future will go.

If instantaneous velocity is after all the sort of property that is included in a momentary holographic state of the world, then a possible world with a disk that is rotating now and a possible world with a disk that is not rotating now will *eo ipso* differ with respect to their momentary holographic states. It is not clear to me, however, that 'diskworld' arguments against the supervenience holographic thesis essentially depend on the partial other-timelyness of velocity. Kripke asks us to consider a pair of possible worlds: in one a (lonely) disk made of uniform continuous matter rotates, in the other, it just sits there. In the Cornell lectures Kripke did not explicitly require that the disk start rotating (in the world in which it rotates); he allowed that the disk might be rotating 'from all eternity' (in the world in which it rotates). So let H_0 be a possible world containing nothing but an eternally rotating disk made of uniform continuous matter. In an alternative possible world the same disk, made of the same continuous matter, just sits there forever. In such an alternative possible world, just as the whole disk never moves, the disk's parts never move: wherever they ever are, they always are. But where is that? There are myriad possibilities. The parts might be permanently located just where they are (temporarily) located at time t in world H_0. They might instead be permanently located just where they are (temporarily) located at time $t + 1$ second. (Assume that the disk does not have an angular velocity of 1 revolution per second.) And so on. Corresponding to these myriad possibilities, we have a series of possible worlds $H_1 \ldots H_n$. If instantaneous velocity turns out to be a 'holographic' rather than a 'non-holographic' property, then it will turn out that H_0 will be discernible from each member of the series of possible worlds $H_1 \ldots H_n$ with respect to its holographic states. Be that as it may, it still seems that different members of that series will be indiscernible from each other with respect to their holographic states, which is enough to sink the holographic supervenience thesis. If we grant Kripke that the same lonely disk, made of the same uniform continuous matter and the same parts, that rotates forever in one world H just sits there forever in another world H', then, as far as I can see, Kripke can run his argument against the holographic supervenience thesis without insisting that instantaneous motion is a non-holographic property.

I have touched on just two of the themes of Kripke's Cornell lectures. This is not because other themes of those lectures are less important or interesting. It is only because Kripke's views on the stem–plant case and the revolving–non-revolving disk case are so well known that I could discuss them without making public what Kripke might not want made public. In view of the wealth of fascinating material in the lectures, one can only hope that Kripke will some day publish them in some form, so that those not lucky enough to have got hold of a typescript of them can benefit from them.

4

The Mental and the Physical

Towards the end of the third lecture of *Naming and Necessity* Kripke notes that materialists have often endorsed some or all of the following claims:

> Persons are (identical with) their bodies.
> Sensations are (identical with) neural events.
> Types of sensations are (identical with) types of neural events.

In *Naming and Necessity* and 'Identity and Necessity' Kripke sets out a number of broadly Cartesian arguments against each of the above claims. Although he stops short of endorsing any of those arguments, he suggests that they are intuitively plausible and show that the identifications at issue are highly problematic. Moreover, Kripke avers, the modal considerations on which the arguments turn tell heavily, not just against the identifications in question, but also against materialism as such, inasmuch as materialism requires that mental facts are ontologically dependent upon physical facts in the sense of following from them by necessity ('N & N', 342). In this section I shall focus on the arguments against identifying persons with their bodies.

> Descartes, and others following him, argued that a person or mind is distinct from his body, since the mind could exist without the body. He might equally well have argued the same conclusion from the premise that the body could have existed without the mind. Now the one response which I regard as plainly inadmissible is the response which cheerfully accepts the Cartesian premise, while denying the Cartesian conclusion. Let 'Descartes' be a name, or rigid designator, of a certain person, and let '*B*' be a rigid designator of his body. Then if Descartes were indeed identical to *B*, the supposed identity, being an identity between two rigid designators, would be necessary, and Descartes could not exist without *B* and *B* could not exist without Descartes ... A philosopher who wishes to refute the Cartesian conclusion must refute the Cartesian premise, and the latter task is not trivial. ('N & N', 334–5)

> All arguments against the identity theory which rely on the necessity of identity, or on the notion of essential property, are, of course, inspired by Descartes' argument for his dualism ... The simplest Cartesian argument can perhaps be restated as follows: let '*A*' be a *name* (rigid designator) of Descartes' body. Then Descartes argues that since he could exist, even if *A* did not, $\Diamond \sim(\text{Descartes} = A)$, hence $\sim(\text{Descartes} = A)$. Those who have accused

him of a modal fallacy have forgotten that 'A' is rigid. His argument is valid, and his conclusion is correct, provided its (perhaps dubitable) premiss is accepted ('I & N', n. 19).

In the first of these passages Kripke starts by setting out two different modal arguments against the identification of Descartes with his body. The first has the following structure:

> Descartes might have existed without his body.
> ———————————————————————
> So Descartes ≠ his body.

The second is the argument Kripke says Descartes might just as well have offered in support of the distinctness of Descartes from his body:

> Descartes's body might have existed without Descartes.
> ———————————————————————
> So Descartes ≠ his body.

In the first passage Kripke avers that it is 'plainly inadmissible' to accept the Cartesian premiss (presumably, *Descartes might have existed without his body*) and deny the Cartesian conclusion (*Descartes is not his body*). This suggests that he regards each of the above arguments as valid. In the second passage cited, though, he suggests that some philosophers have thought that Descartes's argument involves a modal fallacy, and (re)states that argument in a way that, he thinks, makes it evident that it is not fallacious.

Why might the Cartesian argument be thought fallacious? I take it Kripke has something like this in mind: it might be thought that its logical structure is:

> $\diamondsuit(t$ exists without $t')$
> ———————————————————————
> So $t \neq t'$,

where t' is non-rigid.

This argument form, it might be said, is invalid: even if there is some possible world with respect to which (the non-rigid designator) t' designates something that t (sometimes) exists without at that world (so that the premiss is true), it could still be that t and t' designate the same thing in the actual world (so that the conclusion is false).

Kripke's response is to suggest that we think of the Cartesian argument as involving two rigid designators—one for Descartes, and one for Descartes's body.[228] Then, whether we think of Descartes's argument as having the structure just mentioned, or the slightly more complicated structure envisaged in the second passage—to wit:

> $\diamondsuit(t$ exists without $t')$
> So $\diamondsuit (t \neq t')$
> ———————————————————————
> So $t \neq t'$,

---

[228] In both of the passages cited Kripke suggests we think of the argument as involving a name of Descartes's body. But we could equally well think of the argument as involving an impure demonstrative (*this body*), or a rigidified definite description *(the thing that is actually) Descartes's body*.

that argument will be valid, given the rigidity of all the singular terms appearing therein. After all, $\Diamond(t \text{ exists without } t')$ entails $\Diamond(t \neq t')$, and Kripke defines rigidity in such a way that, whenever t and t' are rigid, $\Diamond(t \neq t)$ entails $t \neq t'$.[229]

If we construe the Cartesian argument in the way Kripke suggests (as involving only rigid designators), then there is no gainsaying the validity of that argument—at least as long as we construe rigidity in Kripkean fashion.[230] The same goes for the argument Kripke says Descartes might just as well have offered for the distinctness of Descartes from his body. Assuming a Kripkean construal of rigidity, the only way to resist the Cartesian conclusion about the distinctness of Descartes from his body is to deny the Cartesian premiss that Descartes might have existed without his body B (along with the premiss that his body B might have existed without him). Because Kripke takes it to be initially plausible that Descartes could exist without his body, and that Descartes's body could exist without Descartes, he concludes that the argument Descartes did offer, and the argument he might just as well have offered, constitute a serious challenge to the identification of Descartes with his body.

Both of the modal arguments for the distinctness of Descartes from his body we started with have temporal analogues—to wit:

> Descartes once existed or will exist without (his body) B.
> _____
> So Descartes $\neq$ (his body) B.

and

> (Descartes's body) B once existed or will exist without Descartes.
> _____
> So Descartes $\neq$ (his body) B.

Each of these arguments is clearly valid. Moreover, as long as the modal arguments of which they are the analogues are valid, if either temporal argument is sound, so is the corresponding modal one (since the premiss of the temporal argument is true only if the premiss of the corresponding modal argument is true). Thus someone who accepted the truth of the premiss of the second temporal argument might take this to establish both the

---

[229] There is a complication here. As we have seen (cf. 'I & N', 137), Kripke is non-committal about whether, when t and t' are rigid, and designate the same thing, '$t = t'$' is true in worlds in which t (and t') do not exist: he allows that in a world in which t (and t') do not exist, it might be either indeterminate or false that $t = t'$. Thus he allows that in a world in which t and t' do not exist, it might be either indeterminate or true that $t \neq t'$. (I assume here that '$t \neq t'$' just means '$\sim(t = t')$'.) If $t \neq t'$ is true in possible worlds in which t and t' do not exist, even when t and t' are (non-strongly) rigid designators that designate the same thing, then it won't, after all, be true that, whenever t and t' are rigid, $\Diamond(t \neq t')$ entails $t \neq t'$. But it will still be true that, whenever t and t' are rigid, $\Diamond(\,(t \text{ exists}) \,\&\, (t \neq t')\,)$ entails $t \neq t'$, which is enough for the Cartesian argument to go through.

[230] Of course, a philosopher such as David Lewis, who rejects an identitarian account of modal predication, will construe rigidity in a different, non-Kripkean way.

distinctness of Descartes from his body and the soundness of the second modal argument.

In 'Identity and Necessity' Kripke endorses the validity of the second temporal argument, but is neutral on its soundness:

provided that Descartes is regarded as having ceased to exist upon his death, Descartes ≠ A can be established without the use of a modal argument; for if so, no doubt A survived Descartes when A was a corpse. Thus A had a property (existing at a certain time) which Descartes did not. ('I & N', n. 19)[231]

In the corresponding passage from *Naming and Necessity* Kripke appears to endorse tentatively the truth of the second temporal argument's premiss:

Of course, the body *does* exist without the mind and presumably without the person, when the body is a corpse. This consideration, if accepted, would already show that a person and his body are distinct. ('N & N', n. 74)

Similarly, in the Cornell lectures Kripke averred that the premiss of the second temporal argument is initially plausible, and that philosophers such as Fred Feldman who do not accept it need to provide an argument to the effect that we should not accept it.

Kripke made reference to Feldman in the Cornell lectures because Feldman had earlier argued that neither the modal arguments nor the temporal arguments discussed thus far constitute a serious challenge to the identification of persons with their bodies. According to Feldman, the clear-headed proponent of (what Feldman calls) the person–body identity thesis will—and should— say that such arguments, though valid, have undefended and indefensible premisses.[232] She, should, for example, insist that Descartes goes on existing exactly as long as his body does.

Given that Descartes's body existed after Descartes's death, won't this commit the champion of the person–body identity thesis to the claim that Descartes goes on after death?

Distinguo, says Feldman. The person–body materialist who accepts that Descartes's body does not go out of existence when Descartes dies must say that Descartes goes on existing after his death, but needn't say that Descartes goes on living after death. And, as Feldman sees it, the claim that most animals and most persons will end up as (end their days as) dead animals or dead persons is not just something the person–body materialist will have to accept, but also something that is independently plausible.

---

[231] In order to establish the distinctness of Descartes from his body, we don't need the (stronger) supposition that Descartes will not exist after his death, as long as we have the (weaker) supposition that Descartes isn't after his death in the place that his body is after his death. (As far as establishing the distinctness of Descartes from his body is concerned, it makes no odds whether, immediately after Descartes's death, Descartes is no more, or is in purgatory, as long as he isn't (at that time) where his body is (at that time).)

[232] See F. Feldman, 'Kripke on the Identity Theory', *Journal of Philosophy*, 71 (1974), 665–76, esp. 667.

Why suppose that most persons and animals will go on existing after their death? Bernard Williams writes:

Aristotelian enthusiasts will point out that, leaving aside immortality, when Jones (or his body) dies, Jones ceases to exist ('he is no more'), while his body does not. There may be something here: but it surely cannot be pressed too hard. For, taken strictly, it should lead to the conclusion that 'living person' is a pleonasm, and 'dead person' a contradiction; nor should it be possible to see a person dead, since when I see what is usually called that, I see something that exists. And if it is said that when I see a dead person I see the dead but existent body which was the body of a sometime person, this rewriting seems merely designed to preserve the thesis from the simpler alternative that in seeing a dead body I see a dead person because that is what it is.[233]

The gist of Williams's argument is the following: when I see a dead person, what I see—a dead person—exists. But a dead person is a person that is dead, and a person that is dead is a person. So when I see a dead person, someone now in existence is both a person and dead. Moreover, a dead K is a K that was once alive, but is no longer alive. So, when I see a dead person, I see someone who existed before his death, and continues to exist after it. Thus a person (typically) does not go out of existence when he dies; he simply goes from being a living person to being a dead person, just as a person does not go out of existence when she falls asleep, but simply goes from being an awake person to a sleeping person.

As Williams notes, 'Aristotelian enthusiasts' will object that the dead person I see is not a person, any more than a shoe tree is a tree, or a papier mâché lion is a lion; the dead person I see is *proprie loquendo* the dead body of a person who is no more. Although Williams suggests that this move is unmotivated, it does not seem so to me. As long as they have been well and truly shredded, shredded documents are not documents, puréed carrots are not carrots, and fossilized leaves are not leaves. In all these cases, what we call an F-ish K is not a K that is F, but rather what is left of a K after it goes out of existence (sometimes as a result of being F-ed, sometimes as a result of something that happened earlier). So it does not seem ad hoc to suppose that 'dead person' means 'what is left of a person after her death'. Or rather, in the sentence 'I saw a dead person in a funeral casket' it means that. In a sentence like 'For all those years, she has (unwittingly) been writing letters to a dead person', it means 'person who has ceased to live (in the right sort of way, and has not come back to life)'.[234]

Feldman finds this unpersuasive.[235] After all, he says, in biology classes schoolchildren dissect frogs. Now you can write to someone who does not

---

[233] B. Williams, 'Are Persons Bodies?', in Williams, *Problems of the Self* (Cambridge: Cambridge University Press, 1973), 74.

[234] I say 'ceased to live in the right sort of way' because, if there are non-human persons, some of them might cease to live without dying (if, say, they undergo fission).

[235] See F. Feldman, *Confrontations with the Reaper* (New York: Oxford University Press, 1992), 93–5. Although the frogs example is Feldman's and the burial example is Rosenberg's via Feldman, I have not presented the arguments in quite the form Feldman does. But the gist of the arguments is 'Feldmanian'.

exist at that time, but you cannot dissect someone (or something) that does not exist at that time. The frogs the schoolchildren dissect accordingly exist—in an ex-animate state—when they are dissected. The same would be true if aliens dissected human beings, or human persons.

Also, Feldman says, suppose that I die on Thursday and I am buried on Saturday. I cannot be buried on Saturday if I don't so much as exist on Saturday. So I—this person—exist (in an ex-animate state) on Saturday.

Considerations of this sort do, I think, (at least initially) increase the plausibility of the view that animals and persons don't (typically) go out of existence when they die. But, *pace* Feldman, I don't think they 'decisively establish' that it is contrary to common sense to suppose that persons or animals (typically) cease to exist when they die;[236] I accordingly don't think they refute the premiss of the second temporal argument for the distinctness of Descartes from his body.[237]

Feldman argues from the kinds of things we ordinarily say about dead animals or dead persons to the thesis that animals and persons exist post mortem as dead animals or dead persons. But I think the propriety (if not, in every case, the literal truth) of what we ordinarily say can be accounted for, even if the Aristotelian view is true, and animals and persons cease to be when they cease to live. Moreover, I think that we can motivate a construal of what we ordinarily say about dead animals or dead persons on which what we say is compatible with the Aristotelian view.

Papier mâché lions are not (real) lions. Suppose, though, that you are looking at a bunch of different papier mâché animals in a shop window. You might tell the sales assistant you wanted to buy the lion and the giraffe. You don't want to buy a (real) lion, but there's no impropriety in what you said, since the context makes it clear it's a papier mâché lion, rather than a real one, that you want.

The Aristotelian might say that in the frogs example something similar is going on. The things the schoolchildren are dissecting aren't really frogs; they're dead frogs (and pickled frogs). But in a normal context it is perfectly proper for the teacher to say that the children in her class are dissecting frogs, since it is common knowledge that what children dissect are dead frogs.

Williams or Feldman would object that, unless we have already bought into the Aristotelian view, there is no motivation for saying that the frog the schoolchildren dissected is like the lion in the shop window. But suppose that you want to find out how many cats there are in Twickenham. You will no doubt count the sleeping cats as well as the cats that are awake. But are you really going to count the dead cats as well as the living ones? Or suppose that

---

[236] Ibid. 95.

[237] Some of the arguments set out here appear in a slightly different form in my 'On the Real Distinction Between Persons and Their Bodies', in M. Marsonet (ed.), *The Problem of Realism* (Aldershot: Ashgate, 2002).

a child's only pet is a gerbil. One day, the gerbil dies, and the child buries him in her back garden. Suppose that, the week after, someone asks the child, 'Do you have a gerbil?' I take it that she would not answer in the affirmative. (She would naturally reply: 'I used to, but it died'.[238]) This suggests that the dead cats in Twickenham aren't cats, and that the dead gerbil in the garden isn't a gerbil. Someone who identifies animals with their bodies can of course say that, when we say *There are n cats in Twickenham* or *I have a gerbil*, we mean *There are n living cats in Twickenham* or *I have a living gerbil*. But this is just the sort of move the Aristotelian makes in the case of *the schoolchildren dissected frogs today* and the Feldmanian deems ad hoc.

As for the burial case, suppose that Jones ate fish. This might have happened in a number of ways. Perhaps Jones caught a very small fish and ate it alive, in one gulp. Perhaps Jones caught a good-sized fish cooked it, boned it, and cut it up in small pieces before he started eating. In the latter case I take it that even someone who thinks that a fish goes on existing after death as an ex-animate piscine body will agree that at the time Jones starts to eat, the fish is no more. (If the body of the fish still exists even after its parts are sundered, when does it go out of existence?)

If Smith says 'Jones caught a fish and ate it', he does not thereby commit himself on whether Jones cut up his fish before he started eating. Suppose Brown said to Smith, 'Look, if Jones caught a fish and ate *it*, he couldn't have cut the fish up first, since you can't eat a thing at a time unless it exists at that time.' Smith would quite properly accuse Jones of pedantry.

The moral is that one could naturally and properly (relative to non-pedantic standards of propriety) say that someone caught a fish and ate it, even if what he ate was not *sensu stricto* a fish, but something left behind after the fish (and, in this case, after the fish's body) went out of existence. *Pari ratione*, it might be said, one could naturally and properly say that the murderer killed White and buried him in the back garden, even if what the murderer buried was not strictly speaking White, but something left behind after White went out of existence.[239]

Again, Williams or Feldman might object that we need to be given a reason to think that *The murderer killed White and buried him in the back garden* is like *Jones caught a fish and ate it* (in not entailing that the thing killed or caught is still in existence when it is buried or eaten). But are such reasons so hard to find?

Suppose you live with Bob, who had been planning to go to Tibet last Thursday, but went into a coma on Wednesday (one that the doctors are confident he will come out of). Suppose that a friend of Bob knew that Bob

---

[238] Suppose, on the other hand, that someone who needed a dead gerbil to use as a prop in a play asked the child whether she had a *dead* gerbil. It seems she could naturally answer: 'Yes; it's buried in the garden.'

[239] This strategy for blocking the 'burial case' argument was suggested to me by an example of Feldman's ('The fish you eat today, last night slept in Chesapeake Bay') which he uses to argue for the idea that (most) animals (sooner or later) exist as dead bodies (see *Confrontation with the Reaper*, 95).

was planning to go to Tibet, but wasn't sure exactly when, and knows nothing about the coma. The friend calls Bob's home on Thursday, and, when you pick up the phone, says, 'I need to talk to Bob: is he still in England?' You might naturally answer:

Yes, but he's in a coma.

Now fill in the story as before, but suppose that on Wednesday, Bob died. When the friend says, 'I need to talk to Bob; is he still in England?', would you really answer:

Yes, but he's dead,

even if you knew that Bob's (dead) body still exists, and is still in England? If not, that suggests you think that Bob goes on being in England as long as, but only as long as, he goes on being alive in England. And if you do think that, you have a motivation for saying that what is buried at Bob's funeral is what Bob left behind (as a minister might say, Bob's earthly remains).

Again, suppose you are at Bob's funeral, sitting next to an open casket, with Bob's dead body in plain view. If the person giving the eulogy says, 'Bob is no longer here, but...', you probably won't be surprised; unless you're a 'Feldmanian', you probably won't look at the casket, and say to yourself: 'That can't be true.' If, on the other hand, the person giving the eulogy says, 'Bob's dead body is no longer here' you will no doubt be very surprised: you will say to yourself: 'How can that be? I can see Bob's dead body in the casket.' This suggests that you think that what the person giving the eulogy would say, were she to say that Bob is no longer here, is true, and that what the person giving the eulogy would say, were she to say that Bob's dead body is no longer here, is false. Why isn't this a motivation for thinking that what is buried at Bob's funeral is not strictly speaking Bob, but something Bob left behind?

Although I am inclined to think that persons or animals do not go on existing after their death as corpses, I am not sure to what extent I could support that view by appeal to things we ordinarily say about dead animals and dead persons. As we have seen, some things we say about dead animals or persons initially seem consonant with a Feldmanian view of dead animals and persons, and other things we say initially seem consonant with an Aristotelian view thereof. I suppose I think animals and persons don't exist as dead bodies after their death, because it just seems very plausible to me—though not completely beyond doubt—that the dead body that will still exist after my death is not *me*—is not this person or this human animal. It isn't, in its lumpish inanimateness, the right kind of thing to be me.[240]

---

[240] One reason we might be tempted to think that a (recently) ex-animated human body is a (real) person is that it looks so much like a (real) person. But imagine a machine whose exterior is made of a highly conductive metal with a very high melting point, and whose innards are made of metal with a very low melting point. If, as a result of being subjected to intense heat, the machine's 'innards' are fused, though its exterior is unaffected, I want to say that what is left is not the machine we started

If the reader does not resonate to this intuition, she may come to by think-ing about when human persons come into existence. If I am an entirely corporeal being, it seems at least initially natural to suppose (as Feldman does) that I came into existence at conception.[241] (After all, one might say, how can *I* be conceived at a time when *I* do not yet exist?) But considerations involving (the prehistory of) embryonic development have led some people to think otherwise.[242]

Before the second week the cells 'descended from' the fertilized ovum adhere to each other only loosely, and have, as it were, separate lives, metabo-lizing and dividing independently of each other. The cells are functionally speaking interchangeable, and most will develop into the placenta or other structures that support the embryo, rather than the embryo itself. Separating the clump of cells into two clumps will result in identical twins, although, if the two clumps are put back together (within the period before cell special-ization), we will end up with only one human being. Rearranging the cells in the clump will have no effect on future development.

At about sixteen days, though, the cells in the clump undergo specialization and begin to grow and function in a coordinated way. The clump of cells will begin to exhibit bilateral symmetry around the ancestor of a spinal cord (the so-called 'primitive streak'), and splitting the clump of cells into two smaller clumps would result in death, rather than in two embryos.

According to many embryologists, it is at this point that I come into existence. As A. McLaren puts it:

One can trace back directly from the newborn baby to the foetus, and back further to the origin of the individual embryo at the primitive streak stage in the embryonic plate at sixteen or seventeen days. If one tries to trace back further than that there is no longer a coherent entity. Instead, there is a larger collection of cells, some of which are going to take part in the subsequent development of the embryo and some of which aren't . . . To me the point at which I began was at the primitive streak stage.[243]

According to McLaren, before the two-week point we haven't got me, just a 'collection of cells', some but not all of which will take part in the development of the embryo at a later time. Why don't the pre-embryonic cells—or at least the pre-embryonic cells that will subsequently take part in the development of

with, no matter how much it looks like it (from the outside). Similarly, I want to say, a day-old corpse with disabled and decaying 'micro-innards' is not the person who died a day ago, however much the two might resemble each other from a 'macro' point of view (cf. E. Olson, *The Human Animal: Personal Identity Without Psychology* (Oxford: Oxford University Press, 1997, 152). Moreover, at the phenome-nological level I think people find (recently ex-animated) bodies 'creepy' precisely because they think: 'It looks so much like a person—but it isn't.'

[241] If I consist simply of an immaterial soul, or consist of an immaterial soul together with a human body, then there is no obvious reason to suppose that I begin to exist at the moment of conception.

[242] See the very helpful discussion of these matters in Olson, *The Human Animal*, 89–93. In what follows I draw on Olson.

[243] A. McLaren, 'Prelude to Embryogenesis', cited in Olson, *The Human Animal*, 174 n. 10.

the embryo—constitute me even before embryogenesis? After all, I, as I am now, 'grew from' those cells, just as I, as I am now, grew from the cells that constituted me when I was 6 years old.

A natural thought here is that the pre-embryonic cells from which I grew do not constitute me before embryogenesis because (as we have seen) they are not sufficiently coordinated in their activities to be regarded as taking part in my life.[244] In this respect, the pre-embryonic cells are very unlike the cells that constituted my body when I was 6, and more like the sperm and egg pair from which I came, which obviously never took part in my life. (Even though I 'grew from' the sperm and egg pair, few would be tempted to suppose that the sperm cell and egg cell jointly constituted me prior to my conception.)

Suppose that the pre-embryonic cells from which I grew did not constitute me before embryogenesis because they do not (then) take part in my life. Then, it would seem, neither do the cells of my dead body constitute me after my death. After I die, various things may happen to the cells in my body. Some or all may die along with my body, in which case they won't take part in anyone's life. Some may be transplanted into someone else's body (as parts of a transplanted organ) in which case they will take part in someone's else life. There are other possibilities. But whatever happens, they won't be taking part in my life after my death.

Now, if the cells in my body after I die won't constitute me then, then I will not exist as an ex-animate body; and if nothing other than cells of my body will ever constitute me (no soul, no ethereal entelechy, no computer hardware), then I will not exist after my death, just as I did not exist before embryogenesis.

Feldman thinks that his view of when animals and persons go out of existence is the common-sense view. If he were right, he would be right to hold that the proponent of the person–body identity thesis can dismiss the second temporal argument on the grounds that its premiss is false. But, for the reasons adduced above, I think he is wrong.

Nevertheless, we can see why Kripke stops short of endorsing the soundness of the second temporal argument. Whether or not Feldman ultimately makes a convincing case for the claim that animals and persons exist (exactly) as long as their bodies do, it is not as though there is just no case to be made. Indeed, a quite different—non-Feldmanian—case might be made for that claim.

Feldman and Kripke consider it obvious that, after Descartes dies, his body goes on existing, as a no-longer-living body. Aquinas would not agree.[245] As he sees it, when a person dies, the dead body he leaves behind is not identical—*numero* or even *specie*—with the body he had while he lived.[246] There is the man's (living) body, and there is the man's dead body (which isn't

---

[244] The connection between being a part or constituent of a living organism and taking part in that organism's life is emphasized by Peter van Inwagen; see his *Material Beings*, 91.

[245] At least, he would not agree if what he means by the Latin 'corpus' is what we mean by the English 'body'.   [246] See *Quodlibetum* II, q. 1, a. 1.

really a body at all), and there is the matter that first constituted the living body and then constituted the dead body. But there is nothing which *is* a body, and first was a living body, and then was a dead body or corpse, although there is something (a parcel of matter) that first *constituted* a living body and then constituted a dead body or corpse.

This view may seem less strange when we see that it is a counterpart of a commonly held view about persons. Suppose that a living person becomes a dead person (in the sense of 'becomes' in which Lot's wife became a pillar of salt, and in the 'dead-person-I-just-saw' sense of dead person). Then, non-Feldmanians will typically say, there is the (live) person, and there is the dead person (which isn't really a person at all), and there is the body that first constituted the live person and then constituted the dead person, but there is nothing which *is* (as opposed to constitutes) a person, and was first alive, and then dead.

Someone might object here that dead bodies surely are bodies; as often as not, when we talk about bodies without qualifying them as dead or alive, we have dead ones in mind. But, a defender of Aquinas might say, our willingness to say things like 'There's a body in the road' when there's a corpse in the road no more shows that dead bodies are bodies than our willingness to say 'The children are dissecting frogs today' shows that dead frogs are frogs.

Still, couldn't I truly say, pointing to a dead body, 'This body, which is now dead, was once alive'?

Call the (essentially living) body Aquinas thinks I have my *body*$_L$ (where 'L' stands for living). Call the (accidentally, and probably only temporarily, living) body Kripke and Feldman think I have my *body*$_{DOA}$ (where 'DOA' stands for 'dead or alive'). Aquinas thinks that there are no bodies$_{DOA}$; by his lights, the belief that there are bodies$_{DOA}$, as well as or instead of bodies$_L$ and bodies$_D$ (that is, corpses), is like the belief that there is 'wineorvinegar'—something that is accidentally and temporarily wine, and accidentally and temporarily vinegar—as well as or instead of wine that turns into (and is replaced by) vinegar.

Whatever the attractions of this view, it is not one that everyone who identifies persons with their bodies must hold. Someone who thinks that persons are their bodies can hold that both bodies$_L$ and bodies$_{DOA}$ are perfectly respectable entities, that 'my body' is ambiguous between 'my body$_L$' and 'my body$_{DOA}$', and that I am my body (that is, my body$_L$).

It may be worth underscoring that this construal of person–body materialism does not trivialize the doctrine. A philosopher who believed in bodies$_L$ might have various reasons for thinking they are distinct from the persons that have them. She might, for example, think that two persons could exchange bodies$_L$, so that persons and their bodies$_L$ were discernible with respect to modal properties. Or she might be a dualist and a believer in the afterlife, in which case she would deny that persons and their bodies$_L$ exist at all the same times.

Someone might object to the construal of person–body materialism under discussion in either of two ways. He might deny that there are such things as bodies$_L$. Alternatively, he might allow that there are or might be such things, but deny that the word 'body' can be used to pick them out.

Why shouldn't bodies$_L$ exist? Eric Olson reports that neither of the German words for 'body' (*Körper* and *Leib*) can be applied to corpses (*Leichen*).[247] (According to the *OED*, *Leib* is in fact derived from the Middle High German for 'life'.) If Olson is right, neither *Körper* nor *Leib* means 'body$_{DOA}$', and it is at least initially plausible that those terms mean 'body$_L$'. So why aren't bodies$_L$ perfectly respectable entities, unambiguously picked out by certain German sortals? Also, it is not as though belief in bodies$_L$ clearly involves a problematic increase in our ontological commitments. It is at least arguable that as long as we've got pigs in our ontology, we've already got porcine bodies$_L$, inasmuch as a pig just is a porcine body$_L$ (and more generally, an organism just is its body$_L$; cf. the *OED* definition cited below).

Of course, it might be that the English term 'body' never means 'body$_L$', just as the German term *Leib* never means 'body$_{DOA}$'. In that case—even on the assumption that there are such things as bodies$_L$, and that persons are identical to them—it won't be true that persons are their bodies.

In fact, though, I am doubtful that 'body' can *never* mean 'body$_L$'. A dead frog (I think) isn't really a frog at all. In the same way, I can see someone saying, a dead body isn't really a body at all—in the relevant (biological) sense of 'body';[248] it is too unlike the (living) things that are uncontroversially bodies (in the biological sense) to count as a body (in that sense). (See note 212.) If someone did say that, I wouldn't respond, 'You are simply using the term "body" incorrectly.' In this connection, it is interesting to note that under the very first heading for the word, the *OED* defines 'body' as 'the whole material organism viewed as an organic entity'.[249]

At this point, someone might object:

> Suppose that 'body' can mean 'body$_L$'. It still won't be true that persons are their bodies, because bodies$_L$ sometimes outlast the persons that had them. Suppose someone's cerebral cortex is completely destroyed, so that her higher brain functions are lost for ever. Suppose also that the rest of her brain and body are more or less intact. In such a case, the person goes out of existence, but her body$_L$ does not.

---

[247] Cf. Olson, *The Human Animal*, 151.

[248] There is a sense of 'body' according to which things that are not and never have been alive can be bodies (e.g. celestial bodies). A corpse may perfectly well be a body, in a non-biological sense of 'body'.

[249] That 'body' at least used to mean something like 'organism'—and not that long ago—is evident from the (current) meaning of the term 'antibody'.

The objection raises large issues, to which I cannot do justice here.[250] But I would answer it by denying that, in the case at issue, the person goes out of existence when the cerebral cortex is destroyed.

Whether or not I can go on existing after my life has ended, I cannot start existing after my life has already begun. But, I want to say, my life began long before my higher mental functions came into operation.[251] (My life was under way at the primitive streak stage (roughly a fortnight after conception), and my cerebrum did not come into existence until at least the sixth week; it did not become operational as an organ of sensation or thought until much later than that.) So it certainly looks as though I existed as a not yet thinking being.[252] If so, why couldn't I exist as a no longer thinking being after my cerebral cortex had been destroyed? If my becoming a *res cogitans* is not the beginning of my life, why should my ceasing to be a *res cogitans* be the end of my life? Perhaps *being a person* entails *being a thinking being*, so that I was not always a person, and (maybe) will not always be a person. But the claim that persons are their bodies$_L$ is perfectly consistent with the claim that bodies$_L$ (and persons) weren't always persons, and won't always be persons.

I have been arguing that Kripke's take on the second temporal argument is better than Feldman's. If this is right, then Kripke's take on the second modal argument is likewise better than Feldman's, inasmuch as the second modal argument is sound if the second temporal argument is.[253] But what about the first (and more narrowly Cartesian) modal argument? Does it (as Kripke suggests) pose a serious challenge to the person–body identity thesis, inasmuch as its premiss is at least initially plausible? Or is it (as Feldman suggests) a valid argument with a question-begging premiss?

A champion of the person–body identity thesis might ask why anyone should suppose that Descartes might have existed without his body. Why is it any more possible for Descartes to exist without his body than it is for a pair of earrings to exist without one or the other earring, or for a snowball to exist without the snow it is made of?

I imagine that those who think Descartes could have existed without his body would respond along these lines:

> A pair of earrings is not the sort of thing that could exist apart from the earrings in it. That is to say, a pair of earrings is the sort of thing that could exist only where and when the earrings in it do. Similarly, a snowball is not the sort of thing that could exist apart from the snow it is made

---

[250] For a discussion of some of those issues, see my 'Identità personale e entità personale', in A. Bottani and N. Vassallo (eds), *Identità personale: un dibattito aperto* (Naples: Loffredo, 2001).

[251] As do many embryologists (cf. the passage from McLaren cited earlier).

[252] For a vigorous and to my mind persuasive defence of the claim that I existed as a not yet 'enminded' fetus, see Olson, *The Human Animal*, ch. 4.

[253] Again, assuming a Kripkean understanding of rigidity and an identitarian account of modal predication.

of. But whether or not Descartes ever did or ever will exist apart from his body, he is the sort of being that could exist apart from his body (could exist where and/or when his body did not).

Those who think Descartes might have existed without (his body) B do so because they think that (i) Descartes might have existed apart from B, (ii) B might not have existed, and (iii) *B does not exist* does not imply *Descartes does not exist apart from B.*

Notice, though, that (i) is sufficient to show the distinctness of Descartes from B, whether or not (ii) and (iii) are true. If Descartes might have existed apart from B, then Descartes is not the same as B, since B could not have existed apart from B. It doesn't matter whether or not B's non-existence is possible, or compossible with Descartes's separation from B. The opponent of the person–body identity thesis accordingly doesn't need the stronger premiss that Descartes might have existed without B, but only the weaker premiss that Descartes might have existed apart from B.

Still, why accept even the weaker premiss? The Cartesian finds it obvious that Descartes's existing apart from B is at least possible, and concludes that Descartes and B are distinct. But, the defender of the person–body identity thesis could say, Kripke has provided us with the materials to see the incogency of arguing from the possibility of Descartes's existing without B to the distinctness of Descartes from B.

As we have seen in Chapter 2, Kripke adjudicated the dispute between Quine and Marcus by agreeing with Quine that *Hesperus is Phosphorus* is empirical, and agreeing with Marcus that it is necessary. Along with this adjudication came a distinction between 'epistemic mights' and 'metaphysical mights' (cf. 'N & N', 332 and n. 72). Before it was known whether Hesperus was Phosphorus, someone could have said truly that Hesperus might be different from Phosphorus. But the possibility at issue here is epistemic, not metaphysical; it reflects a gap in knowledge, rather than a genuine (metaphysical) possibility. Similarly, before the four-colour theorem was proven, someone could have said truly that the four-colour theorem might turn out to be false. 'Obviously', Kripke concludes, 'the "might" here is purely "epistemic"—it merely expresses our present state of ignorance, or uncertainty' ('N & N', 307).

So, the person–body materialist could say, *Descartes's existing apart from his body is possible* is ambiguous. It could mean that Descartes's existing apart from his body is epistemically possible (either in the (stronger) sense that it is true for all we know, or in the (weaker) sense that it is true for all we know a priori; cf. Chapter 2, section 2). Or it could mean that Descartes's existing apart from his body is genuinely (metaphysically) possible. Moreover, the person–body materialist can say, an argument from the epistemic possibility of Descartes's existing apart from B to the distinctness of Descartes from B will be fallacious; and an argument from the metaphysical possibility of Descartes's existing apart from B to the distinctness of Descartes from B will be question-begging. Either way, we won't get a serious challenge to the person–body identity thesis.

But doesn't Descartes's existing apart from B look genuinely possible, and not just possible for all we know? Perhaps. Be that as it may, Kripke himself has stressed that statements that we know are non-contingent may nevertheless—at least initially—look like statements whose truth, and whose falsehood, is genuinely (metaphysically) possible. For instance, early on in 'Identity and Necessity' Kripke says:

> We could venture this conclusion: that whenever 'a' and 'b' are proper names, if a is b, that it is necessary that a is b. Identity statements between proper names have to be necessary if they are going to be true at all . . . According to this view, whenever, for example, someone makes a correct statement of identity between two names, such as, for example, that Cicero is Tully, his statement has to be necessary if it is true. But such a conclusion *seems* plainly to be false. ('I & N', 140–1)

Elsewhere Kripke says that one has 'the illusion' that *Heat is the motion of molecules* is contingent ('I & N', 160), and speaks of 'the illusion' that water might not have been hydrogen hydroxide ('N & N', 337).

In both 'Identity and Necessity' and *Naming and Necessity* Kripke makes various suggestions about the factors that might give rise to this kind of illusion. One is a failure clearly to distinguish epistemic from metaphysical modalities ('I & N', 154): if we don't distinguish what might (for all we know) turn out to be false from what might not have been the case, we shall misclassify a posteriori necessities as contingencies.

The failure to distinguish epistemic from metaphysical modalities can lead us to mistake necessarily true (or false) statements for contingently true (or false) statements, whether those statements are identities or non-identities (e.g. *Water is made of hydrogen and oxygen*). But, Kripke suggests, other factors may explain in particular our inclination to mistake necessarily true (or false) identity statements for contingent statements. We may mistake a necessarily true (or false) identity statement of the form $m = n$ for a contingent one, because we misascribe to $m = n$ the modal status of the statement *The* F $=$ *the* G, where 'the F' and 'the G' are (respectively) reference-fixing descriptions for m and n:

> Let 'R_1' and 'R_2' be two rigid designators which flank the identity sign. Then '$R_1 = R_2$' is necessary if true. The references of 'R_1' and 'R_2', respectively, may well be fixed by nonrigid designators 'D_1' and 'D_2'; in the Hesperus and Phosphorus cases these have the form 'the heavenly body in such-and-such a position in the sky in the evening (morning)'. Then although '$R_1 = R_2$' is necessary, '$D_1 = D_2$' may well be contingent, and this is often what leads to the erroneous view that '$R_1 = R_2$' might have turned out otherwise.[254] ('N & N', 333–4; see also 338)

---

[254] Note that in a case where we don't have only contingently co-referential reference-fixing descriptions, the illusion of contingency does not arise. Consider a case in which a person 'renames' or 'nicknames' herself. If Philippa introduces a nickname for herself via the formula 'From now on, Philippa will (also) be *Pippa*', she will not (then) be subject to the illusion that Philippa might not have been Pippa.

The inclination to misascribe the modal status of $D_1 = D_2$ to $R_1 = R_2$ may in turn be due to a failure to distinguish reference-fixing descriptions from 'meaning-providing' descriptions—that is, a tendency to see a name and its associated reference-fixing description as synonymous ('I & N', 157).[255]

There is a riddle that goes: 'If "leg" meant "tail-or-leg", how many legs would a horse have?' People often answer 'Five', at which point the riddler says 'No, four: calling a tail a leg doesn't make it so.' If 'leg' meant 'tail-or-leg', then we could truly say the words, 'Horses have five legs'; but horses would still have just four legs.

Just as there are possible circumstances in which we can truly say the words 'Horses have five legs', there are possible circumstances in which we can truly say the words 'Hesperus is different from Phosphorus'. It doesn't follow that there are possible circumstances in which Hesperus is not Phosphorus, or horses have five legs:[256] as Kripke emphasizes, we use our actual words, with their actual meanings, to describe counterfactual possible situations. But, as the riddle indicates, and Kripke warns us, one can get muddled about this:

> it could have been the case that Venus did indeed rise in the morning in exactly the position in which we saw it, but that on the other hand, in the position which is in fact occupied by Venus in the evening, Venus was not there, and Mars took its place . . . Now one can also imagine that in this counterfactual other possible world, the earth would have been inhabited by people and that they should have used the names 'Phosphorus' for Venus in the morning and 'Hesperus' for Mars in the evening. Now this is all very good, but would it be a situation in which Hesperus was not Phosphorus? Of course, it is a situation in which people would have been able to *say*, truly, 'Hesperus is not Phosphorus'; but we are supposed to describe things in our language, not in theirs. ('I & N', 155)

Another factor which Kripke thinks may incline us to see statements which are necessarily true (or false) as contingent is a failure clearly to distinguish states of affairs involving an individual's having a certain property from states of affairs involving (what Kripke calls) an *epistemic counterpart* of that individual's having that property. Someone could be under the illusion that this table might have been made of ice, because she seemed to be able to imagine that being the case. Thinking harder, though, she should be able to see that what she actually can imagine is there being another table—a table which

---

[255] As Kripke notes, failure to distinguish reference-fixing for meaning-giving can also give rise to illusions of necessity: 'The interesting fact is that the way the reference is fixed seems overwhelmingly important to us in the case of sensed phenomena . . . The fact that we identify light in a certain way seems to us to be *crucial*, even though it is not necessary; the intimate connection may create an *illusion* of necessity' ('N & N', 331).

[256] Given that *Horses have five legs* means something like *Normally, horses have five legs*, it is at least doubtful whether *Horses have five legs* could have been true (even if, say, some deformed horse could have five legs).

resembles this table in certain ways—being made of ice ('I & N', 160–1).[257] Similarly, an ancient astronomer who doesn't know whether Hesperus is Phosphorus might think it is (not merely epistemically) possible that Hesperus is not Phosphorus, because he can imagine Hesperus being a different heavenly body from Phosphorus. But what he actually can imagine is an epistemic counterpart of Hesperus (something that looked the same as Hesperus actually looks, was in the same place in the evening as Hesperus actually is, and so on) being a different heavenly body from Phosphorus:

> If someone protests, regarding the lectern, that it *could* after all have *turned out* to have been made of ice, and therefore could have been made of ice, I would reply that what he really means is that *a lectern* could have looked just like this one, and have been placed in the same position as this one, and yet have been made of ice . . . I have argued that the same reply should be given to protests that Hesperus could have turned out to be other than Phosphorus, or Cicero other than Tully. Here, then, the notion of 'counterpart' comes into its own. For it is not this table, but an epistemic 'counterpart' which was hewn from ice; not Hesperus–Phosphorus–Venus, but two distinct counterparts thereof, in two of the roles Venus actually plays (that of Evening Star and Morning Star), which are different. ('I & N', 157, n. 15)

If our ancient astronomer does not clearly distinguish between Hesperus' being F, and some epistemic counterpart of Hesperus's being F, he may well think that there is a (metaphysically) possible world, which is, for all he knows, the actual world, in which Hesperus (and not just some epistemic counterpart thereof) is different from Phosphorus. In other words, he may think that Hesperus' being different from Phosphorus is metaphysically possible, as well as epistemically possible in the stronger sense. And even if he subsequently learns that Hesperus = Phosphorus, that won't stop him from thinking that there is a (metaphysically) possible world, which is, for all he knows, a priori the actual world, in which Hesperus is different from Phosphorus. That is, it won't stop him from thinking that Hesperus' being different from Phosphorus is metaphysically possible, as well as epistemically possible in the weaker sense.

Returning to the question under discussion (the cogency or otherwise of the first modal argument for the distinctness of Descartes from (his body) B): someone who identifies Descartes with B may want to deny that Descartes's existing apart from B looks genuinely (metaphysically) possible. Some person–body materialists think that Descartes's existing apart from B won't look genuinely possible to you, unless you have already bought into a dualistic account of Descartes. In fact, it is not even as though all *dualists* have

---

[257] Compare: someone might think there could have been a barber who shaved all and only those who don't shave themselves, because she thought she could imagine a barber who shaves all and only those who don't shave themselves. Upon reflection, she could come to see that what she could imagine was there being a barber who shaves all and only those persons *different from the barber* who don't shave themselves.

the intuition that Descartes could have existed apart from B. Aquinas, for example, thinks of a human being as constituted by a material body and an immaterial rational soul, which soul separates from the body at death. But, since he thinks that a (created) human being is essentially a human being, and that nothing can be a human being without having a human body and a human soul, he does not think that Descartes himself (as opposed to an immaterial part of Descartes) could continue to exist after separation from B.

In any case, a person–body materialist needn't deny that Descartes's existing apart from B looks genuinely possible. She can say that Descartes's existing apart from B, like Hesperus' existing apart from Phosphorus, or gold's being a compound, at least initially looks genuinely possible. And she can explain (or explain away) this apparent possibility along Kripkean lines.

Thus she can say that, since it is a posteriori that Descartes = B, and since there is, in Kripke's words, 'a very strong feeling which leads one to think that, if you can't know something by a priori ratiocination, then it's got to be contingent' ('N & N', 306), there is, initially at least, a strong feeling that *Descartes does not exist apart from B* is at most contingently true. Moreover, she could say, there is a possible world, which could, for all we know, a priori be the actual world, in which some epistemic counterpart of Descartes exists apart from B (or any epistemic counterpart thereof). That epistemic counterpart of Descartes might be an initially embodied, but subsequently disembodied, soul; or a being that is initially constituted by an immaterial soul and a material body, and subsequently undergoes disembodiment (without ceasing to exist). Given our tendency not to distinguish clearly an individual from its epistemic counterparts, this gives rise to the illusion that there is a (metaphysically) possible world, which, for all we know, a priori is the actual world, in which Descartes exists apart from B.

To forestall some possible misunderstandings: none of these considerations show that the first modal argument for the distinctness of Descartes from B is unsound. Even if not all epistemic possibilities are metaphysical possibilities, and even if some a posteriori (metaphysical) impossibilities look like (metaphysical) possibilities, it perfectly well still might be that Descartes's existing apart from (and indeed without) B not only looks, but is, metaphysically possible.

Nor do the considerations just set out show that the first modal argument is incogent. It seems consistent with all those considerations that we should accept the premiss of that argument, and thus its conclusion. An analogy: suppose we have an argument that moves from the claim that mental events cause physical events, together with ancillary premisses, to the conclusion that mental events are physical events. An epiphenomenalist will reject the main premiss of the argument. Suppose the epiphenomenalist grants that mental events *appear* to cause physical events, and has an explanation of why they would appear to cause physical events, even if (as she supposes) they do not. (Such an explanation would not, on the face of it, be so hard to provide.)

The argument for the identity of mental events with physical events might still be a cogent as well as a sound one, resting on premises that we ought to accept (even if the epiphenomenalist does not in fact accept them).

On the other hand, the fact that someone who identifies persons with their bodies could tell a Kripkean story about why *Descartes exists only where and when B does* looks contingent, even though it isn't, has implications for the question set a few pages earlier. That question was: Does the first modal argument pose a severe challenge to the person–body identity thesis? It now looks as though the answer is: no—*as long as* there are good reasons for identifying Descartes with B in the first place. If the person–body materialist has good reasons to identify Descartes with B, just as astronomers have good reasons to identify Hesperus with Phosphorus, then the person–body materialist has good reasons to suppose that the first modal argument is unsound.[258] The person–body materialist needn't, but is perfectly able, to concede that the premise of the first modal argument (at least initially) *seems* to be true, since he can tell a Kripkean story about why it would seem to be true, even though it is not.

This point is perhaps worth underscoring, because it brings out a difference between the thesis that persons are their bodies and the thesis that mental properties are physical properties. As we shall see, Kripke seems to grant that there are some apparently compelling considerations in favour of identifying mental properties with physical properties (cf. 'N & N', 355 n. 77). But, he thinks, the identification of mental with physical properties remains highly problematic, because (i) mental properties and the physical properties with which type-identity theorists want to identify them certainly *appear* to be separable, and (ii) the materialist has no account of how mental properties and physical properties could so much as *appear* to be separable if they are in fact identical (and thus inseparable):

The identity theorist, who holds that pain is the brain state, also has to hold that it necessarily is the brain state. He therefore cannot concede, but has to deny, that there would have been situations under which one would have had pain but not the corresponding brain state . . . He has to hold that we are under some illusion in thinking that we can imagine that there could have been pains without brain states. And the only model I can think of for what the illusion might be, or at least, the analogy the materialists themselves suggest, namely, heat and molecular motion, simply does not work in this case. So the materialist is up against a very stiff challenge. He has to show that these things we think we can see to be possible are in fact not possible. He has to show that these things which we can imagine are not in fact things we can imagine. And that requires some very different philosophical argument from the sort which has been given in the case of heat and molecular motion. And it would have to be a deeper and subtler argument than I can fathom and subtler than has ever appeared in any materialist literature that I have read. ('I & N', 163)

---

[258] Assuming that 'Descartes might have existed apart from (or without) B' is understood as making a claim about what is metaphysically possible. (If it is understood as making a claim about what is epistemically possible, then the first modal argument is invalid.)

As Kripke sees it, the crucial difficulty facing the type-identity thesis is not that there are no considerations in its favour. It is instead that, if we accept the type-identity thesis, we are committed to saying that we are subject to the illusion that *Someone is in pain if and only if she is in such-and-such a neural state* is contingent, and we have no 'model' of how such an illusion might arise. If I am right, the difficulties faced by the person–body identity thesis are strongly disanalogous.[259] There may well be difficulties in motivating the claim that Descartes = B in the first place, but if there are good reasons to suppose that Descartes and B are identical and thus inseparable, there do not seem to be insuperable difficulties about how the illusion of Descartes's separability from B might arise.

A proponent of the first modal argument might not be too bothered about this. He might say: of course, if there were good evidence in favour of the identity of Descartes with B, that would be a reason to think that the first modal argument is unsound. But there is no such evidence. And, in the absence of evidence for the identity of Descartes with B, we should regard the first modal argument as sound. After all (he might say), it looks as though Descartes's existing apart from (or without) B is genuinely (metaphysically) possible, and there is no evidence that the appearances are deceptive. So, just as it is reasonable to believe it is metaphysically possible for there to be dogs without cats, given that it looks metaphysically possible, and there is no reason to think it isn't, so it is reasonable to believe that Descartes could exist apart from (or without) B, given that it looks metaphysically possible, and there is no reason to think it isn't. The first modal argument is accordingly both sound and cogent, and even the person–body materialist should admit that it should be regarded as sound, until or unless evidence turns up for the identity of Descartes with B.

To say what I think is wrong with this reasoning, I shall draw an analogy between it and a more obviously wrong-headed bit of reasoning.

Suppose that, observing the sky in the evening, two philosophically inclined astronomers discover an unfamiliar celestial object, which they baptize *New Hesperus*. Observing the sky the next morning, they again come across a celestial object that they cannot identify, and baptize it *New Phosphorus*. At this point, let us suppose, the astronomers have no reason to suppose that New Hesperus is New Phosphorus, but also have no reason to suppose that New Hesperus isn't New Phosphorus. Imagine the following dialogue between the astronomers:

A: I think New Hesperus' existing apart from New Phosphorus is possible.
B: Epistemically, yes.
A: That's not what I mean: I mean that New Hesperus' existing apart from New Phosphorus is metaphysically possible.

---

[259] Though he is not explicit about it, Kripke might agree, judging from what he says (and does not say) at 'N & N', 341 (bottom paragraph).

B: Why do you think that?

A: Well, New Hesperus' existing apart from New Phosphorus looks like a genuine possibility to me. I realize that things can look genuinely possible without being genuinely possible; but in this case, there's no reason to think appearances are deceptive. That's why I think New Hesperus' existing without New Phosphorus is metaphysically, and not just epistemically, possible.

B: But doesn't it look to you like a genuine possibility that New Hesperus = New Phosphorus?

A: Suppose it does. Is that a problem?

B: Well, you accept, don't you, that, if New Hesperus = New Phosphorus, then it is (metaphysically) necessary that New Hesperus = New Phosphorus?

A: Yes; that's been shown by Kripke.

B: So you presumably also accept that, if New Hesperus ≠ New Phosphorus, then it is necessary that New Hesperus ≠ New Phosphorus.

A: In which case . . . ?

B: In which case, you must also accept that, if it is possible that New Hesperus = New Phosphorus, then New Hesperus = New Phosphorus.[260]

A: All right.

B: And you agree that, if New Hesperus = New Phosphorus, then New Hesperus' existing apart from New Phosphorus is (metaphysically) impossible.

A: Yes.

B: So you have just as good reason to suppose that New Hesperus' existing apart from New Phosphorus is metaphysically impossible as you do to suppose that New Hesperus' existing apart from New Phosphorus is metaphysically possible—which means you don't have good reason to suppose the latter.

A: Run that by me again.

B: When I asked why you thought that New Hesperus' existing apart from New Phosphorus was metaphysically possible, you said 'it looks metaphysically possible, and there's no reason to think it isn't.' But New Hesperus' being identical to New Phosphorus looks no less metaphysically possible than New Hesperus' existing apart from New Phosphorus, and there's no more reason to think that New Hesperus' being New Phosphorus is, contrary to appearances, metaphysically impossible than there is to think that New Hesperus' existing apart from New Phosphorus is, contrary to appearances, metaphysically impossible. Since New Hesperus = New Phosphorus only if New Hesperus' existing apart from New Phosphorus is metaphysically impossible, that means

---

[260] From ~(New Hesperus = New Phosphorus) ⊃ □~(New Hesperus = New Phosphorus) we get ~□~(New Hesperus = New Phosphorus) ⊃~~(New Hesperus = New Phosphorus), and thus ◇(New Hesperus = New Phosphorus) ⊃ (New Hesperus = New Phosphorus).

that you have no better reason to think that New Hesperus' existing apart from New Phosphorus is metaphysically possible than you do to think that it is metaphysically impossible.

A: Something has gone wrong.

B: Indeed. Suppose you know that something is true, and you know that, if it is true, then something else is metaphysically impossible. Then you obviously shouldn't believe that that something else is metaphysically possible, even if you, as it were, encounter no resistance in trying to imagine it—even if it looks metaphysically possible to you. If you know that Hesperus = Phosphorus, you shouldn't believe that Hesperus (metaphysically) might have existed apart from Phosphorus, even if Hesperus' existing apart from Phosphorus looks (metaphysically) possible to you. Or it might be better to say: even if it initially looked (metaphysically) possible to you; perhaps once you've taken on board that Hesperus = Phosphorus, Hesperus' existing apart from Phosphorus stops looking (metaphysically) possible to you, and only looks for-all-you-knew possible.[261] Now suppose that something might perfectly well, for all you know, be true, and suppose (you know) that, if it is true, then something else is metaphysically impossible. Then once again you shouldn't believe that that something else is metaphysically impossible—even if you encounter no resistance when you try to imagine it—even if it looks metaphysically possible to you. If it might perfectly well for, all you know, be that New Hesperus = New Phosphorus, and you know that, if New Hesperus = New Phosphorus, then New Hesperus' existing apart from New Phosphorus is (metaphysically) impossible, then you shouldn't believe that New Hesperus (metaphysically) might have existed apart from New Phosphorus—even if New Hesperus' existing apart from New Phosphorus looks (metaphysically) possible to you. Or, it might be better to say, even if it initially looked (metaphysically) possible to you; perhaps once you're attending to the fact that New Hesperus might perfectly well, for all you know, be New Phosphorus, New Hesperus' existing apart from New Phosphorus stops looking (metaphysically) possible to you, and only looks for-all-you-know possible. There is an analogy with perceptual judgements here. If you know that something is true, and you know that, if it is true, then the thing you're looking at isn't orange, then you obviously shouldn't believe that the thing you're looking at is orange, even if it looks orange to you. But it's equally true that, if there's something that might perfectly well, for all you know, be true, and you know that, if it's true, then the thing you're looking at isn't orange, then you shouldn't believe that the thing you're looking at is orange, even if it looks orange to you.

---

[261] There is an interesting issue here about the extent to which judgements about what looks (metaphysically) possible are modular, in the way that perceptual judgements about, say, whether two lines look equal in length are modular.

A: So my judgement that New Hesperus (metaphysically) might have existed apart from New Phosphorus was (ultima facie) unjustified, not because there was a 'rebutting defeater' for that judgement (as there would have been if I had known that New Hesperus = New Phosphorus), but because there was an 'undercutting defeater' for that judgement?

B: Something like that.

To return to the point at issue: Kripke notes that people think, though they may be wrong, that they can imagine totally disembodied creatures ('I & N', 161). I think that those who intuit that Descartes's existing without B is (metaphysically) possible often do so because, when they try to imagine Descartes undergoing total disembodiment—say, at the moment of his death—without ceasing to exist, they encounter no resistance. Imagining Descartes without his body seems to them as unproblematic as imagining dogs without cats. From this they conclude that what they seem to be able to imagine is genuinely possible, so that Descartes's existing apart from or without B is genuinely possible.

But are there some propositions that might perfectly well for all we know be true, and are such that, if they are true, Descartes's being embodied one moment and disembodied the next is impossible? Perhaps. Following Snowdon, let us call a sort an *abiding sort* if necessarily nothing could come to fall under it, or cease to fall under it, without coming to be or ceasing to be.[262] According to a venerable tradition—going back at least as far as Aristotle—animality is an abiding sort. According to the same tradition, Descartes is an animal of a certain sort (a member of the species *Homo sapiens*). Suppose that Descartes is an animal, and animality is an abiding sort. Then Descartes could continue existing in bodiless form only if something could be an animal at a time without having a body at that time. And how could animality and bodilessness be compatible properties? So if, necessarily, Descartes goes on existing only if Descartes goes on being an animal, and, necessarily, Descartes is an animal only if Descartes has a body, then Descartes's being embodied one moment and disembodied the next is, after all, metaphysically impossible.

Now I am inclined to believe that Descartes, and the rest of us, are animals, and that animality is an abiding sort. But others (and I) have defended these theses elsewhere, and I shall not rehearse those arguments here.[263] My point is just that, if we are initially strongly inclined to regard Descartes's continuing to exist in bodiless form as (metaphysically) possible, the strength of that inclination may well be a result of failing to attend to our total evidence. Once we bring to mind the hypothesis that Descartes is an animal, that hypothesis

---

[262] Cf. P. Snowdon, 'Persons, Animals, and Ourselves', in C. Gill, ed., *The Person and the Human Mind* (Oxford: Clarendon Press, 1990), 87.

[263] See as well as Snowdon's piece, Olson's *The Human Animal*, and my 'Identità personale e entità personale'.

is likely to strike us as not so far-fetched.[264] And once we take seriously the hypothesis that Descartes is an animal, this will call into question our initial inclination to think Descartes could go on existing in bodiless form (are (any) animals really, like souls, the sorts of things that could exist in either an embodied or a non-embodied state)?

To be fair, not everyone who thinks that Descartes's existing without B is metaphysically possible does so because they think that Descartes could exist in a bodiless state. Perhaps they think that Descartes, unlike B, could survive the envatment of his brain and the destruction of the rest of B.[265] Perhaps they think that Descartes, unlike B, could survive a very gradual but ultimately complete 'de-organification.' (Imagine replacing part of Descartes's heart with an electronic pacemaker, his hip joint with an artificial hip joint, and keep on going in this way until all the organic parts of Descartes have been replaced by inorganic substitutes.) Perhaps they think that Descartes could survive his being 'resited' in another body via an instantaneous transfer of information from B's brain to the brain in a different ('blank-brained') body B′, and the subsequent destruction of B.[266] Again, though, I think that, if we are initially strongly inclined to think that Descartes could exist without B in either the envatment scenario, or the de-organification scenario, or the re-siting-in-a-new-body-cum-destruction-of-the-old-body scenario, that may well be because we are not attending to the hypothesis that Descartes is an animal.

It may helpful at this point to take stock. The first modal argument for the distinctness of Descartes from B is not an argument that someone who identifies persons with their bodies can dismiss as question-begging. She needs to say something about why it would be a mistake to regard Descartes's existing without B as a metaphysical, and not just an epistemic or 'merely conceptual', possibility.[267] As Kripke says ('N & N', 335), making this case is not a trivial task,

---

[264] Philosophers like Shoemaker who endorse a 'psychologistic' (broadly Lockean) conception of our nature and persistence conditions often hold that we are *not* animals (in the straightforward sense of being identical to some animal), although we share a body with some animal. But Shoemaker admits that the thesis that we are animals (in the straightforward sense) is initially at least very intuitively plausible. For various arguments to the effect that each of us is identical to some animal, see Olson, *The Human Animal*, and my 'Identità personale e entità personale'.

[265] Note that someone who wants to resist this conclusion could hold either that in the case under discussion what survives envatment is not Descartes, but only a part thereof, or that what survives envatment is not just a part of B, but B itself (in a drastically 'pared-down' form). Presumably B can survive some paring down, though it's far from clear that it could survive being pared down to a brain.

[266] For a defence of the possibility of resiting via information transfer (especially against 'duplication-based' objections), see S. Shoemaker, 'Personal Identity: A Materialist's Account', in S. Shoemaker and R. Swinburne, *Personal Identity* (Oxford: Basil Blackwell, 1984). For a critique of that defence, see my 'Identità personale e entità personale'.

[267] If a state of affairs is epistemically possible in the weak sense (since we don't know a priori that it does not obtain), but not in the strong sense (since we know it does not obtain), we could call it a 'merely conceptual' possibility. Hesperus' existing apart from Phosphorus is a merely conceptual possibility, at least according to ordinary, not especially exigent, standards for knowledge; New Hesperus' existing apart from New Phosphorus is a not merely conceptual, but also epistemic, possibility—what we could call a 'live epistemic possibility'.

but I do not think it is an impossible one. For reasons I hope are clear from what I have said so far, I suspect that we can make a case for the metaphysical impossibility of Descartes's existing without B only if we accept a broadly Aristotelian (broadly animalist) account of Descartes's nature and persistence conditions. Obviously, if we think of Descartes's essence in a Cartesian way (so that Descartes persists if and only if his immaterial mind does), it will be difficult to defend the metaphysical impossibility of Descartes's existing without B. Suppose, on the other hand, we think of Descartes's essence in a broadly Lockean way. That is, suppose we think that Descartes will continue to exist if and only if his mental life will go on, and that his mental life will go on if and only if there is an appropriately causally connected sequence of 'total mental states', starting from Descartes's current total mental state, the adjoining members of which sufficiently overlap in their contents (and there aren't two or more ('branching') sequences of this type). Then it will be possible for a mental life to start off in one brain and end up in another, and it will again be difficult to defend the metaphysical impossibility of Descartes's existing without B.

The second modal argument for the distinctness of Descartes from B is in certain respects less problematic than the first one. Unlike the first modal argument, it doesn't have a premiss that is in tension with animalism, and it has a premiss whose truth can be defended by appeal to what actually happens, and not what we can apparently imagine happening. (If bodies can outlast persons, bodies do (at least in some cases) outlast persons. So even if perchance B did not outlast Descartes, we can say more in favour of the possibility of B's outlasting Descartes than that we can imagine its happening; we can say that it is the sort of thing that actually does happen all the time.) Still, as Kripke says, there is room for resisting the second modal argument. Indeed, there is perhaps more room than Kripke concedes. As well as saying that human persons are human persons$_{DOA}$, and, like bodies—like the bodies they in fact are—(typically) end their days as corpses, the opponent of the second modal argument can say that human bodies are bodies$_L$, and, like humans—like the humans they in fact are—(necessarily) cease to exist at death.

A last remark on modal and temporal arguments for the distinctness of Descartes from B: in at least two places Kripke compares the second temporal argument for the distinctness of Descartes from B with Wiggins-type temporal arguments for the distinctness of a statue from the matter of which it is composed ('N & N', 354 n. 73; 'I & N,' 163 n. 19). The arguments in question are obviously structurally similar. But I think that the temporal argument for the distinctness of a statue from its matter is much closer to conclusive than the temporal argument for the distinctness of Descartes from B. As we have seen, one can resist the conclusion of the second temporal argument either by insisting that Descartes is a body$_{DOA}$, or by insisting that (i) Descartes's body is, as the *OED* puts it, a 'whole material organism viewed as an organic entity', and (ii) organisms go out of existence at death (in other words, by insisting that Descartes's body is an organism$_L$). While it is unclear just how much plausibility either of these moves has, it is, I think, clear that the corresponding moves in attempting to resist the argument for the distinctness of the statue from its matter are very implausible.

The move that corresponds to the insistence that Descartes is a body$_{DOA}$ would be the insistence that the bronze statue, like the bronze, will continue to exist even after it has been melted down (or melted down and subsequently poured into a different mould). To take this line is to suppose that 'statue' is a phase-sortal, and that the abiding sort to which a bronze statue belongs is something like *bit of bronze*—in much the way 'child' is a phase-sortal, and the abiding sort to which a human child belongs is *human being*. But on the face of it this analogy cannot be drawn. Suppose some Tibetan parents leave their child at a monastery when the child is 5 years old so that he can become a lama. Suppose they come back fifteen years later and say to the monk who receives them, 'Fifteen years ago, we left our child with you; where is he now?' The monk could truthfully answer, pointing at a 20-year-old, 'Over there.' By contrast, suppose I loan you my bronze statue of (the dragon) Smaug. You melt the bronze down and subsequently pour it into a new mould, ending up with a statue of Jerry Garcia. Some time later I knock on your door and say, 'A while ago, I loaned you my statue of Smaug. Where is it now?' You surely could not truthfully answer, pointing to the statue of Jerry Garcia, 'Over there.' (Suppose that, when I loaned you the statue, I said, 'I'll lend you my statue of Smaug if you promise to return it,' and you said 'I do.' If you then give me the statue of Jerry Garcia, you could hardly claim to have kept your promise!)

The move that corresponds to the insistence that Descartes's body is an organism$_L$ would be the insistence that the bronze of which a statue is made, like the statue made of the bronze, did not exist before it was poured into the mould in which it acquired its current shape, and will not survive the loss of its current shape. This again seems simply wrong. The bronze statue was made *from* the bit of bronze it is now made *of*, and x cannot be made from y, unless y was there before x.

So the identification of persons with their bodies is, at least in some ways, more defensible than the identification of artefacts with their matter. Still, the materialist who wants to defend the identification of Descartes with his body has her philosophical work cut out for her.

But why must she take on the job? Isn't the thesis that persons are their bodies surplus to materialist requirements? In his Cornell lectures on identity through time Kripke averred that someone could be 'quite a materialist' and accept that a person is distinct from his body. A materialist is presumably committed to saying that I am a (purely) material being (at least in the sense of having a purely material constitution). Perhaps a materialist is committed to identifying me with some body—where 'body' means '(purely) material substance'.[268] But why can't she just say that each person is (identical to) some (material) body, and leave it at that?[269]

It might be objected that, if I am a material substance, I must be some particular kind of material substance, and the only plausible candidate for the kind of material substance I am is *human body*. But why couldn't the kind be *human person*, or perhaps (if 'person' is a phase-sortal) *human animal*? Someone who thinks I am a human animal is not thereby committed to thinking I am a human body; if she believes in neither animals$_{DOA}$ nor bodies$_L$, she is in fact committed to denying that I am a human body.

So if a materialist wants to embrace a thesis about persons which is robustly materialist, but less contentious than the thesis that persons are their bodies, a natural candidate would be:

Persons are (identical with) (material) bodies,

where this means nothing more or less than

Persons are (purely) material things.

This last thesis is, however, mute on how a person is related to *her* body. What should the materialist say about this? Kripke suggests that a materialist might say that a person is nothing over and above her body, in much the way that a statue is nothing over and above the stuff of which it is made. But, Kripke thinks, the materialist who makes this move is not yet out of the woods:

One might say that [a statue] is 'nothing over and above' [the hunk of matter of which it is composed]; and the same device might be tried for the relation of the person and the body. The difficulties in the text would not then arise in the present form, but analogous difficulties would appear . . . ('N & N', 354 n. 73)

We might have expected the 'analogous difficulties' to take the form of a modal argument to the effect that Descartes could not have had the same constitution as B—say:

> If all of Descartes's parts were parts of B, then all of Descartes's parts were material parts (since B had only material parts). Nothing all of whose parts are material could exist without having any material parts. (In other words, nothing with a purely material constitution could have

---

[268] According to the *OED*, 'body' can mean 'a separate portion of matter, large or small, a material thing; something that has physical existence or extension in space'. When people speak of celestial bodies, or discuss whether there can be two bodies in the same place at the same time, it is presumably this (not specifically biological or organic) sense of 'body' they have in mind.

[269] Bernard Williams writes: 'Why should we not say that persons form a class of material bodies? . . . For the rest of this paper, I shall consider four leading objections to the view that persons are material bodies; what I say will . . . I hope, be discouraging, if no more, to the objections' ('Are Persons Bodies?', 70). In fact, though, many of the objections Williams considers (that Jones's body will outlast him, that 'Jones' and 'Jones's body' are not intersubstitutable *salva veritate* in certain contexts) are not objections to the view that persons are or form a class of material bodies; they are objections to the view that persons are *their* bodies. In the article Williams does not distinguish the two views.

a purely immaterial constitution.) Thus B could not have existed without having any material parts. But, whether or not Descartes ever *did* exist without having any material parts, he is the sort of being that *could* have existed (in a disembodied state) without having any material parts. So Descartes cannot have a purely material constitution, and cannot have the same constitution as B.[270]

In fact, the 'analogous difficulties' Kripke has in mind here are of a different sort:

A theory that a person is nothing over and above his body in the way that a statue is nothing over and above the matter of which it is composed, would have to hold that (necessarily) a person exists if and only if his body exists and has a certain additional physical organization. Such a thesis would be subject to modal difficulties similar to those besetting the ordinary identity thesis. ('N & N', 354 n. 73)

If I have not misunderstood Kripke, he is supposing that, if a statue S is nothing over and above the matter M of which it is composed, then

(1) Necessarily: S exists if and only if M exists and has a certain (additional) physical organization.

Similarly, if a person P is nothing over and above the body B that constitutes her, then

(2) Necessarily: P exists if and only if B exists and has a certain (additional) physical organization.[271]

But, Kripke seems to be saying, there are worries about whether (2) is true.

Kripke doesn't say what the worries are, and I'm not sure what he has in mind. Perhaps he is thinking of the (alleged) possibility that, say, Descartes could exist in a completely bodiless state, even after B had gone out of existence; perhaps he is thinking of the (alleged) possibility that Descartes could exist as a brain in a vat, even after B had gone out of existence; perhaps he is thinking of the (alleged) possibility that Descartes and Arnauld could switch bodies, even though their bodies remained in existence and retained the same physical organization they had before the exchange, and Arnauld could subsequently cease to exist; perhaps he is thinking of the (alleged) possibility that Descartes could go out of existence, even if B remained in existence and retained its (additional) physical organization. As I have already indicated, I have my doubts about all of these possibilities (construed as genuine, rather than merely conceptual or epistemic, possibilities); but I agree that the kind of

---

[270] The reservations expressed earlier in this chapter about the first modal argument for the distinctness of Descartes from B obviously also apply to this argument. Other worries about it are voiced by Shoemaker in 'On an Argument for Dualism', in S. Shoemaker and C. Ginet (ed.), *Knowledge and Mind* (Oxford: Oxford University Press, 1983).

[271] What might the additional physical organization be? Perhaps whatever physical organization is necessary and sufficient for B to be a living body.

(alleged) metaphysical possibilities that would rule out the identification of Descartes with B, should they be genuine, would also rule out the thesis that Descartes is nothing over and above B, where this last claim is understood as equivalent to (2).

What I find puzzling about the passage just cited is that it's not clear why the claim that a statue is nothing over and above its matter should be thought to entail (1). It is, I take it, true that a statue is nothing over and above its matter. But it is far from obvious that (1) is true.

Suppose that someone makes a statue out of bronze by pouring some liquid bronze into a mould and letting it harden. After many years the statue is melted down. The melted bronze comes to constitute some other object or objects, but is eventually melted down once again, and is then, by an amazing coincidence, recast by someone other than the person who made the original statue into a statue with exactly the same shape as the original statue (by being poured into a different mould which, purely by accident, has exactly the same size and shape as the mould used to cast the original statue). The statue that exists after the second casting is, we may suppose, a perfect duplicate of the one that existed after the first casting; but is it the very same statue?[272] That seems doubtful.

If, however, the statues in question are duplicate statues, rather than the same statue, then (1) is false. (Where S_1 is the original statue, S_2 is the statue produced by the second casting, t_1 and t_2 are (respectively) the moment S_1 begins to exist and the moment S_2 begins to exist, and M is the matter constituting both statues, S_1 *exists if and only if M exists and has such-and-such a physical organization* is true at t_1 and false at t_2, while S_2 *exists if and only if M exists and has that physical organization* is true at t_2 but false at t_1: hence neither biconditional is a necessary truth.)

Suppose, though, that in the case just described there is just one (twice-made) statue. It still seems very doubtful that (1) is true. For a statue is (I think, and I imagine Kripke thinks) essentially an artefact, made in a certain way. If the very same matter that actually constitutes the Statue of Liberty had come together randomly, with the exact same physical organization as the Statue of Liberty, the Statue of Liberty would not thereby have existed, even though a perfect duplicate of it would have.

So far we have been considering worries about whether a statue's matter's existing and having a certain physical organization is a sufficient condition for that statue's existing. But inessentialists about the material origin of artefacts, at least, will have doubts about whether a statue's matter's existing and having a certain physical organization is a necessary condition for that statue's existing. (If, as inessentialists about the material origin of artefacts maintain, the *QE II* could have built from a different batch of metal, the same presumably goes for the Statue of Liberty.) Even those who hold that artefacts have their

---

[272] For a discussion of this question, see Aquinas, *Commentary on the Sentences* IV, d. 44. 1. 2, *ad* 4.

material origin essentially should be wary of the idea that a statue's matter's existing and having a certain physical organization is necessary for that statue to exist. To return to the sort of example discussed in Chapter 2, suppose I made a statue out of wood, and subsequently artificially petrified it. It seems natural to say that in such a case there is one statue, which is first made entirely of wood, subsequently made partly of wood and partly of calcite, and finally made entirely of calcite. If this is right, then there is no bit of matter whose existence (at a time) is a necessary condition of the existence of the statue (at that time). (Even if none of the calcite matter had ever existed, the statue could still have come into existence, made of wood; even if all the wooden matter of the statue had been annihilated after the statue had been petrified, the statue could have still gone on existing.) Thus condition (1) is not satisfied by a hypothetical statue made first of wood and then of calcite. In that case, it would seem, it isn't satisfied by *actual* wooden statues either, since actual wooden statues could have survived petrification, even if they never in fact underwent it.

In short: although (2) may well be true, (1) looks false. If the claim that a statue is nothing over and above its matter entails (1), then it is false that a statue is nothing over and above its matter. I take it, though, that there is some sense in which a statue is nothing over and above its matter. So the materialist can say that in that same (non-(1)-entailing) sense, a person is nothing over and above her body.

What might that sense be? If a statue is nothing over and above the bronze constituting it, and a forest is nothing over and above the trees constituting it, then there is nothing more to the statue than the bronze, and nothing more to the forest than the trees. So we might say that *a* is nothing over and above *b* just in case only parts of *b* are parts of *a*. Thus construed, the nothing-over-and-above relation differs from identity in that an individual can bear that relation to different things at different times. For example, a forest may now be nothing over and above these trees, even though 300 years from now it will be nothing over and above those trees.

If we construe the nothing-over-and-above relation in these (compositional) terms, and construe materialism about persons as the claim that persons are never anything over and above their bodies, then materialism about persons is the claim that the only parts persons ever have had or ever will have are parts of their bodies. Thus understood, materialism about persons entails that I neither am nor have an immaterial soul; but it is compatible with the claim that Descartes is not his body, and the claim that Descartes's body's existing and having a certain physical organization is neither necessary nor sufficient for his existing. Thus understood, materialism about persons does not commit one to any thesis as contentious as animalism. And thus understood, materialism about persons does not rule out its being metaphysically possible for Descartes to exist as an envatted brain, even though B has ceased to exist; does not rule out its being metaphysically possible for Descartes to exist as a no-longer-organic being, even though B has ceased to exist; and does not even rule out its being metaphysically possible for Descartes to exist as a no-longer-material thing, even though B has ceased to exist.[273]

Kripke considers two materialistic accounts of the person–body relation: one according to which a person is the same as her body, and another according to which a person is 'nothing over and above' her body. He suggests that the first account is problematic, inasmuch as it is subject to certain modal and temporal counter-arguments; and he suggests that the second account is subject to the same sort of difficulties that beset the first. While this may be true of the second account *as formulated by Kripke*, there is another (to my mind, more natural) way of formulating the second account (of spelling out the idea that persons are 'nothing over and above' their bodies). If the second account is formulated in the way suggested above, it is—as best I can see—not subject to any insuperable, or even pressing modal (or temporal), difficulties.

2, TOKEN-IDENTITY THEORIES AND TYPE-IDENTITY THEORIES

As I noted at the beginning of the last section, Kripke thinks that there are powerful modal arguments, of a broadly Cartesian sort, against the claim that mental events are (identical to) physical events, and against the claim that mental properties are (identical to) physical properties.[274] In both 'Identity and Necessity' and *Naming and Necessity* Kripke devotes most of his attention to Cartesian arguments against the identification of (certain sorts of) mental properties with (certain sorts of) physical properties. In *Naming and Necessity* (and in 'Identity and Necessity'), though, he briefly sets out the following argument against the identification of (particular) mental events with (particular) physical events:

Let 'A' name a particular pain sensation, and let 'B' name the corresponding brain state, or the brain state some identity theorist wishes to identify with *A*. *Prima facie*, it would seem that it is at least logically possible that B should have existed (Jones' brain could have been in exactly that state at the time in question) without Jones feeling any pain at all, and thus without the presence of A. ('N & N', 335)

If X is a pain and Y is the corresponding brain state . . . it *seems* clearly possible that . . . the brain state should have existed without being felt as pain. ('I & N', 162 n. 17)

---

[273] For a very neat discussion of how a materialist can allow that Descartes might some day be a non-material being, see Shoemaker, 'On an Argument for Dualism'.

[274] The view that mental events are (identical to) physical events is often called the token-identity theory; the view that mental properties are (identical to) physical properties is often called the type-identity theory.

If, however, it is possible that B should have existed without A, then A and B cannot be identified. So, Kripke concludes, there is a valid argument from initially plausible premises, whose conclusion is that a particular sensation cannot be identified with any particular cerebral event.

Also, Kripke avers,

> Just as it seems that the brain state could have existed without any pain, so it seems that the pain could have existed without the corresponding brain state. ('N & N', 336)

> If X is a pain, and Y is a corresponding brain state . . . it seems clearly possible that X should have existed without the corresponding brain state. ('I & N', 162 n. 17)

The first argument goes: Jones's brain state B could have occurred, even if (Jones had felt no pain, and thus even if) Jones's pain A had not occurred. So B only accidentally co-occurs with A; so B ≠ A.

A proponent of the identity of pains with neural events might ask: Why *does* it prima facie clearly seem possible for the brain state B to occur without the pain A? I take it Kripke would respond to this challenge along these lines: B occurs when certain neurons in Jones's brain fire in a certain way at a certain time. We certainly appear to be able to imagine those neurons firing that way at that time, even though Jones feels no pain. Also, because pains are essentially pains,[275] and are essentially the pains of whomever they are actually the pains of,[276] a pain that is actually Jones's pain could not have occurred without being Jones's pain. And, because *being a pain (of Jones)* implies *being felt as a pain (by Jones)*,[277] no pain that is actually a pain of Jones could occur without being felt as a pain by Jones. So A couldn't occur without Jones's feeling A as a pain; so A couldn't occur without Jones feeling pain. Thus, if there are possible circumstances in which Jones's neurons fire in the right way at the right time but Jones feels no pain, then there are circumstances in which B occurs but not A. And it certainly *looks* as though there are such circumstances.

The second argument goes: the pain A could have occurred, even if the brain state B had not occurred. So A only accidentally co-occurs with B; so A ≠ B.

Again, the proponent of the identity of pains with neural events might ask: Why does it seem prima facie clearly possible for the pain A to occur without the brain state B? I take it Kripke would respond along these lines:

---

[275] 'If X is a pain . . . then *being a pain* is an essential property of X' ('I & N', 162 n. 17).

[276] I don't know of any place where Kripke explicitly says that pains are essentially of whomever they are actually of. But without this premiss, it seems that the argument under discussion at 'N & N', 336, won't go through. (Suppose that the pain A could occur without being Jones's pain. Then we could not infer from the fact that Jones felt no pain, and thus did not feel A as a pain, that A was not felt as a pain, and thus was not a pain, and thus did not occur.) The argument under discussion at 'I & N', 162 n. 17, on the other hand, does not seem to depend on the assumption that pains are essentially 'owned' by whoever actually 'owns' them. Changing the variables to accord with the 'N & N' passage, that argument goes: B could occur without being felt as pain. But A could not occur without being felt as pain. So A ≠ B.   [277] See 'I & N', 162 and n. 18.

A occurs when Jones feels a certain way at a certain time. We certainly appear to be able to imagine Jones's feeling that way at that time, even though Jones's neurons were not firing in the way they would have to be firing in order for B to occur. (It is assumed here that, if B actually involves neurons firing in such-and-such way, then B could not have occurred without neurons firing in that way.[278]) Since Jones's feeling the right way at the right time is sufficient for the occurrence of A, and apparently compatible with the non-occurrence of B, it again looks as though there are circumstances in which A occurs but not B.

The intuitions driving the two arguments for the distinctness of A from B under consideration could be put this way: A is an event with a phenomenological essence. That is, A's presence in the world by itself guarantees that something feels a certain way. B is an event with a physical-structural essence: its presence in the world by itself guarantees that neurons are firing a certain way (that brain cells are configured in a certain way). But A does not have the physical-structural essence of B, and B does not have the phenomenological essence of A; so A and B must be distinguished (twice over).

A defender of the identity of A with B could resist this line of thought either by arguing that A's presence in the world does not by itself guarantee anything's feeling a certain way, or by arguing that B's presence in the world does by itself guarantee something's feeling a certain way.

As Kripke observes, identity theorists have often taken the first line. They have often supposed that *being a pain* is an accidental property of a physical state—say, the property of having a certain causal role (of having certain typical causes and effects), which that physical state might not have had. But he thinks such a view is completely indefensible:

The difficulty [in identifying *A* with *B*] can hardly be evaded by arguing that although *B* could not exist without *A*, *being a pain* is a merely contingent property of *A*, and that therefore the presence of *B* without pain does not imply the presence of *B* without *A*. Can any case of essence be more obvious than the fact that *being a pain* is a necessary property of each pain? The identity theorist who wishes to adopt the strategy in question must even argue that *being a sensation* is a contingent property of *A*, for *prima facie* it would seem logically possible that *B* could exist without any sensation with which it might plausibly be identified. Consider a particular pain, or other sensation, that you once had. Do you find it at all plausible that *that very sensation* could have existed without being a sensation, the way a certain inventor (Franklin) could have existed without being an inventor? ('N & N', 335)[279]

---

[278] See 'N & N', 336: 'Note that *being a brain state* is evidently an essential property of *B* (the brain state). Indeed, even more is true: not only being a brain state, but even being a brain state of a specific type is an essential property of *B*. The configuration of brain cells whose presence at a given time constitutes the presence of *B* at that time is essential to *B*, and in its absence *B* would not have existed.'

[279] Also see 'I & N', 162 n. 17: 'that a certain pain *X* might have existed, yet not have been a pain . . . seems to me self-evidently absurd. Imagine any pain: is it possible that *it itself* could have existed, yet not have been a pain?'

Here Kripke does not exactly argue for the view that pains are essentially pains. I imagine he thinks there is no point in doing so, since the premises of such an argument would be no more plausible than its conclusion, which conclusion is the denial of something 'self-evidently absurd'. This makes it hard to know how to adjudicate the dispute between Kripke and philosophers like Lewis, who are quite happy with the idea that pains are only accidentally pains in virtue of only accidentally having the causal properties that make a thing a pain.

For what it is worth, the idea that a pain might not have been a pain does not strike me as self-evidently absurd. Perhaps a sensation is a pain only if it has some feature that it might have been present without having. Suppose that a sensation is a pain only if it surpasses a certain threshold of a certain kind of (felt) intensity. And suppose that a sensation that actually has a certain degree of intensity of that kind might still have occurred, even if it had been slightly less intense (in that way) than it actually was. If both of these things are true, it seems there could be a sensation which was 'just barely' a pain (was just barely intense enough (in the right way) to count as a pain) and might have occurred without being a pain (if it had been occurred, but been a little less intense (in that way)).

It might be objected here that the slightly less intense sensation that might have occurred is numerically as well as qualitatively different from the pain (just barely) that actually occurred. Perhaps it is. It's hard to say. We do sometimes talk as though one and the same sensation could increase or decrease in intensity, just as one and the same headache can get worse or better. ('That strange sensation has almost disappeared', we might say.) And if the intensity of a particular sensation can vary across time, it presumably can also vary across possible worlds. On the other hand, if someone were to say that, in the kind of situation in which we might say that a strange sensation is gradually disappearing, we really have a series of numerically as well as qualitatively different sensations, ordered by decreasing intensity, that wouldn't obviously seem like a misdescription. Compare:

(a) The man next door is gradually losing his hair.
(b) The sensation (or the pain) is gradually becoming less intense.
(c) The distance between the earth and the sun is gradually shrinking.

It would be daft to think that when, as we might say, the man next door is gradually losing his hair, we have a series of numerically as well as qualitatively different men, ordered by decreasing hairiness: the very same man next door who had a little more hair yesterday has a little less hair today. On the other hand, it would not be daft to say that when, as we might say, the distance between the earth and the sun is gradually shrinking, we have a series of numerically different distances, ordered by the less-than relation: it is not the case that the very same distance between the earth and the sun that was *m* kilometres yesterday is *m* 2 *n* kilometres today.[280] (*b*) seems to me a case intermediate between (*a*) and (*c*): it is not immediately obvious whether we should assimilate it to (*a*) or to (*c*).

Also, consider a case in which (to put it neutrally) I first have an intensely painful sensation, then a moderately painful sensation, then a no longer painful but almost painful sensation, and finally no sensation at all. (Those who have trouble imagining this sequence need only grab hold of some nettles, as I once did shortly after arriving in England.) Perhaps one could say that, as the sensation caused by grabbing hold of the nettles decreased in intensity, it got less and less painful, until it wasn't painful at all, and became a 'mere tingle', before disappearing altogether. But, once it stopped hurting—once I no longer had any sensation that could be described as a pain (rather than, say, a tingling)—I would naturally say 'The pain has gone' (or, 'The pain has disappeared'). I would not naturally say, 'The pain I had a minute ago is still present, though it is not a pain any more.' If, however, there were just one sensation, which started out as an intense pain and subsequently became (a moderate pain, and a mild pain, and then) an ex-pain, then it would be true that the pain I had before is still present, though it is currently a tingle and not a pain, in the same way that it is true that the child we had in 1984 still exists, though she's currently an adult and not a child. All this suggests that the pain caused by the nettles is absent, unless it is a pain, as Kripke maintains.[281]

Whether or not something that was a pain (but just) might have been not a pain (though almost), it does seem, if not self-evident, at least initially plausible that a pain is essentially a sensation. Supposing that this very pain could have been present without being a pain or any other kind of sensation is, on the face of it, rather like supposing that the very walk you took could have happened without being a walk, or a trot, or a stumble, or . . .

The idea that a pain might not have felt bad or indeed any way at all is a strange one. But defenders of the identity of A with B needn't embrace it. Rather than saying that the one event that is both A and B has a physical-structural essence but not a phenomenological essence, they could say that that one event has an essence with two aspects—one physical-structural and one phenomenological. The mere presence of A—and thus the mere presence of B—is sufficient to guarantee that something feels a certain way; the

---

[280] If the earth was *m* kilometres away from the sun yesterday, and is *m* − *n* kilometres away from the sun today (for *n* ≠ 0), then the earth is at a *different* distance from the sun today than it was yesterday (is not at the *same* distance from the sun today as it was yesterday). One could hardly say: 'The earth is at the same distance from the sun today as it was yesterday, though that distance has shrunk.'

[281] On the other hand, it is at least arguable that 'bit of ice' is a phase-sortal for something whose abiding sort is *bit of water* (*bit of* H_2O). If this is so, when a bit of ice melts, it stops being ice, but it doesn't stop being (full stop). Even so, after it has melted, I might say, 'The ice in my drink is gone.'

mere presence of B—and thus the mere presence of A—is sufficient to guarantee that neurons fire a certain way (or that brain cells are configured in a certain way).

Granted, the champion of the identity of A with B might say,

> It is not immediately obvious that A has a physical-structural essence, or that B has a phenomenological essence. If we have fixed the reference of A by saying, A shall be the (painful) feeling I am having now, then it will be a priori for us that, if A is present, something feels a certain way; by contrast, it will be a posteriori that, if B is present, something feels a certain way. Similarly, if we have fixed the reference of 'B' by saying, B shall be the neural event of this type going on now, then it will be a priori for us that, if B is present, then neurons fire, although it will be a posteriori that, if A is present, then neurons fire. Because we don't always clearly distinguish what might, for all we know (or might, for all we know, a priori), be true from what might (metaphysically) be true, we are subject to the illusion that it is (metaphysically) possible for A to be present without neurons firing, or for B to be present without anything feeling any way.
>
> Also, for the sort of reasons Kripke has brought out, we are subject to the illusion that the falsehood of $A = B$ is metaphysically as well as epistemically or conceptually possible. Where R_1 and R_2 are co-referential rigid designators, associated (respectively) with reference-fixing descriptions D_1 and D_2, and $D_1 = D_2$ is contingent, this often creates the illusion that $R_1 = R_2$ is contingent (cf. 'N & N', 333–4): and this is just the sort of situation we get with A and B, if their reference is fixed in the way suggested above. (Although one and the same thing is both B and the painful feeling I am having now, something else might have been the painful feeling I am having right now, in which case it would be false that the neural event of this type = the (painful) feeling I am having right now.)
>
> Furthermore, if I am having the sensation A now, there is a possible world in which I am in the same (qualitative) epistemic situation I am actually in (everything, as it were, looks just the same to me as it actually does), and an epistemic counterpart of A is present, even though neither B nor any epistemic counterpart thereof is present. This gives rise to the illusion that the falsehood of $A = B$ is a metaphysical, as well as an epistemic or a conceptual, possibility.

In the previous section of this chapter I suggested that, if someone has good reasons to suppose that Descartes = (his body) B, she needn't be worried about the appearance that Descartes could have existed without B (or vice versa), since she can explain (away) that appearance in much the way that Kripke explains (away) the appearance that Hesperus might have existed without Phosphorus (or vice versa). It now looks as though, if someone has good reasons to suppose that (the pain) A = (the brain state) B, she again needn't be worried about the appearance that A could have existed without B (or vice versa), since she can explain (away) that appearance along Kripkean lines. The point may be worth emphasizing, for two reasons. Although Kripke does not, so far as I know, think that there are compelling considerations in favour of the identification of persons with their bodies, he seems to concede that there are (at least some) compelling motivations in favour of the identification of pains with neural events.[282] And, although Kripke does not explicitly say that someone who identifies persons with their bodies will be unable to explain away the apparent separability of a person from his body (or vice versa) along Kripkean lines, he does express doubt that someone who identifies mental events with physical events will be able to explain away the apparent separability of the former from the latter (or vice versa) along those lines.[283]

Here someone might object that the strategy for explaining away the apparent possibility of A without B breaks down at a crucial point. He might say:

As Kripke says at 'N & N', 339, 'the notion of an epistemic situation qualitatively identical to one in which the observer had a sensation S simply *is* one in which that observer had that sensation'. So it's no good saying that we are subject to the illusion that the sensation A could be present without the neural event B, because we mistake that impossible state of affairs for a (metaphysically, as well as epistemically or concep-tually) possible state of affairs in which we are in a qualitatively identical epistemic situation, and an epistemic counterpart of A is present without B. If in an epistemic situation qualitatively identical to one in which an observer has a sensation A, an epistemic counterpart of the sensation A is present, so is A itself.

Suppose Kripke held that, if I am actually having a particular pain, then in any possible world in which I am in an epistemic situation qualitatively identical to the one I am actually in, and have an epistemic counterpart of the particular pain I am actually having, I am having that particular pain. Then, it seems, he would be mistaken. Suppose that I become consciously aware of a pain that seems to have been in my experience for some time before

---

[282] See 'N & N', 355 n. 77: 'identity theorists have presented positive arguments for their view, which I certainly have not answered here. Some of these arguments seem to me to be weak or based on ideological prejudices, but others strike me as highly compelling, arguments which I am at present unable to answer convincingly.' It is not entirely clear whether by 'their view' Kripke has in mind a type-identity theory or a token-identity theory, but it makes no difference for present purposes, since the identification of mental properties with neural properties implies the identification of mental events with neural events.

[283] Cf. 'N & N', 341: 'Suffice it to say that I suspect that the theorist who wishes to identify various particular mental and physical events will have to face problems fairly similar to those of the type–type theorist; he too will be unable to appeal to the standard alleged analogues.'

I attended to it. I baptize that pain A. I subsequently become distracted, and cease to have A 'in the front of my mind'; I also forget that I baptized my pain A. A while later I become aware of a pain that seems to me to have been in my experience for some time before I attended to it, and I baptize that pain A'. This new baptismal act jogs my memory, and I remember that I had earlier baptized a pain I was having A. At this point, I might wonder whether A = A'. Perhaps there was just one ongoing pain, which I baptized A shortly after attending to it, subsequently 'disattended to', and then reattended to, and rebaptized A'. Perhaps A stopped shortly after I stopped attending to it, and A' is a brand new pain that just happens (as best I can remember) to feel exactly the way A felt. If A is not in fact A', then it is metaphysically impossible that A = A'. But even if A ≠ A', there is still a possible world in which I am in an epistemic situation qualitatively identical to the one I am actually (currently) in, in which I am having a pain that is an epistemic counterpart of A. So a pain can, after all, be an epistemic counterpart of a particular pain in an epistemic situation qualitatively identical to the actual one, without being that particular pain.

In fact, though, I don't think Kripke is committed to the view that, if I am actually having a particular pain, then in any possible world in which I am in an epistemic situation qualitatively identical to the one I am actually in, and have an epistemic counterpart of the particular pain I am actually having, I am having that particular pain. An expression like 'the experience of X' or 'the sensation of Y' is ambiguous; it can denote either a token or a type. On one understanding of 'experience', if you are different from me, then I can't have your experiences, even if my experience is exactly like yours.[284] And (probably) I can't have any of my experiences more than once. On another understanding of 'experience', if you tell me about a curious experience you had yesterday, and I tell you that I have never had that experience, what I say is not a trivial truth; and if I tell you that I have had that experience on many different occasions, what I say is not a trivial falsehood. In the same way, if I say, 'Ugh: that's the sensation I get whenever I go on a roller coaster', I say something whose truth does not imply that I will only get on a roller coaster once in the course of my life.

When Kripke says, 'the notion of an epistemic situation qualitatively identical to one in which the observer had a sensation S simply *is* one in which that observer had that sensation', one might think he has in mind sensations as tokens. But the context of the passage makes it look as though what Kripke has in mind here is sensations as types. Kripke is committed to the idea that any epistemic counterpart of the sensation type pain present in an epistemic situation qualitatively indiscernible from the actual one just is the sensation

---

[284] Unless perhaps you and I can share a mind, as some philosophers think: see e.g. Lewis's postscript A to 'Survival and Identity'.

type pain.[285] But that does not (in any obvious way) commit him to the view that any epistemic counterpart of a given token of pain present in an epistemic situation qualitatively identical to the actual one is that token of that type.

Thus someone who finds it plausible that any epistemic counterpart of pain is pain might think that Kripke has put his finger on a good argument against the identity of sensation types with brain-state types, which, however, has no force against the identity of sensation tokens with brain-state tokens. Philosophers inclined to something like anomalous monism might take this to be grist for their mill.

On the other hand, there is a resistible but not initially implausible line of thought that would take us from the distinctness of sensation types from brain-state types to the distinctness of sensation tokens to brain-state tokens. Sensation tokens and brain-state tokens are particular events. Particular events are changes (inconstancies) or 'unchanges' (constancies). (An object's accelerating is a change; an object's continuing to move (or not move) at the same velocity is an unchange.) Now, it is natural to suppose that a change consists in an individual's coming (or ceasing) to be a certain way at a certain time, and that an unchange consists in an individual's continuing to be a certain way at a certain time. And it is initially, at least, not implausible to suppose that, if one and the same individual comes to be (or continues to be, or ceases to be) two different ways at the same time, we have two different changes (or two different unchanges), and thus two different events.[286] Suppose, then, that at one and the same time Jones acquires both a phenomenological property and a certain neural property. If the phenomenological property (*having this type of sensation*) and the neural property (*having a brain in such-and-such condition*) are identical, then nothing prevents us from identifying the change which is Jones's coming to have that sensation with the change that is Jones's coming to have a brain in such-and-such a condition. If, however, the phenomenological property Jones acquires at the relevant time is different from the neural property he acquires at that time, there seems to be a case for saying that Jones's acquiring the phenomenological property is one change (one event) and Jones's acquiring the neural property is another change (another event).[287]

---

[285] See 'I & N', 163 n. 18: 'For a sensation to be *felt* as pain is for it to *be* pain.' Assuming that an epistemic counterpart of pain feels like pain, this implies that any epistemic counterpart of the sensation type pain is pain. Note that in this sentence what Kripke means by 'a sensation' is clearly a sensation type, rather than a sensation token, since a sensation token couldn't be pain (as opposed to *a* pain).

[286] See J. Kim, 'Causation, Nomic Subsumption, and the Concept of an Event', *Journal of Philosophy*, 70 (1973), 217–36, for a conception of events along these lines.

[287] The case seems especially strong if the phenomenological property and the neural property are each determinate (rather than determinable) properties, and the properties are mutually independent (that is, either could have been instantiated without the other). As we shall see, Kripke argues for the distinctness of phenomenological properties from neural properties from the mutual independence of phenomenological properties and neural properties.

Returning to our main topic: as I indicated earlier, the argument for the non-identity of sensation types with brain-state types is the anti-materialist argument that Kripke sets out in the greatest detail and shows the greatest sympathy for. That argument has the following structure:

> Suppose that pain is identical to some neurophysiological state—say, the stimulation of C-fibres. In that case, 'Pain = C-fibre stimulation' will be either contingently true or necessarily true. Various philosophers have supposed it is only contingently true. But this cannot be right, since 'pain' and 'C-fibre stimulation' are both rigid and identity statements flanked by rigid designators are necessarily true or necessarily false. So if 'Pain = C-fibre stimulation' is true, it is necessarily true. But it certainly looks as though someone's C-fibres could be being stimulated, even though she is not in pain. And it looks as though someone could be in pain, even though her C-fibres are not being stimulated. If, however, 'pain' and 'C-fibre stimulation' rigidly designate one and the same state type, then it is impossible for someone's C-fibres to be stimulated, even though she is not in pain, or for someone to be in pain, even though her C-fibres are not being stimulated. So, it would seem, 'Pain = C-fibre stimulation' is no more necessarily true than it is contingently true.

How do we know that 'pain' and 'C-fibre stimulation' are rigid? Kripke simply assumes that 'C-fibre stimulation' is rigid, but notes that nothing hangs on the assumption. For suppose that 'C-fibre stimulation' non-rigidly designates a physical state that pain is identical to. Then we can introduce a term T that rigidly designates what 'C-fibre stimulation' non-rigidly designates, and run the above argument to show that pain is not T, and thus is not C-fibre firing ('N & N', 337).

How do we know that 'pain' is rigid? Well, Kripke says, it seems absurd to suppose that pain might have been some phenomenon other than the one it actually is ('N & N', 333). 'Pain' is rigid, just like 'heat', or 'light', or other names of physical types.

Some materialists disagree. As we have noted, Lewis thinks that what makes a particular pain a pain is its (accidentally) being apt (in certain circumstances) to have certain sorts of causes and certain sorts of effects. Similarly, Lewis holds, what makes the *type* pain the type pain is its (accidentally) having certain typical causes and effects:

The concept of pain . . . is the concept of a state that occupies a certain causal role, a state with certain typical causes and effects. It is the concept of a state apt for being caused by certain stimuli and apt for causing certain behavior. Or better, a state apt for being caused in certain ways by stimuli plus other mental states and apt for combining with other mental states to jointly cause certain behavior.[288]

---

[288] Lewis, 'Mad Pain and Martian Pain', in Lewis, *Philosophical Papers*, i. 124.

Because the concept of pain is a causal-role concept, Lewis thinks, 'pain' is non-rigid:

> If the concept of pain is the concept of a state that occupies a certain causal role, then whatever state does occupy that role is pain. If the state of having neurons hooked up in a certain way and firing in a certain pattern is the state properly apt for causing and being caused, as we materialists think, then that neural state is pain. But the concept of pain is not the concept of that neural state. The concept of pain, unlike the concept of that neural state which is in fact pain, would have applied to some different state if the relevant causal relations had been different. Pain might not have been pain . . . Something that is not pain might have been pain.[289]

> Kripke vigorously intuits that some names for mental states, in particular 'pain', are rigid designators: that is, it is not contingent what their referents are. I myself intuit no such thing, so the non-rigidity imputed by the causal-role analysis troubles me not at all.[290]

Why doesn't Lewis hold that 'pain' rigidly designates some neural state— say, the firing of C-fibres? Presumably because he intuits that

(1) It might have been that: (C-fibre-less) aliens were in pain

and

(2) Pain might have been a state of (C-fibre-less) aliens,

neither of which could be true if 'pain' rigidly designated C-fibre firing (and both of which could be true if 'pain' non-rigidly designated C-fibre firing).

But if 'pain' designates C-fibre firing (albeit non-rigidly), it could not be true that

(3) Pain is a state that (C-fibre-less) aliens might have been in.

(2) will be true (or at least have a true reading) for the same reason that

> The current president of the United States might have been Dan Quayle

has a true reading; (3) will have no true reading for the same reason that

> The current president of the United States is a person that might have been Dan Quayle.

has no true reading. This seems (to me) an unwelcome result, inasmuch as (2) and (3) seem (to me) have the same truth-conditions: pain might have been found in C-fibre-less aliens just in case pain is a (kind of) state that C-fibre-less aliens might have been in.

Also, suppose that 'pain' non-rigidly designates C-fibre firing. Then, even though it will be true (on one reading) that pain might have been a state that (C-fibre-less) aliens were in, it will not be true (on any reading) that

[289] Ibid. 125.   [290] Lewis, 'Reduction of Mind', 41.

(4) Pain might have been a state that (C-fibre-less) aliens as well as creatures exactly like us were in.

((4) will be false for the same sort of reason as 'The current president of the United States might have been both Dan Quayle and Bill Clinton'.) This, too, seems an unwelcome result. I can see how someone might hold that (since 'pain' rigidly designates C-fibre-firing) pain couldn't have been a state C-fibre-less aliens were in. And I can see how someone might hold that pain could have been a state C-fibre-less aliens were in. But it seems odd to hold that pain could have been a state of C-fiber-less aliens were in if, but only if, it weren't a state that creatures just like us were in.

Here is another way to put essentially the same point. While I can say that *x* might have been F, even if I know that *x* is not in fact F, I cannot say that *x* might *be* F under those circumstances. (If I know that you are out, I can say 'You might have been home right now', but not 'You might be home right now'.) Lewis's account allows us to say that

(5) Pain might have been a state that creatures radically unlike us physically were in.

But it doesn't allow us to say that

(6) Pain might be (may be) a state that creatures radically unlike us physically are in.

For, Lewis would say, we know that pain is a (neural) state of us. So, he would have to say, we know that pain is not a state of any creature too physically unlike us to have neural states. Again, though, it seems very odd to accept (5) and reject (6). I should have thought that one of the main attractions of not supposing that 'pain' rigidly designates C-fibre firing is that it allows us to say that we may some day find out that creatures very unlike us (say, Lewis's Martians) are in pain.

I have offered some (indirect) arguments against Lewis's account of pain as non-rigid. Kripke argues more directly (and perhaps as or more persuasively) that 'Pain might have been something else (some other phenomenon)' is obviously false. All these arguments depend on intuitions that, as we have seen, Lewis does not share. How are we to know that Kripke's intuitions (and mine) are better than Lewis's?

I don't know to how to *show* that Lewis's intuitions are less good than Kripke's or mine, any more than I would know how to show that someone who does not intuit the falsehood of 'Richard Nixon might have been some other man' has bad intuitions.[291] This worries me, but I take comfort from the

---

[291] I am tempted to say something like: 'pain' is a name of a state; names are rigid; so 'pain' is rigid. But Lewis denies that (all) names are rigid; he explicitly characterizes 'pain' as 'a *contingent* name—that is, a name with different denotations in different worlds' ('An Argument for the Identity Theory', in

fact that Lewis does not claim to intuit the non-rigidity of 'pain' (he only claims not to intuit its rigidity), and the fact that (as Lewis concedes) the view that 'pain' (like 'heat', 'light', and so on) is rigid is widely regarded as intuitively plausible (by materialists and non-materialists alike).[292] It would seem at least somewhat less surprising if relatively few philosophers failed to intuit rigidity where it was than it would if a great many philosophers intuited rigidity where it wasn't.

Suppose, though, that Lewis is right about the non-rigidity of 'pain'. It doesn't look as though that it is sufficient to block the argument under consideration against identifying (certain) mental states with neural states. For that argument could simply be reformulated as follows:

> Pain is a (kind of) feeling. Suppose that that (kind of) feeling is identical to C-fibre stimulation. Then *That (kind of) feeling = C-fibre stimulation* is either contingently true or necessarily true. It cannot be contingently true, because 'that (kind of) feeling', like 'C-fibre stimulation', is rigid (demonstratives rigidify). So it must be necessarily true. But it certainly looks as though someone's C-fibres could be stimulated, even though she is not having that (kind of) feeling. And it looks as though someone could have that (kind of) feeling, even though her C-fibres are not being stimulated. If, however, 'that (kind of) feeling' and 'C-fibre stimulation' rigidly designate one and the same state type, then it is impossible for someone's C-fibres to be stimulated, even though she is not having that (kind of) feeling, or for someone to have that (kind of) feeling, even though her C-fibres are not being stimulated. So, it would seem, *That (kind of) feeling = C-fibre stimulation* is no more necessarily true than it is contingently true.

Those who would defend the truth of *That (kind of) feeling = C-fibre stimulation* in the face of this argument will have to insist that that identity statement is necessarily true. This might not look too hard to do. In the argument the only reason provided for thinking that *That (kind of) feeling = C-fibre stimulation* is not necessarily true is that it looks as though we could have that kind of feeling without C-fibre stimulation, or have C-fibre stimulation without that kind of feeling. But *That (kind of) feeling = C-fibre stimulation* is a theoretical identification, like *Hesperus = Phosphorus*, or *Water = H₂O*, or *Heat = molecular motion*. In the last three cases, Kripke argued, the identity statements are necessarily true, even though we are subject to the illusion that 'either' of the things identified could be present

Lewis, *Philosophical Papers*, i. 101). Against this, I would want to say that it is very strange to call an expression a name—even if it looks like one syntactically—if it is non-rigid. To use an example of Lewis's, 'Miss America' is a non-rigid designator that looks like many genuine names (e.g. 'Miss Brody'). But we wouldn't say that 'Miss America' was a name of anyone, just because it is (temporally and modally) non-rigid.

[292] Cf. Lewis, 'Events', in Lewis, *Philosophical Papers*, ii. 253 n. 8.

without 'the other'. So why can't the (type-)identity theorist say the same thing about the first case?

Kripke grants that the (type-)identity theorist might be able to show that *That (kind of) feeling = C-fibre stimulation* is a necessary truth that doesn't look like a necessary truth. But, he thinks, the prospects for doing so are not bright ('N & N', 337–8). And, Kripke tries to show, the (type-)identity theorist will not be able to tell the same story about why *That (kind of feeling) = C-fibre stimulation* is necessary, though it appears not to be, that Kripke has told about why, say, *Hesperus = Phosphorus*, or *Water = H$_2$O*, or *Heat = molecular motion* is necessary, though it appears not to be.

As we have seen, Kripke has an account of why (*a*) it is necessary that Hesperus = Phosphorus, even though (*b*) it does not (initially, at any rate) look necessary that Hesperus = Phosphorus. It may not hurt to say a bit more here about the Kripkean defence-cum-explanation of the necessity of *Hesperus = Phosphorus*. It goes like this: we know that 'Hesperus' and 'Phosphorus' are rigid (since we know that there is no true reading of 'Hesperus might not have been Hesperus' or 'Phosphorus might not have been Phosphorus'). Whence we may infer that *Hesperus = Phosphorus* is necessarily true, if true. And it is true. (How do we know it is true? Kripke does not say explicitly, but the following description of one way in which we could know it is true seems consonant with Kripke's views:[293] We know that Hesperus is a celestial body and that Phosphorus is a celestial body. Also, given what we know about where Hesperus and Phosphorus are at certain times, and about how celestial bodies move, we can infer that Hesperus and Phosphorus are in some of—indeed, in all of—the same places at all the same times.[294] And we know that different celestial bodies don't occupy all the same places at all the same times. So we know that Hesperus and Phosphorus are the same celestial body, and thus that *Hesperus = Phosphorus* is true.)

Having established the necessity of *Hesperus = Phosphorus*, Kripke goes on to explain why *Hesperus = Phosphorus* (initially, at least) does not look like a necessary truth. Since I have already said a good bit about how the explanation goes in the previous section of this chapter, I shall not say more here.

The (parallel) Kripkean account of the necessity and apparent non-necessity of *Water = H$_2$O* goes like this: Since we know that 'water' and 'H$_2$O' are rigid, we know that *Water = H$_2$O* is necessarily true, if true. And it is true. (How do we know it is true? Again, Kripke is not very explicit on this matter. But an answer that appears consonant with Kripke's views would be: We know (through empirical investigation) that water is a substance with a particular

---

[293] I do not know whether the way I have suggested one could know that Hesperus = Phosphorus is the way in which astronomers actually came to know that Hesperus = Phosphorus.

[294] We can imagine one astronomer working out exactly where Hesperus was throughout a certain interval, and another astronomer working out exactly where Phosphorus was throughout that same interval. Comparing notes, they discover that, throughout that interval, Hesperus and Phosphorus are in exactly the same places at exactly the same times; whence they infer that Hesperus = Phosphorus.

chemical composition—the substance whose chemical name is 'H_2O'. So we know that water = H_2O.) Moreover, the same sorts of factors that give rise to the illusion that *Hesperus = Phosphorus* is not necessarily true give rise to the illusion that *Water = H_2O* is not necessarily true.

The Kripkean account of the necessity and apparent non-necessity of *Heat = molecular motion* follows familiar lines. Given that 'heat' and 'molecular motion' are rigid, we know that *Heat = molecular motion* is necessarily true, if true. And it is true. (How do we know it is true? An answer apparently consonant with Kripke's views would be: We know (through empirical investigation) that heat is a natural phenomenon with a particular 'intrinsic constitution' and particular causes and effects—the natural phenomenon whose (physical) name is 'molecular motion'. So we know that heat = molecular motion.) And, for the usual reasons, we are subject to the illusion that *Heat = molecular motion* is not necessarily true.

A (type-)identity theorist who wants to give a Kripkean account of the necessity and apparent non-necessity of *That (kind of) feeling = C-fibre stimulation* would presumably have to argue that we have good reason to suppose that that (kind of) feeling—i.e. pain—is a natural phenomenon with a certain 'intrinsic constitution' and certain causes and effects—the natural phenomenon whose (neurophysiological) name is 'C-fibre stimulation'. So we know that that (kind of) feeling = C-fibre stimulation. To complete the account, the identity theorist would have to show how the illusion of the non-necessity of *That (kind of) feeling = C-fibre stimulation* arises.

One might have various doubts about whether a good case can in fact be made for the truth of *Water = H_2O*, or *Heat = molecular motion*, or *That (kind of) feeling = C-fibre stimulation*. Indeed, as I shall try to show, the doubts— which, as far as I can see, do not arise in the case of *Hesperus = Phosphorus*— get more pressing as we move from *Water = H_2O* to *Heat = molecular motion* to *That (kind of) feeling = C-fibre stimulation*.

How do we know that water is a substance with a particular chemical composition—the substance whose chemical name is 'H_2O'? Someone might think that there is no particular problem about how we could find this out. We can identify samples of water and investigate their chemical composition. On that basis, we can ascertain that the stuff water has a particular sort of hydrogen and oxygen composition, and is in fact the (chemical) substance whose chemical name is 'H_2O'.

Notice, though, that, even if all (and only) the samples of a stuff or substance S have the kind of composition that is definitive of samples of a stuff or substance S' (the kind of composition that is both necessary and sufficient for being a sample of substance S'), it is still perfectly possible that S ≠ S'. Suppose that the only kind of wood in the world were balsa. Then all and only samples of balsa would have the kind of composition the having of which is both necessary and sufficient for being samples of wood. But it would still be true that wood ≠ balsa. After all, wood is something that there could be, even if there were no balsa; balsa is not something there could be, if there were no balsa.

Similarly, suppose that (i) all (and only) bits of the stuff water have the kind of composition the having of which is necessary and sufficient for being a sample of the (chemical) substance H_2O. This by itself does not guarantee that (ii) water = H_2O. For (i) is, but (ii) is not, compatible with its being genuinely possible for there to have been bits of water that were not bits of H_2O, or vice versa.

Indeed, not a few philosophers have thought that, even if all the bits of water there actually are have the kind of composition necessary and sufficient for being bits of H_2O, there still could have been bits of water that weren't bits of H_2O, because they had a chemical composition that no bit of H_2O could have.[295] For such philosophers, empirical research has favoured the claim that all the bits of water—or, at least, all the bits of water around here—are bits of H_2O.[296] But it hasn't favoured the claim that water = H_2O, any more than research could favour the claim that wood = balsa. As those philosophers see it, *Water = H_2O*, like *Wood = balsa*, is necessarily false.

This brings to light a difference between the case for *Hesperus = Phosphorus* and the case for *Water = H_2O*. Someone can clearly make a case for the identity of Hesperus with Phosphorus without having a view on any contentious modal issues. (If one can show that Hesperus is a planet and Phosphorus is a planet, and Hesperus and Phosphorus are in all the same places at all the same times, then one has made the case for the identity of Hesperus with Phosphorus.) It is not clear, though, that one can make the case for the identity of water with H_2O without having a view on the (at least mildly) contentious modal issue of whether there could have been bits of water that weren't bits of H_2O.

Of course, Kripke, along with Putnam and Albritton, persuaded most people (or, at least, most analytic philosophers) that those philosophers who think that there could have been bits of water that had the wrong sort of composition to be H_2O are mistaken (cf. 'N & N', 323). Upon reflection, most of us (myself included) are inclined to agree with Kripke that bits of stuff on another planet or in another possible world couldn't be water, however much they resembled bits of water in certain 'superficial' ways, unless they had the right sort of composition to be bits of H_2O. And, upon reflection, most of us (myself included) are inclined to agree with Kripke that bits of stuff on another planet or in another possible world would (have to) be bits of water, however much their 'superficial' properties differed from the superficial properties of bits of water around here, as long as they had the right sort of composition to be bits of H_2O.

---

[295] The (plausible) assumption here is that no bit of H_2O could have been composed of anything but H_2O molecules.

[296] Where the 'is' is the 'is' of identity, or perhaps (more weakly) the 'is' of (co)-constitution.

If Kripke is right about these matters, then a certain sort of case for the non-identity of water with H_2O (one that rests on the possibility of bits of water that aren't bits of H_2O) cannot be made. It also seems that a case for the identity of water with H_2O can be made. For, if Kripke is right, then necessarily water is present where and when, and only where and when H_2O is present.[297] And it is hard to see what grounds there could be for distinguishing two necessarily compresent stuffs. It is much easier to see how different stuffs could actually be present at all the same places and times than it is to see how different planets actually could be present at all the same places and times (remember the possible world in which all the bits of wood are bits of balsa); but it is, on the face of it, no easier to see how different stuffs could necessarily be present at all the same places and times than it is to see how different planets could necessarily be present at all the same places and times. In the case at issue, if water and H_2O are necessarily compresent, with respect to what properties might they be discernible?

Just as (permanent) compresence is not a sufficient condition for the identity of stuffs, neither is it a sufficient condition for the identity of natural phenomena.[298] Suppose that the only sort of oxidation there was was rusting. Then oxidation and rusting would be present at all the same places and times. But they are different phenomena, inasmuch as oxidation, but not rusting, is something that there could have been instances of, even if there hadn't been any metals.

Similarly, even if heat and molecular motion are always found together in the actual world, heat $\neq$ molecular motion if there could have been instances of heat that weren't instances of molecular motion (or vice versa). If, on the other hand, heat and molecular motion are necessarily compresent phenomena, then it is at least arguable that heat $=$ molecular motion, inasmuch as it is unclear with respect to what sort of properties heat and molecular motion could be discernible, if they were necessarily compresent.

Kripke intuits that we couldn't have heat without molecular motion, or vice versa. Any (instance of a) phenomenon present that was present in the absence of (an instance of) molecular motion couldn't be (an instance of) heat, however much it resembled heat, even if, say, it had many of the effects heat actually has, including the effects on our sense organs. Similarly, any (instance of a) phenomenon present in the absence of (an instance of) heat couldn't be (an instance of) molecular motion ('N & N', 326; 'I & N', 158–60).

---

[297] Depending on your views about what sort of thing stuffs are, you may be unhappy with the idea that stuffs—as opposed to bits of stuff—are present in space and time.(You might, for example, think that stuffs are properties, and that properties are aspatial and atemporal.) If so, take 'a stuff is present at this place and time' to be elliptical for 'a stuff has some instance present at this place and time' and take 'necessarily compresent stuffs' to mean 'stuffs that necessarily have their instances at exactly the same places and times'.

[298] Again, if you think that instances of natural phenomena are present in space and time, but not natural phenomena, take 'necessarily compresent phenomena' to mean 'phenomena that necessarily have instances at all the same places and times'.

Perhaps Kripke is right about this. Myself, I have no strong intuitions about this question, and, given my ignorance of the relevant bits of physics, I wouldn't be entitled to intuitions, if I had any. I do suspect that the intuition that any instance of a phenomenon that is present in the absence of an instance of molecular motion couldn't be an instance of (genuine) heat is less strong and widely shared than the intuition that any bit of stuff that is where no bit of H_2O is couldn't be a bit of (genuine) water.[299]

Be that as it may, what seems quite clear is that, even if in fact that (kind of) feeling (pain) and C-fibre stimulation always go together, there is no strong or widely shared intuition that an instance of a phenomenon that was present in the absence of an instance of C-fibre stimulation couldn't be an instance of pain. This point has been brought out very nicely by George Bealer.[300]

Imagine an interplanetary explorer landing on a hitherto unexplored planet. Shortly after getting out of her ship, she comes across a liquid that looks very like water. Before drinking a bit of that liquid, she subjects it to chemical analysis, and discovers that it doesn't have the right sort of composition to be H_2O. Would she conclude—should she conclude—that the bit of liquid is not a bit of water (and that the liquid (at least some of) whose bits are present where no bit of H_2O is present is not water)? Yes, most of us judge.

Imagine that, shortly after testing the sample of liquid, the explorer runs into some alien inhabitants of the planet. Suppose she learns that the term in the alien's language for the liquid a sample of which she tested is 'hydor'. The explorer should conclude that the word 'hydor' does not mean 'water'. ('Hydor' means that liquid (the liquid a sample of which the explorer tested) and that liquid isn't water (since it isn't H_2O).) Suppose that the explorer subsequently learns that the aliens have a term 'algos' that a radical interpreter would be initially inclined to translate into English as 'pain'. When the aliens bump up (hard) against (hard) objects, or have their skin broken by sharp objects, they grumble about the *algos* this brings about. When the aliens are operated on, they are first given a substance that they say is meant to prevent or at least mitigate *algos*. The aliens say that *algos* is typically the result of events that damage or endanger an organism, and typically leads to aversive behaviour (behaviour that takes the organism away from the source of the damage or danger). If the aliens have learned some English, and they want to get across to English speakers what 'algos' means, they tell them, '*Algos* is a kind of feeling—an unpleasant one.'

---

[299] Kripke notes that some people have been inclined to think that, although the only actually present phenomenon that is heat is molecular motion, there might have been a kind of heat different from 'our heat', and different from molecular motion; he adds that a similar suggestion has been made for gold. Though he registers his disinclination to accept this view of heat or gold, he does not pursue the matter further (see 'N & N', 353 n. 68). I have also heard people who know more physics than I do say that the identification of heat with molecular motion is problematic, even if we limit our attention to 'our heat' (to actual instances of heat).

[300] See Bealer, 'Mental Properties', *Journal of Philosophy*, 91 (1994), 185–208.

Now suppose that our explorer becomes familiar with alien anatomy, and learns that the aliens have no C-fibres. Would she conclude—should she conclude—that, since the aliens are not mistaken when they judge that they have *algos*, and have no C-fibres to be stimulated, their word 'algos' does not mean 'pain'? Would she conclude—should she conclude—that, since the aliens mean by 'algos' what we mean by 'pain', and pain = C-fibre stimulation, the aliens are mistaken when they judge they have *algos*? No and no. We don't feel any intuitive pressure to say that the explorer would or should draw either of these conclusions, because we don't have the intuition that, if the phenom-enon the aliens call 'algos' (the phenomenon that is the referent of the alien term 'algos') is present in the absence of C-fibre stimulation, then that phenomenon could not really be pain. So there is a striking disanalogy between the water–*hydor* case and the pain–*algos* case. In the former case we have (at least, most of us have) the intuition that, if the stuff the aliens call 'hydor' (the stuff that is the referent of the alien term 'hydor') is present in the absence of H_2O, then that stuff can't be water. The corresponding intuition concerning *algos* is simply absent.[301]

The type-identity theorist holds that just as water = H_2O and rusting = a certain kind of oxidation, pain = C-fibre stimulation. It seems plausible that rusting is to a certain kind of oxidation as water is to H_2O. Imagine our explorer coming across what looks like an instance of rusting. If she finds out that the phenomenon she has encountered is present in the absence of oxidation, she seems entitled to conclude that that phenomenon is not really rusting, even if it looks quite a lot like rusting, in just the way she is entitled to conclude that the stuff she encountered shortly after getting out of the spaceship is not water, even if it looks quite a lot like water.[302] But, for the reasons just given, it does not seem initially plausible that pain is to C-fibre stimulation as water is to H_2O. If pain is to be compared to a stuff, it looks less like a (chemical) stuff such as water, or wood, than it does like a (non-chemical) stuff such as glue. If our explorer comes across a sample of some homoeomer-ous stuff, then that sample, which obviously does not have the right compos-ition to be a sample of H_2O, or a sample of (mostly) cellulose, is not a sample of water, or wood.[303] But if our explorer comes across a sample of some homoeomerous sticky stuff that the aliens produced with the intention

---

[301] See Bealer, 'Mental Properties', *Journal of Philosophy*, 91 (1994), 198–9.

[302] Or perhaps 'rusting' is ambiguous, and can pick out either a phenomenon that cannot be present in the absence of oxidation, or a phenomenon of a sort that typically occurs when metals are oxidized, but may occur even when oxidation does not. See the *OED* entry for 'rust': 'a red, orange, or tawny coating formed upon the surface of iron or steel by oxidation, esp. through the action of air or moisture: also, by extension, a similar coating formed upon any other metal by oxidation *or corrosion*' (my emphasis). The *OED* similarly allows that there is a sense of 'fish' that applies to aquatic mammals such as whales, and a sense that does not.

[303] Or at least, I want to say, there is one sense of 'wood' in which the homoeomerous stuff is not wood. Perhaps in one sense of 'tree' homoeomerous things could be trees, and in a corresponding sense of 'wood', the homoeomerous hard stuff homoeomerous trees are made of could be wood.

of sticking things together, and actually use to stick things together, then our explorer has come across a sample of glue. The homoeomerous water-like substance or wood-like substance would be, as it were, 'fool's water' or 'fool's wood', but the homoeomerous sticky stuff would be genuine glue. Before she came across this weird homoeomerous glue, the following was true about all the bits of glue the explorer had come across: they weren't (and, presumably, they couldn't have been) present in the absence of atomic matter. This bit of stuff is and a fortiori could have been present in the absence of atomic matter (i.e. occupies a place at a time that is occupied by no atomic matter at that time). But that is not a good reason to doubt that it is glue: it is still glue, as long as it has the right origin and the right function. Before she came across (instances of) weird alien pain, the following was true about all the instances of pain the explorer had come across: they weren't (and, perhaps, they couldn't have been) present in the absence of C-fibre stimulation. The instance of the mental phenomenon the explorer has come across in the aliens is and a fortiori could have been present in the absence of C-fibre firing. But that is not a good reason to doubt that it is an instance of pain; it is still an instance of pain, as long as it has the right phenomenological and causal properties.

As Bealer points out, in the water–H_2O case we initially seem subject to a conflict of intuitions. We have the intuition that (some or all) water could have turned out not to be H_2O, so that we could have had water in the absence of H_2O. This tells against the identification of water with H_2O. But we also have the intuition that anything present in the absence of H_2O couldn't be (genuine) water, and anything present in the absence of (genuine) water couldn't be H_2O; this, I have suggested, tells in favour of the identification of water with H_2O. By contrast, in the pain–C-fibre case we have anti-identification intuitions, but appear to lack pro-identification intuitions. The (type-)identity theorist accordingly cannot assume that, if there is a good case for the truth (and, via rigidity, the necessary truth) of *Water = H_2O*, then there is a similar and similarly good case for the truth (and, via rigidity, the necessary truth) of *Pain = C-fibre stimulation* (or *That (kind of) feeling = C-fibre stimulation*). The (type-)identity theorist needs to make a new case for the truth (and necessary truth) of *Pain = C-fibre stimulation*. That case will have to take account of the differences brought to light by Bealer between our intuitions concerning the possibility or otherwise of the presence of water in the absence of H_2O, and our intuitions concerning the possibility or otherwise of the presence of pain in the absence of C-fibre stimulation.

In light of all this, one might have thought Kripke would argue that the (type-)identity theorist cannot (successfully) model her account of the necessity and apparent non-necessity of *Pain = C-fibre firing* on Kripke's account of the necessity and apparent non-necessity of *Water = H_2O*, or *Heat = molecular motion*, because there is no good case to be made for the necessity of *Pain = C-fibre stimulation*. But Kripke does not take this line.[304] Instead he argues that the (type-)identity theorist cannot offer a Kripkean account of the necessity and apparent non-necessity of *Pain = C-fibre firing*, because the part of that account that explains why *Pain = C-fibre firing* appears not to be a necessary truth (if Kripkean) will entail that *Pain = C-fibre firing* not only appears not to be, but actually is not, a necessary truth.

Given that *Heat = molecular motion* is a necessary truth, why doesn't it look like a necessary truth? We have already seen that, for Kripke, the illusion of contingency can arise when we conflate a possibility involving an epistemic counterpart of a thing with a possibility involving that thing itself. To refresh our memory:

If someone protests, regarding the lectern, that it *could* after all have *turned out* to be made of ice, and therefore could have been made of ice, I would reply that what he means is that *a lectern* could have looked just like this one, and have been placed in the same position as this one, and yet have been made of ice. In short, I could have been in the *same epistemological situation* in relation to *a lectern made of ice* as I actually am in relation to *this* lectern. In the main text, I have argued that the same reply should be given to protests that Hesperus could have turned out to be other than Phosphorus, or Cicero other than Tully. Here, then, the notion of 'counterpart' comes into its own. For it is not this table, but an epistemic 'counterpart', which was hewn from ice; not Hesperus–Phosphorus–Venus, but two distinct counterparts thereof, in two of the roles Venus actually plays (that of the Evening Star and Morning Star), which are different. ('I & N', 157 n. 15)

'Heat is the motion of molecules' will be necessary, not contingent, and one only has the *illusion* of contingency in the way one could have the illusion of contingency in thinking that this table might have been made of ice. We might think one could imagine it, but if we try, we can see on reflection that what we are really imagining is just there being another lectern in this very position here which was in fact made of ice. The fact that we may identify this lectern by being the object we see and touch in such and such a position is something else. (I & N, 160–1).

Suppose I don't know whether this lectern is made of ice. Then there will be a possible world **H**—a possible world that might, for all I know, be the actual world—in which I am in an epistemic situation qualitatively indistin-guishable from the one I am actually in, and where *a* lectern made of ice—one which, for all I know, is this lectern—has all the actual features of this lectern whereby I identify it. In such a case we can call the ice lectern in **H** (the one that has the identifying features of this lectern, the one that might, for all I know, be this lectern) an epistemic counterpart (in **H**) of this lectern. In the case as described, I know it is genuinely possible that something that, for all I know, is this lectern is made of ice; and I know it is, for all I know, possible that this lectern is made of ice. But I don't know whether it is genuinely

---

[304] This may be connected with the already mentioned fact that Kripke appears to hold that some arguments may provide us with compelling reasons to think that some statement of the form 'Pain is neural state N' is true, and thus, given rigidity, necessary (cf. 'N &N', 355 n. 77).

possible for this lectern (as opposed to an epistemic counterpart thereof) to be made of ice.

Now suppose that I learn that this lectern is not made of ice. Presuming (I know) that, if this lectern is not made of ice, it could not have been made of ice, I should conclude that this lectern could not have been made of ice. But, if I am insufficiently reflective, I may still be inclined to think that this lectern could have turned out to be, and thus could have been, made of ice. Even after I have learned that this lectern is not made of ice, I am aware that there is a possible world in which I am in an epistemic situation qualitatively indistinguishable from the one I was in before I learned the lectern was not made of ice, and in which I am in the same epistemic situation in relation to an ice lectern with the identifying features of this lectern (to an ice epistemic counterpart of this lectern) that I was actually in, in relation to this lectern. If I don't clearly distinguish the impossibility that this lectern is made of ice from the possibility that an epistemic counterpart of this lectern is in a qual-itatively indistinguishable epistemic situation is made of ice, I will be inclined to think that *This very lectern is made of ice* is true in a possible world that was once epistemically possible (i.e. is compossible with everything I knew then), even if it is no longer epistemically possible (compossible with everything I know now).

Similarly, suppose that I don't know whether heat is molecular motion or some other phenomenon—say, caloric. Then there will be a possible world **H**—a possible world that might, for all I know, be the actual world—in which I am in an epistemic situation qualitatively indistinguishable from the one I am actually in, and where a 'calorical' phenomenon—one which, for all I know, is heat—has all the actual features of heat whereby I identify heat. I know it is genuinely possible that something that, for all I know, is heat is caloric; and I know it is, for all I know, possible that heat is caloric. But I don't know whether it is genuinely possible for heat (as opposed to an epistemic counterpart thereof) to be caloric.

Suppose that I subsequently learn that heat is not caloric. Presuming (I know) that, if heat is not caloric, it could not have been caloric, I should conclude that heat could not have been caloric. But, if I am insufficiently reflective, I may still be inclined to think that heat could have turned out to be, and thus could have been, caloric. Even after I have learned that heat is not caloric, I am aware that there is a possible world in which I am in an epistemic situation qualitatively indistinguishable from the one I was in before I learned heat was not caloric, and in which I am in the same epistemic situation in relation to a calorical phenomenon with the identifying features of heat (to a calorical epistemic counterpart of heat) that I was actually in, in relation to heat. If I don't clearly distinguish the impossibility that heat is caloric from the possibility that an epistemic counterpart of heat is caloric, I will be inclined to think that *Heat is caloric* is true in a possible world that was once epistemically possible, even if it is no longer epistemically possible.

Now suppose the (type-)identity theorist attempts to explain (away) the apparent non-necessity of *Pain = C-fibre stimulation* along the same lines that Kripke explains (away) the apparent non-necessity of *Heat = molecular motion*. That is, suppose she says:

> Since Pain = C-fibre stimulation (and 'pain' and 'C-fibre stimulation' are rigid), there is no possible world in which pain ≠ C-fibre stimulation, and no possible world in which pain is present in the absence of C-fibre stimulation. But we may well be initially inclined to think there is such a world. For we didn't always know that pain = C-fibre stimulation. So there is a possible world in which we are in an epistemic situation qualitatively indistinguishable from the one we were in before we learned that pain = C-fibre stimulation, and in which we are in the same epistemic situation in relation to a phenomenon with the identifying features of pain (to an epistemic counterpart of pain) that we were actually in, in relation to pain, even though that phenomenon is not C-fibre stimulation, and is in fact present in the absence of C-fibre stimulation. If we don't clearly distinguish the impossibility that pain is present in the absence of C-fibre stimulation from the possibility that an epistemic counterpart of pain is present in the absence of C-fibre stimulation (or any epistemic counterpart thereof), we shall be inclined to think that *Pain is present in the absence of C-fibre stimulation*, and consequently *Pain ≠ C-fibre stimulation*, are true in a possible world that was once epistemically possible (i.e. is compossible with everything we knew then).

Suppose that in an alternative world we are in the same epistemic situation we were in before we found out that pain was C-fibre stimulation, and an epistemic counterpart of pain is present in the absence of C-fibre stimulation (or any epistemic counterpart thereof). In that alternative world the epistemic counterpart of pain has the (actual) feature of pain whereby we (actually) identify pain. But Kripke thinks, 'the way we identify pain is by feeling it' ('I & N', 163 n. 18); we identify pain by its having a certain characteristic 'immediate phenomenological quality' ('N & N', 340). Moreover, for Kripke, necessarily, any state that has that immediate phenomenological quality *is* pain.[305] So if there is an alternative possible world in which we are in the same epistemic situation we were in before we found out that pain was C-fibre stimulation, and in which an epistemic counterpart of pain is present in the absence of C-fibre stimulation, then there is an alternative possible world in which pain itself is present in the absence of C-fibre stimulation. In that case, we obviously cannot appeal to an alternative possible world in which we are in a situation qualitatively indistinguishable from the one we were in before we learned that pain was C-fibre stimulation, and an epistemic counterpart of pain is present in the absence of C-fibre stimulation, in order to explain why

---

[305] See 'I & N', 163 n. 18: 'for a state to be *felt* as pain is for it to *be* pain'.

we are subject to the *illusion* that *Pain = C-fibre stimulation* is non-necessary, since the existence of such a world implies that *Pain = C-fibre stimulation is non-necessary.* The case is quite different from the case of heat and molecular motion. There one can appeal to an alternative possible world in which we are in an epistemic situation qualitatively indistinguishable from the one we were in before we learned that heat was molecular motion, and an epistemic counterpart of heat is present in the absence of molecular motion, in order to explain why we are subject to the illusion that *Heat = molecular motion* is non-necessary. For in that case the presence of an epistemic counterpart of heat in the absence of molecular motion (in the alternative world) does not entail that heat itself is present in the absence of molecular motion (in the alternative world, or in any possible world).

Alternatively, suppose the (type-)identity theorist says:

> Since pain = C-fibre stimulation (and 'pain' and 'C-fibre stimulation' are rigid), there is no possible world in which pain ≠ C-fibre stimulation, and no possible world in which pain is absent in the presence of C-fibre stimulation. But we may well be initially inclined to think there is such a world. For we didn't always know that pain = C-fibre stimulation. So there is a possible world in which we are in an epistemic situation qualitatively indistinguishable from the one we were in before we learned that pain = C-fibre stimulation, and in which we are in the same epistemic situation in relation to a phenomenon with the identifying features of pain (to an epistemic counterpart of pain) that we were actually in, in relation to pain, even though that phenomenon is not C-fibre stimulation, and is in fact absent in the presence of C-fibre stimulation. If we don't clearly distinguish the impossibility that pain is absent in the presence of C-fibre stimulation from the possibility that an epistemic counterpart of pain is absent in the presence of C-fibre stimulation (and any epistemic counterpart thereof), we shall be inclined to think that *Pain is absent in the presence of C-fibre stimulation,* and consequently *Pain ≠ C-fibre stimulation,* are true in a possible world that was once epistemically possible (i.e. is compossible with everything we knew then).

If in an alternative possible world C-fibre stimulation is present in the absence of an epistemic counterpart of pain, then in that world C-fibre stimulation is present, even though no state is present that has the feature whereby we actually identify pain. The feature whereby we actually identify pain is, for Kripke, the way it feels. So if there is an alternative possible world in which we are in an epistemic situation qualitatively indistinguishable from the one we were in before we discovered that pain is C-fibre stimulation, and C-fibre stimulation is present in the absence of an epistemic counterpart of pain, then there is an alternative possible world in which C-fibre stimulation is present, even though nothing feels the way pain actually feels. But, Kripke thinks, if no state feels the way pain actually feels, then pain cannot be present: 'if a C-fiber

stimulation would have occurred without our feeling any pain, then the
C-fiber stimulation would have occurred, without there *being* any pain' ('I & N',
163 n. 18). So, if there is a possible world in which we are in an epistemic situation qualitatively indistinguishable from the one we were in before we discovered that pain is C-fibre stimulation, and C-fibre stimulation is present in
the absence of an epistemic counterpart of pain, then there is a possible world
in which C-fibre stimulation is present in the absence of pain. So the (type-)
identity theorist cannot appeal to a possible world in which C-fibre stimulation is present in the absence of an epistemic counterpart of pain, in order to
explain why we are subject to the illusion that *Pain = C-fibre firing* is non-
necessary, since the existence of such a world implies that *Pain = C-fibre stimulation* is non-necessary. The case is once again different from that of heat and
molecular motion. There one can appeal to a possible world in which we are
in an epistemic situation qualitatively indistinguishable from the one we were
in before we learned that heat was molecular motion, and molecular motion
is present in the absence of an epistemic counterpart of heat, in order to
explain why we are subject to the illusion that *Heat = molecular motion* is
non-necessary. For in that case the absence of an epistemic counterpart of
heat (in an alternative possible world in which we are in an epistemic situation qualitatively indistinguishable from the one we were in before we learned
that heat was molecular motion) does not entail that heat itself is absent in
that world.[306]

The basic intuition driving these arguments against a certain sort of attempt
to explain away the apparent non-necessity of *Pain = C-fibre stimulation* is
this: in the case of heat, there is a gap between identifying features and
essence—a gap that allows something to have the identifying features of
heat without being heat, and allows something to be heat without having the
identifying features of heat. In the case of pain, the intuition is, there is no such
gap between identifying features and essence. Nothing can have the identifying
feature of pain (the 'immediate phenomenological quality') without being
pain, and nothing can be pain without having that identifying feature. So,
although there is no difficulty about finding a state in another world that is an
epistemic counterpart of heat (in that world) even though it isn't heat, or is
heat even though it isn't an epistemic counterpart of heat (in that world), there
is no state in another world that is an epistemic counterpart of pain (in that
world) even though it isn't pain, or is pain even though it isn't an epistemic
counterpart of pain (in that world).

We saw earlier that one might think of a particular pain either as having a
non-phenomenological essence, or as having a 'dual-aspect' essence, one aspect

---

[306] If in some possible world heat is present, but lacks the actual features of heat whereby we
actually identify heat, then that world may lack an epistemic counterpart of heat (in the actual world),
in spite of containing heat. (Alternatively, it may contain both heat and some other state that is an
epistemic counterpart of heat (in the actual world).)

of which was phenomenological and one aspect of which was not. Similarly (although this certainly does not recommend itself to untutored intuition), one might think of the type pain as having a non-phenomenological essence, or, less counter-intuitively, think of it as having an essence one aspect of which was phenomenological and one aspect of which was not. A (type-)identity theorist could accordingly grant Kripke that, for a state to be pain, it is required that it feel a certain way. Still, she might say, *pace* Kripke, that is not all that is required: a state that felt the right way, but did not have the right sort of typical causes and effects, would not be pain. If she took this line, she could maintain that a state could feel like pain without having the right sort of causal properties to be pain. This would allow her to maintain that a state could be an epistemic counterpart of pain without being pain. She could say:

> Something with the wrong sort of intrinsic constitution to be heat could nevertheless have the feature(s) we identify heat by (say, producing a certain sensation in us). Likewise, something with the wrong sort of causal profile to be pain could nevertheless have the feature we identify pain by (its immediate phenomenological quality). Such a state could, *pace* Kripke, be an epistemic counterpart of pain without actually being pain. So Kripke hasn't after all shown that we can't explain away the apparent non-necessity of *Pain = C-fibre stimulation* in the way we can explain away the apparent non-necessity of *Heat = molecular motion.*

Alternatively, a type-identity theorist might grant that all that is required for a state to be pain is that it feel a certain way, but insist that the way a state feels is partly a matter of its causal profile (or its 'intrinsic constitution'). David Lewis writes:

Pain is a feeling. To have pain and to feel pain are one and the same. For a state to be painful and for it to feel painful are likewise one and the same. A theory of what it is for a state to be pain is inescapably a theory of what it is like to be in that state, of the phenomenal character of that state. Far from ignoring questions of how states feel . . . I have been discussing nothing else! Only if you believe on independent grounds that considerations of causal role and physical realization have no bearing on whether a state is pain should you say that they have no bearing on how that state feels.[307]

Someone who finds this persuasive could grant that any epistemic counterpart of pain that feels the way pain actually feels is pain, but deny that any epistemic counterpart of pain feels like pain, on the grounds that an other-wordly state might have some intrinsic property (in that other world) that I actually identify pain by, without having the right sort of causal profile or perhaps the right sort of 'intrinsic constitution' to feel like pain (feel the way pain actually feels).

In fact, though, I doubt that either of these strategies is of much use to someone who wants to defend the necessary truth of *Pain = C-fibre stimulation* in

---

[307] 'Mad Pain and Martian Pain', 130.

the face of Kripke's arguments. On either strategy, it is a necessary condition of a state's being pain that it occupy a certain causal role. But, as Kripke notes, it seems not to be a necessary condition of a state's being C-fibre stimulation that it occupy that causal role. If pain = C-fibre stimulation, but C-fibre stimulation might not have occupied the causal role anything would have to occupy in order to be pain, then we are back to a Lewis–Armstrong identity theory, on which *Pain = C-fibre stimulation* is contingently true, because 'pain' non-rigidly designates what 'C-fibre stimulation' designates.

As a number of philosophers have noted,[308] Kripke seems to have overlooked a strategy for explaining away the apparent non-necessity of *Pain = C-fibre stimulation* by recourse to qualitatively indistinguishable epistemic situations and epistemic counterparts. Even if nothing except pain can be an epistemic counterpart of pain, a state that is not C-fibre stimulation can be a counterpart of C-fibre stimulation. So, for all Kripke has said, there is a possible world in which an epistemic counterpart of C-fibre stimulation is present in the absence of pain, and a possible world in which an epistemic counterpart of C-fibre stimulation is absent in the presence of pain. (Think of worlds in which the firing of some other kind of fibre has the features whereby we actually identify C-fibre stimulation.) If there are such worlds, the (type-)identity theorist can say:

> There is a possible world in which we[309] are in an epistemic situation qualitatively indistinguishable from the one we were in before we learned that pain = C-fibre stimulation, and in which we are in the same epistemic situation in relation to a phenomenon with the identifying features of C-fibre stimulation (to an epistemic counterpart of C-fibre stimulation) that we were actually in, in relation to C-fibre stimulation, even though that phenomenon is not pain, and is in fact present in the absence of pain. If we don't clearly distinguish the impossibility that C-fibre stimulation is present in the absence of pain from the possibility that an epistemic counterpart of C-fibre stimulation is present in the absence of pain, we shall be inclined to think that *Pain is absent in the presence of C-fibre stimulation*, and consequently *Pain ≠ C-fibre stimulation*, are true in a possible world that was once epistemically possible (i.e. is compossible with everything we knew then). Similarly, there is a possible world in which we are in an epistemic situation qualitatively indistinguishable from the one we were in before we learned that pain = C-fibre stimulation, and in which we are in the same epistemic situation in relation to a phenomenon with the identifying features of C-fibre stimulation (to an epistemic counterpart of C-fibre stimulation) that we are actually in, in relation to C-fibre stimulation, even though that phenomenon is not

---

[308] See e.g. A. Woodfield, 'Identity Theories and the Argument from Epistemic Counterparts', *Analysis*, 38 (1978), 143.   [309] Or creatures who are epistemic counterparts of us.

pain, and is in fact absent in the presence of pain. If we don't clearly distinguish the impossibility that C-fibre stimulation is absent in the presence of pain from the possibility that an epistemic counterpart of C-fibre stimulation is absent in the presence of pain, we shall be inclined to think that *C-fibre stimulation is absent in the presence of Pain*, and consequently *Pain ≠ C-fibre stimulation*, are true in a possible world that was once epistemically possible. So, in order to explain away the appearance that *Pain = C-fibre stimulation* is not necessary, we needn't suppose that an other-wordly state can be pain without being an epistemic counterpart of pain (in that world) or suppose that an other-wordly state can be an epistemic counterpart of pain (in that world) without being pain. It is enough that an other-wordly state can be C-fibre stimulation without being an epistemic counterpart of C-fibre stimulation (in that world),[310] or that an other-wordly state be an epistemic counterpart of C-fibre stimulation (in that world) without being C-fibre stimulation.

So it begins to look as though the (type-)identity theorist can perhaps explain away the apparent non-necessity of *Pain = C-fibre stimulation* along Kripkean lines. But George Bealer has argued that, for reasons not discussed by Kripke, the attempt to explain away the apparent non-necessity of *Pain = C-fibre stimulation* along Kripkean lines will fail.[311] While I shall not offer exactly the argument Bealer does, my argument will be a transposition of Bealer's.[312]

Consider the following modal argument:

> It might have been that: something was hot, even though it was not made of molecules.
>
> So *being hot* does not imply *having molecules*, and a fortiori does not imply *having molecules in motion*.
>
> So *being hot* is a different property from *having molecules in motion*.

As we have seen, someone who thinks that *being hot* is the same property as *having molecules in motion* has a ready reply. She can say:

> The first premiss of the argument is false. While it might have been that: something had an epistemic counterpart of the property of being hot,

---

[310] Again, someone might be puzzled by the idea that an other-wordly state that is C-fibre stimulation could fail to be an epistemic counterpart of C-fibre firing. But we shouldn't think of *being an epistemic counterpart of* as a two-place relation between (trans-world) individuals. A thing X, as it is in a possible world **H**, is an epistemic counterpart of a thing Y as it is another world—say, the actual world **G**. Even if X = Y, it needn't be true that X, as it is in **H**, is an epistemic counterpart of Y, as it is in **G**. If X does not have the actual identifying features of Y in **H**, and some other thing Z has those features (in **H**), then Z will be the epistemic counterpart of Y in **H**, rather than X, even though X = Y, and Y ≠ Z.

[311] See again Bealer's (very rich) article 'Mental Properties', esp. 201–3.

[312] Bealer's argument turns on (roughly) a distinction between a proposition's being true in a certain possible world and a statement's expressing a true proposition in that possible world; it does not bring in the distinction between states and their epistemic counterparts.

even though it was not made of molecules, nothing could have had the very property *being hot,* unless it was made of molecules (in motion).

Now consider the following analogous argument:

It might have been that: something was a bit of glue, even though it was made of some homoeomerous substance.

So *being a bit of glue* does not imply *being made of atoms.*

So *being a bit of glue* is a different property from *being made of atoms put together in such-and-such a way.*

This seems to me a good argument. Be that as it may, it should be clear that a defender of the identity of *being a bit of glue* with *being made of atoms put together in such-and-such a way* could not effectively counter the argument by suggesting that, although a bit of some homoeomerous substance could have an epistemic counterpart of the property of being a bit of glue, no bit of homoeomerous substance could have the very property *being a bit of glue.* If some other-wordly bit of some other-wordly homoeomerous stuff has all the features whereby we identify something as a bit of glue (if it's a bit of stuff that was produced with the intention of sticking certain kinds of things together, is used that way, and so on), then that bit of stuff just is a bit of glue.

I have it on Bealer's authority that C-fibres have a great many functionally related parts (the number, he speculates, may be over 74 million). And, Bealer suggests, a C-fibre is the kind of thing it is in virtue of having a distinctive morphology, which (necessarily) involves having at least that many functionally related parts.[313] If this is right, then nobody can have stimulated C-fibres without having some parts that have at least that many functionally related parts. Let G rigidly designate the property of having some parts that have at least that many functionally related parts. Consider the following argument:

It might have been that: someone was in pain, but didn't have G.

It couldn't have been that: someone had stimulated C-fibres, but didn't have G.

So *Being in pain ≠ having stimulated C-fibres* (and pain ≠ C-fibre stimulation).

Given assumptions about the rigidity of the terms involved that Kripke and (necessitarian) type-identity theorists share, the argument is valid.[314] The necessitarian type-identity theorist who identifies pain with C-fibre stimulation must accordingly deny either the first or the second premiss. But

[313] Bealer, 'Mental Properties', 201.

[314] A necessitarian type-identity theorist will hold, say, that it is necessarily true that pain = C-fibre stimulation. Lewis and Armstrong are non-necessitarian type-identity theorists.

she is not going to want to deny the second premiss, which is analogous to claims such as 'It couldn't have been that: something was made of water, but didn't have parts that were made of hydrogen.' So she will have to deny the first premiss.

But even if the first premiss is false, as Kripke would say, it certainly *appears* to be true. If it is false, why does it appear to be true? Or, to put the same question a different way, if it is a necessary truth that anyone in pain has G, why does it appear not to be a necessary truth?

At least as long as the (type-)identity theorist accepts the Kripkean intuition that we cannot distinguish between pain and its epistemic counterparts, it looks as though indistinguishable epistemic situations and epistemic counterparts will be of no help to her in answering this question. If she accepts that intuition, she cannot say that we misintuit that someone could have been in pain without having G, because we conflate the impossibility that someone could have the property *being in pain* without having G with the possibility that someone could have an epistemic counterpart of the property *being in pain* without having G. Nor can she say that we misintuit that someone could have been in pain without having G, because we conflate the impossibility that someone could be in pain without having G with the possibility that someone could be in pain without having an epistemic counterpart of G, since there is no plausible story to be told about why it is an epistemic counterpart of G, and not G itself, that it is possible that someone could be in pain without having. So, as Kripke maintains, there seem to be apparent non-necessities concerning the way mental and neurophysical properties are related that cannot be explained away by any straightforward application of the strategy Kripke used to explain away apparent non-necessities whose necessity followed from scientific theoretical identifications such as *Water* = H₂O or *Heat* = *molecular motion*.[315]

Of course, as Kripke concedes ('N & N', 342), this does not imply that no other strategy could be used to explain away the apparent non-necessity of statements like *Anyone in pain has G*.[316]

---

[315] Note that, even if Bealer were wrong about the necessity of *Anyone with stimulated C-fibres has G*, that would not be of much help to the necessitarian type-identity theorist who wants to identify pain with C-fibre stimulation. For there is surely some property P such that (i) it is necessary that anyone with stimulated C-fibres has P, and (ii) the apparent non-necessity of *Anyone in pain has P* cannot be explained away as arising from a failure to distinguish the possibility that someone could be in pain, without having an epistemic counterpart of P, from the impossibility that someone could be in pain, without having P itself.

[316] Cf. Nagel's discussion in 'What Is it Like to Be a Bat?' (*Philosophical Review*, 83 (1974), n. 11): 'A theory that explained how the mind–brain relation was necessary would still leave us with Kripke's problem of explaining why it nevertheless appears contingent. That difficulty seems surmountable, in the following way. We may imagine something by representing it to ourselves . . . perceptually or sympathetically. To imagine something perceptually, we put ourselves in a conscious state resembling the state we would be in if we perceived it. To imagine something sympathetically, we put ourselves in a conscious state resembling the thing itself. (This method can only be used to imagine mental events

So, even if Kripke is right to suppose that there are statements concerning the relation of mental properties to physical properties whose necessity follows from the identity of pain with C-fibre stimulation, and whose apparent non-necessity cannot be explained away in the way that Kripke explains away the apparent non-necessity of statements such as *Anything hot has molecules in motion*, that does not constitute a refutation of the thesis that pain = C-fibre stimulation.

Kripke would agree. In fact, he does not attempt to refute that thesis, and isn't even convinced that it is false. In a passage I have already called attention to, he writes: 'identity theorists have presented positive arguments for their view, which I certainly have not answered here. Some of these arguments seem to me to be weak or based on ideological prejudices, but others strike me as highly compelling arguments which I am at present unable to answer convincingly' ('N & N', 355 n. 77).

By 'their view' Kripke might have in mind here either a type-identity theory or a token-identity theory. But since the footnote occurs at the end of an extended discussion of the type-identity theory, I would guess that 'their view' is the type-identity theory. It is presumably because Kripke thinks that there are apparently compelling considerations both for and against the identification of mental properties with neural properties (and for and against the identification of mental events with neural events) that Kripke concludes (the final foonote of) *Naming and Necessity* by saying 'I regard the mind–body problem as wide open and extremely confusing' ('N & N', 355 n. 77).

The type-identity theory is in a sense a target in Lecture III of *Naming and Necessity*, and in the concluding part of 'Identity and Necessity.' But it is a target in a very different sense from, say, the sense in which the classical descriptivist

---

and states—our own or another's.) When we try to imagine a mental state occurring without its associated brain state, we first sympathetically imagine the occurrence of the mental state: that is, we put ourselves in a state that resembles it mentally. At the same time, we attempt to perceptually imagine the non-occurrence of the associated physical state, by putting ourselves into another state unconnected with the first: one resembling that which we would be in if we perceived the non-occurrence of the physical state. Where the imagination of the physical features is perceptual, and the imagination of mental features is sympathetic, it appears to us that we can imagine any experience occurring without its associated brain state, and vice-versa. The relation between them will appear contingent even if it is necessary, because of the independence of the disparate types of imagination.' On Kripke's strategy for explaining away the apparent non-necessity of certain statements, what is crucial is distinguishing two (easily conflatable) states of affairs. The first is a possible state of affairs in which one is in the same qualitative epistemic situation one is actually in, and an epistemic counterpart of an item X is present (or absent) in the absence (or presence) of the item Y whose identity with X is both necessary and a posteriori. The second is an impossible state of affairs in which one is in the same qualitative epistemic situation one is actually in, and X itself is present (or absent) in the absence (or presence) of the item Y whose identity with X is both necessary and a posteriori. On the strategy suggested by Nagel, what is crucial to explaining away the apparent non-necessity of certain statements is a distinction not between two apparently imaginable states of affairs, but rather between two different ways of imagining a state of affairs (or event). For a development and defence of the Nagelian strategy, see C. Hill, 'Imaginability, Conceivability, Possibility, and the Mind–Body Problem', *Philosophical Studies*, 87 (1997), 64–85.

theory of reference is a target in the first two lectures of *Naming and Necessity*. In those lectures Kripke's aim is to show that 'the whole picture given by [the classical descriptivist] theory of how reference is determined seems to be wrong from the fundamentals' ('N & N', 300). In the last lecture of *Naming and Necessity* his aim is to raise and defend doubts about the type-identity theory (and the token-identity theory).[317] When Kripke gave the *Naming and Necessity* lectures, he thought that three points about the type-identity theory were un-, or at least under-appreciated, namely:

(1) A theory that identifies pain with a neural state (say, C-fibre stimulation) has modal consequences that are, at least at first sight, deeply counter-intuitive.

(2) The type-identity theorist accordingly faces 'a very stiff challenge' ('I & N', 163); he must make the case that we should accept the identity theory, in spite of its apparently deeply counter-intuitive consequences (say, by showing that, upon reflection, those consequences are not so counter-intuitive as they first appear).

(3) The type-identity theorist has not in fact made that case.

Of course, Kripke was not the first to suggest that the identification of pain with a neural state (say, C-fibre stimulation) had modal consequences that were (at least prima facie) deeply counter-intuitive. But, as he says, when this suggestion was made, the typical response was that, although the alleged deeply counter-intuitive consequences of the identification of pain with C-fibre stimulation were indeed deeply counter-intuitive, they were not consequences of that identification. The identity of pain with C-fibre stimulation, it was said, was synthetic, a posteriori, and contingent. So, although it was deeply counter-intuitive to suppose that there couldn't be pain without C-fibre stimulation, or C-fibre stimulation without pain, neither of these suppositions followed from the claim that pain = C-fibre stimulation.[318]

If (as I think) Kripke has made a good case for the rigidity of 'pain', then he has shown that the typical response is indefensible.[319] And he has made a good case for (1) and (2). What about (3)? Here there is room for doubt.

By Kripke's lights, in order to make the case for the identity of pain and C-fibre stimulation, the type-identity theorist needs to overcome the initial

---

[317] See again 'N & N', 355 n. 77, which Kripke begins with 'Having expressed these *doubts* about the identity theory . . .' (my emphasis).   [318] See 'I & N', 144 and 163.

[319] Remember that, given that 'pain' is rigid, we can show that the identification of pain with a neural state that is C-fibre stimulation has counter-intuitive consequences, whether or not that neural state is rigidly designated by the expression 'C-fibre stimulation'. In fact, for reasons already discussed, the identification of the mental state that is pain with a neural state has counter-intuitive consequences, whether or not that mental state is rigidly designated by the expression 'pain'. What is counter-intuitive is that *that* mental state (the one that is pain) necessarily co-occurs with *that* neural state (the one that is C-fibre stimulation). That the identification of pain with C-fibre stimulation has apparently counter-intuitive modal consequences can thus be established without recourse to any claim about rigidity more contentious than the claim that demonstratives rigidify.

presumption that those states are separable, and hence distinct. This, Kripke thinks, the type-identity theorist could do, if she could both tell a plausible story about why pain and C-fibre stimulation should be identified, and explain why pain and C-fibre stimulation would—at least initially—look separable to us, even if they were identical (and thus inseparable).

That is, indeed, one way in which the type-identity theorist could make the case for the identity of pain and C-fibre stimulation. But is it the only way? As I have mentioned, Kripke seems to concede that there are some apparently compelling arguments in favour of the identity of pain with a neural state (let us suppose, C-fibre stimulation). If those arguments were compelling enough, they would presumably give us good reasons to think that pain and C-fibre stimulation are identical, and hence inseparable, even if all of us—including type-identity theorists—were subject to the illusion that pain and C-fibre stimulation are separable.

Of course, if all we had was an argument for identifying pain and C-fibre stimulation, we wouldn't have any explanation of why pain and C-fibre stimulation look separable, even though they aren't. We wouldn't have what Kripke has given us in the case of Hesperus and Phosphorus, or water and H₂O, or heat and molecular motion.

Fair enough. But why would we need that, in order to make the case for pain's being C-fibre stimulation? To be sure, if the type-identity theorist were in a position not just to provide grounds for identifying pain with C-fibre stimulation, but also to explain why pain and C-fibre stimulation (at least initially) look separable, then she would be able to make a stronger case for the identity of pain and C-fibre stimulation than she would be able to make if she could do only the first of those things. After all, if she were in a position to do both of those things, she would be able to explain the fact—agreed on by identity theorists and their opponents—that pain and C-fibre stimulation (initially at least) look separable. Moreover, if the explanation of the apparent separability of pain and C-fibre stimulation were of the right sort, the type-identity theorist would be able not just to explain, but also to explain *away*, the intuition of separability. This is what (Kripke thinks) we are in a position to do in the cases involving Hesperus and Phosphorus, or water and H₂O, or the lectern not made of ice. In all these cases, Kripke thinks, once we have seen how the illusion of non-necessity arises, we cease to be in its grip. For example, once we see that, in the non-ice lectern case, what we are really imagining is *another* lectern's being made of ice, we lose the conviction that we can perfectly well imagine *this* lectern's being made of ice.

So the type-identity theorist is in a better position if she can explain away the apparent separability of pain from C-fibre stimulation than if she can't. It still might, for all that, be true that the type-identity theorist, even without such an explanation-away, is in a better position than her opponent. That is, it might be that, however counter-intuitive the modal consequences of the type-identity theorist's view, they were distinctly less bad than the bad consequences of the 'non-identity theory'. Moreover, the type-identity theorist might be able to show, or at least make it plausible, that such is the case. (Here the 'apparently compelling arguments' Kripke alludes to would have to do the work.) If the type-identity theorist could do that, then, as best I can see, she would have made a good case for the type-identity theory, in spite of the fact that her theory was incompatible with deeply held intuitions that she was in no position to explain away.[320]

It is interesting in this connection to reconsider a passage from Kripke cited earlier:

> The materialist is up against a very stiff challenge. He has to show that these things we think we can see to be possible are in fact not possible. He has to show that these things which we can imagine are not in fact things we can imagine. And that requires some very different philosophical argument from the sort which has been given in the case of heat and molecular motion. And it would have to be a deeper and subtler argument than I can fathom and subtler than has ever appeared in any materialist literature that I have read. ('I & N', 163)

This gives the impression that Kripke has no idea of what a good argument for the impossibility of (say) pain in the absence of C-fibre stimulation might look like. Such an argument, Kripke seems to be saying, would have to be *toto caelo* different from anything he has come across in the literature, and *toto caelo* different from the arguments that can be given for the impossibility of (say) heat without molecular motion.

This is puzzling. Kripke elsewhere seems to grant that in the literature there are apparently compelling arguments in favour of identifying pain with some neural state. At the very least, he does not want to exclude the possibility that there are apparently compelling arguments of that type. Such arguments, if they exist, are in effect arguments for the impossibility of pain without the neural state with which pain is identified. So why does Kripke make it sound as though, in order to defend her view, the type-identity theorist would have to come up with a completely new kind of argument?

The answer may be that here Kripke is implicitly supposing that the type-identity theorist can defend her view only if she can explain why pain in the absence of C-fibre stimulation is apparently possible, though really impossible. I take it that what Kripke did not find in the materialist literature, and what he cannot fathom, is not an argument for the inseparability of pain from

---

[320] Of course, if the type-identity theory had deeply counter-intuitive consequences, and the type-non-identity theory had, say, even more deeply counter-intuitive consequences, the best policy might be to withhold assent to either view. Still, the identity theorist would have made a kind of case for the identity theory, if she could show that it was, though problematic, better than the alternative. I should say, though, that I am in fact very doubtful that the type-identity theorist could show that the denial of her theory has untoward consequences. All the arguments to that effect I am acquainted with depend on (what to my mind are) highly contentious premises about a deeply mysterious thing— causality.

C-fibre stimulation, but rather an explanation-away of the apparent separability of pain from C-fibre stimulation.

I have been suggesting that Kripke makes a convincing case for (1) and (2), but not for (3). For all Kripke has shown, (3) might be false. In fairness to Kripke, though, this may not show a weakness in his arguments. Perhaps his intent in the *Naming and Lectures* was to argue for (1) and (2), and to register his belief in (3),[321] without arguing for it.

By way of summary: Kripke has not refuted, and does not claim to have refuted, the sort of type-identity theory that identifies pain with a neural state such as C-fibre stimulation. But he certainly has shown that such a theory is modally problematic in ways not generally appreciated before the appearance of *Naming and Necessity*. Kripke has expressed his suspicion that token-identity theories, and theories on which a person either is his body or is nothing over and above his body, are also modally problematic in analogous ways. For reasons set out in this chapter, I do not share these suspicions. As far as I can see, the thesis that a person is nothing over and above his body (properly understood) does not have any initially counter-intuitive modal consequences. The thesis that mental events are physical events may have initially counter-intuitive modal consequences. If it does, it is not clear that the 'counter-intuitivity' of those consequences cannot be explained away along broadly Kripkean lines (by recourse to indistinguishable epistemic situations and epistemic counterparts). Again, in fairness to Kripke, he only voices the suspicion that token-identity theories, and theories on which a person is nothing over and above his body, are not less modally problematic than type-identity theories of the sort he argues against in detail. I hope that he some day returns to these topics, as he once suggested he might do,[322] and turns his suspicions into arguments, so that we can have a more informed view of how well-founded they are.

---

[321] As he does at 'N & N', 355.   [322] At 'N & N', 354 n. 73.

BIBLIOGRAPHY

ABBOTT, B., 'A Note on the Nature of "Water" ', *Mind*, 106 (1997), 311–19.

ACKERMANN, D., 'Natural Kinds, Concepts, and Propositional Attitudes', in P. French, T. Uehling, and H. Wettstein (eds.), *Contemporary Perspectives in the Philosophy of Language* (Minneapolis: University of Minnesota Press, 1979).

ADAMS, R., 'Theories of Actuality', *Nous*, 8 (1974), 211–31.

—— 'Primitive Thisness and Primitive Identity', *Journal of Philosophy*, 76 (1979), 5–26.

ALMOG, J., 'Naming Without Necessity', *Journal of Philosophy*, 20 (1986), 210–42.

AUNE, B., 'Determinate Meaning and Analytic Truth', in G. Debrock and M. Hulswit (eds.), *Living Doubt* (Dordrecht: Kluwer Academic Publishers, 1994).

BARCAN, R. C., 'A Functional Calculus of First Order Based on Strict Implication', *Journal of Symbolic Logic*, 11 (1946), 1–16.

BEALER, G., 'Mental Properties', *Journal of Philosophy*, 91 (1994), 185–208.

BECKER, O., 'Zur Logik der Modalitäten', *Jahrbuch für Philosophie und Phänomenologische Forschung*, 11 (1930), 497–548.

BIGELOW, J., and PARGETTER, R., *Science and Necessity* (Cambridge: Cambridge University Press, 1990).

BROWN, J., 'Natural Kind Terms and Recognitional Capacities', *Mind*, 107 (1998), 275–303.

BURGE, T., 'A Theory of Aggregates', *Nous*, 11 (1977), 97–117.

—— 'Other Bodies', in A. Woodfield (ed.), *Thought and Object: Essays in Intentionality* (Oxford: Clarendon Press, 1982).

CHANDLER, H., 'Plantinga and the Contingently Possible', *Analysis*, 37 (1976), 152–9.

CHISHOLM, R., 'Identity Through Possible Worlds: Some Questions', *Nous*, 1 (1967), 1–8; repr. in M. Loux (ed.), *The Possible and the Actual* (Ithaca: Cornell University Press, 1979).

CHOMSKY, N., 'Language and Nature', *Mind*, 104 (1995), 1–61.

DONNELLAN, K., 'Proper Names and Identifying Descriptions', in D. Davidson and G. Harman (eds.), *Semantics of Natural Language* (Dordrecht: D. Reidel, 1972).

—— 'The Contingent A Priori and Rigid Designators', in P. French, T. Uehling, and H. Wettstein (eds.), *Contemporary Perspectives in the Philosophy of Language* (Minneapolis: University of Minnesota Press, 1979).

DUMMETT, M., *Frege: Philosophy of Language*, 2nd edn. (London: Duckworth, 1981).

EVANS, G., 'The Causal Theory of Names', *Proceedings of the Aristotelian Society*, 80 (1976), 187–208.

—— 'Reference and Contingency', *The Monist*, 62 (1979), 161–89.

FELDMAN, F., 'Kripke and the Identity Theory', *Journal of Philosophy*, 71 (1974), 665–76.

—— *Confrontations with the Reaper* (New York: Oxford University Press, 1992).

FEYS, R., 'Les Logiques nouvelles des modalités', *Revue Neoscholastique de Philosophie*, 40 (1937), 517–53.

FINE, K., 'Prior on the Construction of Possible Worlds and Instants', in K. Fine and A. Prior, *Worlds, Times, and Selves* (Amherst: University of Massachusetts Press, 1977).

GEACH, P., *Reference and Generality*, 3rd edn. (Ithaca: Cornell University Press, 1980).

GIBBARD, A., 'Contingent Identity', *Journal of Philosophical Logic*, 4 (1975), 187–221.

GUPTA, A., *The Logic of Common Nouns* (New Haven: Yale University Press, 1980).

HAWLEY, K., 'Persistence and Non-supervenient Relations', *Mind*, 108 (1999), 53–67.

HILL, C., 'Imaginability, Conceivability, Possibility, and the Mind–Body Problem', *Philosophical Studies*, 87 (1997), 64–85.

HINTIKKA, J., 'Modality and Quantification', *Theoria*, 27 (1961), 119–28.

—— *Knowledge and Belief* (Ithaca: Cornell University Press, 1962).

—— 'The Semantics of Modal Notions', in D. Davidson and G. Harman (eds.), *Semantics of Natural Language* (Dordrecht: D. Reidel, 1972).

HOSSACK, K., 'Implicit Definition, Descriptive Names, and the Contingent A Priori', MS, 2000.

HUGHES, C., 'The Essentiality of Origin and the Individuation of Events', *Philosophical Quarterly*, 44 (1994), 26–45.

—— 'Same-Kind Coincidence and the Ship of Theseus', *Mind*, 106 (1997), 53–69.

—— 'Matter and Actuality in Aquinas', *Revue Internationale de Philosophie*, 204 (1998), 269–86.

—— 'Omniscience, Negative Existentials, and Cosmic Luck', *Religious Studies*, 34 (1998), 375–401.

—— 'Bundle Theory from *A* to *B*', *Mind*, 108 (1999), 149–56.

—— 'Identità personale e entità personale', in A. Bottani and N. Vassallo (eds.), *Identità personale: un dibattito aperto* (Naples: Loffredo, 2001).

—— 'On the Real Distinction Between Persons and Their Bodies', in M. Marsonet (ed.), *The Problem of Realism* (Aldershot: Ashgate, 2002).

—— 'Starting Over', in A. Bottani and P. Giaretta (eds.), *Individuals, Essence, and Identity* (Boston: Kluwer, 2002).

—— 'Gli eventi e i loro criteri di identita', in C. Bianchi and A. Bottani (eds.), *Significato e Ontologia* (Milan: FrancoAngeli, 2003).

HUGHES, G. E., and CRESSWELL, M. J., *An Introduction to Modal Logic* (London: Methuen, 1968).

KANGER, S., *Provability and Logic* (Stockholm: Almquist & Wiksell, 1957).

KAPLAN, D., 'Demonstratives', MS, 1977.

—— 'Bob and Carol and Ted and Alice', MS, 1978.

—— 'Trans-world Heir Lines', in M. Loux (ed.), *The Possible and the Actual* (Ithaca: Cornell University Press, 1979).

—— 'Thoughts on Demonstratives', in P. Yourgrau (ed.), *Demonstratives* (Oxford: Oxford University Press, 1990).

KATZ, J., 'Names Without Bearers', *Philosophical Review*, 103 (1994), 787–825.

KEIL, F. C., *Concepts, Kinds, and Cognitive Development* (Cambridge, Mass.: MIT Press, 1989).

KIM, J., 'Causation, Nomic Subsumption, and the Concept of an Event', *Journal of Philosophy*, 70 (1973), 217–36.

—— 'Events as Property Exemplifications', in M. Brand and D. Walton (eds.), *Action Theory* (Dordrecht: D. Reidel, 1975).

—— 'Perception and Reference Without Causality', *Journal of Philosophy*, 74 (1977), 606–20.

KNUUTTILA, S., 'Modal Logic', in N. Kretzmann, A. Kenny, and J. Pinborg (eds.), *The Cambridge History of Later Medieval Philosophy* (Cambridge: Cambridge University Press, 1982).

KRIPKE, S., 'Semantical Considerations on Modal Logic', *Acta Philosophica Fennica*, 16 (1963), 83–94; repr. in L. Linsky (ed.), *Reference and Modality* (Oxford: Oxford University Press, 1977).

—— 'Semantical Analysis of Modal Logic II: Non-normal Modal Propositional Calculi', in J. W. Addison, L. Henkin, and A. Tarski (eds.), *The Theory of Models* (Amsterdam: North Holland Publishing Company, 1965).

—— 'Identity and Necessity', in M. Munitz (ed.), *Identity and Individuation* (New York: New York University Press, 1971).

—— 'Naming and Necessity', in D. Davidson and G. Harman (eds.), *Semantics of Natural Language* (Dordrecht: D. Reidel, 1972).

—— 'Speaker's Reference and Semantic Reference', in P. French, T. Uehling, and H. Wettstein (eds.), *Contemporary Perspectives in the Philosophy of Language* (Minneapolis: University of Minnesota Press, 1979).

—— 'A Puzzle About Belief', in A. Margalit (ed.), *Meaning and Use* (Dordrecht: D. Reidel, 1979).

—— *Naming and Necessity*, slightly rev. edn., with new introd. (Oxford: Basil Blackwell, 1980).

LEWIS, C. I., 'Implication and the Algebra of Logic', *Mind*, 21 (1912), 522–31.

—— and LANGFORD, C. H., *Symbolic Logic* (New York: Dover, 1932).

LEWIS, D., 'An Argument for the Identity Theory', *Journal of Philosophy*, 63 (1966), 17–25; repr. in Lewis, *Philosophical Papers*, vol. i.

—— 'Counterpart Theory and Quantified Modal Logic', *Journal of Philosophy*, 65 (1968), 113–26; repr. in Lewis, *Philosophical Papers*, vol. i.

—— 'Counterparts of Persons and Their Bodies', *Journal of Philosophy*, 68 (1971), 203–11; repr. in Lewis, *Philosophical Papers*, vol. i.

—— *Counterfactuals* (Cambridge, Mass.: Harvard University Press, 1973).

—— 'Survival and Identity', in A. Rorty (ed.), *The Identities of Persons* (Berkeley: University of California Press, 1976); repr. in Lewis, *Philosophical Papers*, vol. i.

—— 'Mad Pain and Martian Pain', in N. Block (ed.), *Readings in the Philosophy of Psychology* (Cambridge, Mass.: Harvard University Press, 1980); repr. in Lewis, *Philosophical Papers*, vol. i.

—— 'Individuation by Stipulation and Acquaintance', *Philosophical Review*, 92 (1983), 3–33.

—— *Philosophical Papers*, 2 vols. (Oxford: Oxford University Press, 1983–6).

—— 'Events', in Lewis, *Philosophical Papers*, vol. ii.

—— *On the Plurality of Worlds* (Oxford: Basil Blackwell, 1986).

—— 'Reduction of Mind', in S. Guttenplan (ed.), *The Blackwell Companion to the Philosophy of Mind* (Oxford: Basil Blackwell, 1998).

LINSKY, L., *Names and Descriptions* (Chicago: University of Chicago Press, 1977).

LOUX, M. (ed.), *The Possible and the Actual* (Ithaca: Cornell University Press, 1979).

LYCAN, L., 'Two—No Three—Concepts of Possible Worlds', *Proceedings of the Aristotelian Society*, 94 (1990), 215–29.

McGINN, C., 'Rigid Designation and Semantic Value', *Philosophical Quarterly*, 32 (1982), 97–113.

MARCUS, RUTH BARCAN, 'A Functional Calculus of First Order Based on Strict Implication', *Journal of Symbolic Logic*, 11 (1946), 1–16.

MASSEY, G. J., *Understanding Symbolic Logic* (New York: Harper & Row, 1970).

MELLOR, H., 'Natural Kinds', *British Journal for the Philosophy of Science*, 28 (1977), 299–312.

MONTAGUE, R., 'Logical Necessity, Physical Necessity, Ethics, and Quantifiers', *Inquiry*, 3 (1960), 159–69.

—— 'The Proper Treatment of Mass Terms in English', in F. J. Pelletier (ed.), *Mass Terms: Some Philosophical Problems* (Dordrecht: D. Reidel, 1979).

NAGEL, T., 'What Is it Like to Be a Bat?', *Philosophical Review*, 83 (1974), 435–50.

NOONAN, H., 'The Closest Continuer Theory of Identity', *Inquiry*, 28 (1985), 195–229.

—— 'Constitution Is Identity', *Mind*, 102 (1993), 133–45.

OLSON, E., *The Human Animal: Personal Identity Without Psychology* (Oxford: Oxford University Press, 1997).

PARSONS, T., 'Essentialism and Quantified Modal Logic', in L. Linsky (ed.), *Reference and Modality* (Oxford: Oxford University Press, 1971).

PLANTINGA, A., *The Nature of Necessity* (Oxford: Clarendon Press, 1974).

—— 'The Boethian Compromise', *American Philosophical Quarterly*, 15 (1978), 129–38.

PUTNAM, H., 'The Meaning of Meaning', in Putnam, *Mind, Language, and Reality* (Cambridge: Cambridge University Press, 1975).

—— *Representation and Reality* (Cambridge, Mass.: MIT Press, 1988).

QUINE, W. V. O., *Word and Object* (Cambridge, Mass.: Mit Press, 1960).

—— *From a Logical Point of View* (New York: Harper Torchbooks, 1961).

—— 'Two Dogmas of Empiricism', in Quine, *From a Logical Point of View*.

—— 'Reference and Modality', in Quine, *From a Logical Point of View*.

—— 'Propositional Objects', in Quine, *Ontological Relativity and Other Essays* (New York: Columbia University Press, 1969).

RUSSELL, B., 'On Denoting', *Mind*, 14 (1905), 479–93.

—— 'Lectures on Logical Atomism', in R. C. Marsh (ed.), *Logic and Knowledge* (London: Allen & Unwin, 1956).

SAINSBURY, M., 'Philosophical Logic', in A. Grayling (ed.), *Philosophy: A Guide Through the Subject* (Oxford: Oxford University Press, 1995).

SALMON, N., *Reference and Essence* (Oxford: Basil Blackwell, 1982).

SEARLE, J., 'Proper Names', *Mind*, 67 (1958), 166–73.

SEGAL, G., *A Slim Book About Narrow Content* (Cambridge, Mass.: MIT Press, 2000).

SHOEMAKER, S., 'Identity, Properties, and Causality', in P. French, T. Uehling, and H. Wettstein (eds.), *Midwest Studies in Philosophy* (Minneapolis: University of Minnesota Press, 1979).

—— 'On an Argument for Dualism', in C. Ginet and S. Shoemaker (eds.), *Knowledge and Mind* (Oxford: Oxford University Press, 1983).

—— 'Personal Identity: A Materialist's Account', in S. Shoemaker and R. Swinburne, *Personal Identity* (Oxford: Basic Blackwell, 1984).

SLOTE, M., 'Causality and the Concept of a Thing', in P. French, T. Uehling, and S. Wettstein (eds.), *Midwest Studies in Philosophy* (Minneapolis: University of Minnesota Press, 1979).

SNOWDON, P., 'Persons, Animals, and Ourselves', in C. Gill (ed.), *The Person and the Human Mind* (Oxford: Clarendon Press, 1990).

SOSA, D., 'The Import of "A Puzzle About Belief" ', *Philosophical Review*, 105 (1996), 376–401.

STALNAKER, R., 'Assertion', *Syntax and Semantics*, 9 (1978), 315–32.

—— 'Anti-Essentialism', in P. French, T. Uehling, and H. Wettstein (eds.), *Midwest Studies in Philosophy* (Minneapolis: University of Minnesota Press, 1979).

—— 'Possible Worlds', in M. Loux (ed.), *The Possible and the Actual* (Ithaca: Cornell University Press, 1979).

—— *Inquiry* (Cambridge, Mass.: MIT Press, 1987).

—— 'Semantics for Belief', *Philosophical Topics*, 15 (1987), 177–90.

STRAWSON, P. F., 'Maybes and Might Have Beens', in A. Margalit (ed.), *Meaning and Use* (Dordrecht: D. Reidel, 1979).

TICHY, P., 'Kripke on Necessity A Posteriori', *Philosophical Studies*, 43 (1983), 225–41.

TRAVIS, C., 'Are Belief Ascriptions Opaque?', *Proceedings of the Aristotelian Society*, 97 (1993), 88–106.

VAN INWAGEN, P., *Material Beings* (Ithaca: Cornell University Press, 1989).

WIGGINS, D., *Sameness and Substance* (Cambridge, Mass.: Harvard University Press, 1980).

WILLIAMS, B., 'Are Persons Bodies?', in Williams, *Problems of the Self* (Cambridge: Cambridge University Press, 1973).

WILLIAMSON, T., 'Bare Possibilia', *Erkenntnis*, 48 (1998), 257–73.

WOODFIELD, A., 'Identity Theories and the Argument from Epistemic Counterparts', *Analysis*, 38 (1978), 140–6.

ZIMMERMAN, D., 'Distinct Indiscernibles and the Bundle Theory', *Mind*, 106 (1997), 305–9.

INDEX